Ecological Studies, Vol. 127

Analysis and Synthesis

Ecological Studies

Volumes published since 1992 are listed at the end of this book.

Springer
Berlin
Heidelberg
New York
Barcelona
Budapest
Hong Kong
London
Milan
Paris
Santa Clara
Singapore
Tokyo

H. Sandermann A.R. Wellburn R.L. Heath (Eds.)

Forest Decline and Ozone

A Comparison of Controlled Chamber and Field Experiments

With 77 Figures, 12 in Color and 22 Tables

Springer

Prof. Dr. Heinrich Sandermann
GSF-Forschungszentrum für Umwelt und Gesundheit GmbH
Institut für Biochemische Pflanzenpathologie
D-85764 Oberschleißheim, Germany

Prof. Dr. Alan R. Wellburn
University of Lancaster
Institute of Environmental and Biological Sciences
Lancaster LA1 4YQ, UK

Prof. Dr. Robert L. Heath
University of California, Riverside
Department of Botany and Plant Sciences
Riverside, CA 92521, USA

ISSN 0070-8356
ISBN 3-540-61321-8 Springer-Verlag Berlin Heidelberg New York

Library of Congress Cataloging-in-Publication Data. Forest decline and ozone: a comparison of controlled chamber and field experiments/Heinrich Sandermann, Alan R. Wellburn. Robert L. Heath, eds. p. cm. – (Ecological studies; vol. 127) Includes bibliographical references and index. ISBN 3-540-61321-8 1. Trees–Effect of atmospheric ozone on–United States. 2. Trees–Effect of atmospheric ozone on–Europe. 3. Forest declines–United States. 4. Forest declines–Europe. I. Sandermann, Heinrich. II. Wellburn, Alan, 1940- . III. Heath, Robert L., 1940 . IV. Series: Ecological studies; v. 127. SB745.F67 1997 634.9′619–dc20 96-30571

Cover design: Design & Production GmbH, Heidelberg

Typesetting: Scientific Publishing Services (P) Ltd., Madras

SPIN: 10468789 31/3137/SPS – 5 4 3 2 1 0 – Printed on acid-free paper

Preface

The idea for this book arose in 1993, after the Free State of Bavaria through its Bayrisches Staatsministerium für Landesentwicklung und Umweltfragen (Bavarian Ministry of Regional Development and the Environment) decided to discontinue both the Bavarian project management (PBWU) for forest decline research and the multidisciplinary field research on the Wank Mountain in the Alps near Garmisch. Forest decline through the action of ozone and other photooxidants was a main topic of the supported research in the Alps and will be a topic of new investigations in the Bavarian Forest. Many interesting results were obtained, but the researchers involved have not had sufficient time to allow reliable conclusions to be drawn. It was therefore decided to ask international experts for contributions in order to summarize the best available evidence of a possible link between ozone and forest decline – a topic which has been studied in the USA since the late 1950s and in Europe since the early 1980s.

The original idea of *Waldsterben* as an irreversible large-scale dieback of forests in Germany was soon recognized to be wrong (Forschungsbeirat 1989). However, the new criteria used for the official German and European damage inventories (loss or yellowing of needles or leaves, tree morphology) indicate that persistently high percentages of damaged spruce and pine remain, and there is an increasing percentage of damaged beech and oak, with a high proportion of biotic disease (Forschungsbeirat 1989; UN-ECE 1995). The enormous volume of damage inventory data in Europe constitutes an operational definition of forest decline. Yet, the symptoms evaluated are nonspecific and therefore difficult to reproduce in laboratory experiments. Much better diagnostics, including molecular stress marker techniques, are needed. Two field guides to diagnose tree diseases have been published: one for California where ozone problems were originally discovered (Scharpf 1993), and one for Germany in the aftermath of the *Waldsterben* concept (Hartmann et al. 1988). Both guides contain excellent color photographs and useful glossaries with current definitions of decline, predisposition, and other terms.

A functional definition of forest decline has been given by Manion (1981). Decline was defined as a multiplicity of tree diseases due to more than a single cause. Manion (1981) also implied that a strict temporal sequence of the abiotic and biotic agents was involved, although this latter aspect may apply only in special cases. The functional definition is close to the classical phytopathological concept of predisposition (Gäumann 1951; Forschungsbeirat 1989; Ayres 1984). In our case, abiotic stress factors are thought to sensitize or desensitize a tree for biotic disease, and vice versa. Forest decline phenomena have also been discussed on the assumption of an abiotic origin (Huettl and Müller-Dombois 1993). However, such a restricted definition does not appear to be justified with respect to ozone since an interaction between ozone exposure and fungal infection, as well as insect attack, has already been established in the San Bernardino Mountains, California (Scharpf 1993).

Atmospheric ozone is a relatively stable product of the photooxidation of hydrocarbons and nitrogen oxides. It is phytotoxic under two basic regimes. Firstly, acute toxicity occurs during short-term exposure of plants to high ozone concentrations, and symptoms are primarily caused by membrane disruption, cell death, and tissue necrosis. Secondly, chronic toxicity occurs during long-term exposure of plants to near-ambient ozone concentrations. In this case, both direct and indirect symptoms occur depending on stomatal apertures and interactions with other environmental abiotic and biotic factors.

Conifers such as spruce (e.g., *Picea abies)* and pine (e.g., *Pinus ponderosa, Pinus sylvestris)* are known to develop visible symptoms primarily on the youngest and the older needle classes under acute and chronic ozone exposure regimes, respectively (Hartmann et al. 1988; Scharpf 1993). Due to moderate levels of ozone in most forests, only the chronic exposure regime seems to be ecologically relevant for conifers. The situation is less clear for deciduous trees and crop plants. Prior to the development of visible symptoms, however, biochemical stress reactions occur, such as the development of ethylene or the induction of defence reactions.

This book will attempt to describe (1) the temporal and causal links between the initial stress reactions and visible ozone symptoms; (2) the possible relationship between ozone effects and complex tree diseases, and (3) the role of ozone in more general forest ecosystem interactions.

At this time, definite proof of a link between ozone and forest decline exists only for California. However, the new data summarized in this book starts to also support a role for ozone in forest decline in eastern USA and Europe. Improved diagnostics for the effects of ambient ozone still need to be developed in order to prove or rule out a link between ozone and forest decline. The present book therefore represents an interim report that illustrates the complexities of ozone research and that hopefully will stimulate more definitive research.

The idea for this volume is largely due to Prof. Dr. Dr. h.c. Hubert Ziegler (Emeritus, Technische Universität München) and Prof. Dr. Ernst-Detlef Schulze (Universität Bayreuth). The constant interest and encouragement by the series editor of Ecological Studies, Prof. Dr. Dr. h.c. Otto L. Lange (Emeritus, Universität Würzburg) are also gratefully acknowledged. Special thanks are due to Drs. P. Miller, M. Arbaugh, and P. Temple who organized an impressive excursion for the editors into the San Bernardino Mountains to see first hand the longest-lasting study site for ozone and the forest.

Oberschleißheim, September 1996 Heinrich Sandermann

References

Ayres PG (1984) The interaction between environmental stress injury and biotic disease physiology. Annu Rev Phytopathol 22:53–75

Forschungsbeirat Waldschäden/Luftverunreinigungen FBW (1989) Dritter Bericht, Kernforschungszentrum Karlsruhe GmbH, Karlsruhe, ISSN 0931-7805

Gäumann E (1951) Pflanzliche Infektionslehre. Birkhäuser, Basel

Hartmann G, Nienhaus F, Butin N (1988) Farbatlas Waldschäden. Diagnose von Baumkrankheiten. E Ulmer, Stuttgart

Huettl RF, Mueller-Dombois D (1993) Forest decline in the Atlantic and Pacific regions. Springer Berlin Heidelberg New York, pp 351–358

Manion PD (1981) Tree disease concepts. Prentice-Hall, Englewood Cliffs, pp 324–339

Scharpf RF (ed) (1993) Diseases of Pacific coast conifers. US Department of Agriculture, Forest Service, Agricultural Handbook 521, Washington DC

UN-ECE (ed) (1995) Forest condition in Europe. United Nations Economic Commission for Europe, Brussels, ISSN 1020–3729

Contents

Contributors

M.J. Arbaugh

Pacific Southwest Research Station, Forest Service,
US Department of Agriculture, 4955 Canyon Crest Drive,
Riverside, CA 92507, USA

J.D. Barnes

Department of Agricultural and Environmental Science,
Ridley Building, University of Newcastle,
Newcastle-upon-Tyne NE1 7RU, UK

A.H. Chappelka

School of Forestry, Auburn University, Auburn, AL, USA

D. Ernst

GSF-Forschungszentrum für Umwelt und Gesundheit GmbH,
Institut für Biochemische Pflanzenpathologie,
Ingolstädter Landstr. 1, 85764 Oberschleißheim, Germany

T.S. Fredericksen

Environmental Resources Research Institute,
The Pennsylvania State University,
University Park, PA 16802, USA

B.A. Hale

Department of Horticultural Science,
University of Guelph,
Guelph, Ontario, Canada N1G 2W1

W.M. Havranek

Forest Research Institute, Rennweg 1, 6020 Innsbruck, Austria

R.L. Heath

Department of Botany and Plant Sciences,
University of California, Riverside, CA 92521, USA

W. Heller

GSF-Forschungszentrum für Umwelt und Gesundheit GmbH,
Institut für Biochemische Pflanzenpathologie,
Ingolstädter Landstr. 1, 85764 Oberschleißheim, Germany

J.L. Innes

Swiss Federal Institute for Forest, Snow and Landscape Research,
Zürcherstr. 111, 8903 Birmensdorf, Switzerland

G. Kramm

Fraunhofer Institute for Atmospheric Environmental Research
(IFU), Kreuzeckbahnstr. 19, 82467 Garmisch-Partenkirchen,
Germany

C. Langebartels

GSF-Forschungszentrum für Umwelt und Gesundheit GmbH,
Institut für Biochemische Pflanzenpathologie,
Ingolstädter Landstr. 1, 85764 Oberschleißheim, Germany

J.A. Laurence

Boyce Thompson Institute for Plant Research, Tower Road,
Ithaca, NY 14853–1801, USA

P.W. Lucas

Institute of Environmental and Biological Sciences,
Division of Biological Sciences, Lancaster University,
Lancaster LA1 4YQ, UK

C. Lütz

GSF-Forschungszentrum für Umwelt und Gesundheit GmbH,
Arbeitsgruppe Expositionskammern, Ingolstädter Landstr. 1,
85764 Oberschleißheim, Germany

T.A. Mansfield

Institute of Environmental and Biological Sciences, Division of Biological Sciences, University of Lancaster, Lancaster LA1 4YQ, UK

R. Matyssek

Department of Forest Botany, University of Munich, Hohenbachernstr. 22, 85354 Freising, Germany

A.R. McLeod

Institute of Terrestrial Ecology, Monk Woods, Abbots Ripton, Huntingdon, PE17 2LS, UK

P.R. Miller

USDA Forest Service, Pacific Southwest Forest and Range Experiment Station, 4955 Canyon Crest Drive, Riverside, CA 92507, USA

V.A. Mohnen

ASRC, State University of New York at Albany, Department of Atmospheric Science, 100 Fuller Road, Albany, NY 12205, USA, and Fraunhofer Institute for Atmospheric Environmental Research (IFU), Kreuzeckbahnstr. 19, 82467 Garmisch-Partenkirchen, Germany

R.C. Musselman

USDA Forest Service, Rocky Mountain Forest and Range Experiment Station, 240 West Prospect Road, Fort Collins, CO 80526, USA

H.-D. Payer

GSF-Forschungszentrum für Umwelt und Gesundheit GmbH, Arbeitsgruppe Expositionskammern, Ingolstädter Landstr. 1, 85764 Oberschleißheim, Germany

A. Polle

Institut für Forstbotanik und Baumphysiologie, Am Flughafen 17, 79085 Freiburg, Germany

H. Rennenberg

Institut für Forstbotanik und Baumphysiologie, Am Flughafen 17, 79085 Freiburg, Germany

M. Reuther

Landesanstalt für Wald und Forstwirtschaft, Hohenbachernstr. 20, 85354 Freising, Germany

H. Sandermann Jr.

GSF-Forschungzentrum für Umwelt und Gesundheit GmbH, Institut für Biochemische Pflanzenpathologie, Ingolstädter Landstr. 1, 85764 Oberschleißheim, Germany

H.-E. Scheel

Fraunhofer Institute for Atmospheric Environmental Research (IFU), Kreuzeckbahnstr. 19, 82467 Garmisch-Partenkirchen, Germany

W. Seiler

Fraunhofer Institute for Atmospheric Environmental Research (IFU), Kreuzeckbahnstr. 19, 82467 Garmisch-Partenkirchen, Germany

G.Selldén

Botaniska Institutionen, Goteborgs Universitet, Avd F Fysiologisk Botanik, Carl Scottbergs Gata 32, 41319 Göteborg, Sweden

L. Skärby

Swedish Environmental Research Institute (IVL), Box 47086, 40258 Göteborg, Sweeden

J.M. Skelly

Department of Plant Physiology, The Pennsylvania State University, 210 Buckhout Laboratory, University Park, PA 16802, USA

W.R. Stockwell

Fraunhofer Institute for Atmospheric Environmental Research (IFU), Kreuzeckbahnstr. 19, 82467 Garmisch-Partenkirchen, Germany

S. Sutinen

Finnish Forest Research Institute, Suonenjoki Research Station, 77600 Suonenyoki, Finland

G.E. Taylor Jr.

Department of Environmental and Resource Sciences, College of Agriculture, University of Nevada, Reno, NV 89557-0013, USA

P.J. Temple

Stateswide Air Pollution Research Center,
University of California, Riverside, CA 92521, USA

M.H. Unsworth

Center for Analysis of Environmental Change,
Oregon State University, Weniger Hall 283,
Corvallis, OR 97331-6511, USA

A.R. Wellburn

Institute of Environmental and Biological Sciences,
Division of Biological Sciences, University of Lancaster,
Bailrigg, Lancaster LA1 4YQ, UK

G. Wieser

Federal Forest Research Institute, Division of Tree Physiology,
Rennweg 1, 6020 Innsbruck, Austria

1 Ozone Formation, Destruction and Exposure in Europe and the United States

W.R. Stockwell, G. Kramm, H.-E. Scheel, V.A. Mohnen and W. Seiler

1.1 Introduction

One of the earliest observations that air pollution could damage vegetation was made in Los Angeles during 1944 (Finlayson-Pitts and Pitts 1986). It was established by Middleton et al. (1950) that plant damage was associated with air pollution episodes. Haagen-Smit et al. (1952) showed that similar plant damage could be caused by synthetic mixtures of alkenes and O_3 or by mixtures of nitrogen dioxide and alkenes that had been exposed to light. In subsequent research, Haagen-Smit and coworkers (Haagen-Smit 1952; Haagen-Smit et al. 1953; Haagen-Smit and Fox 1956) showed that photochemical reactions of nitrogen oxides and organic compounds produced O_3 and other pollutants.

Photochemical air pollution is no longer confined to Los Angeles (EPA 1991b; McKee 1994) and it now affects most major urban locations in the world (Finlayson-Pitts and Pitts 1986). High O_3 concentrations are no longer limited to local hot spots or urban regions. Elevated levels of O_3 appear, especially during the spring and summer months, in rural and even in remote regions of the United States, Canada (Table 1.1), and Europe.

Ozone concentrations are highly variable in the lower atmosphere and here meteorological conditions and photochemical processes greatly affect the O_3 concentrations. Relationships have been established between O_3 concentration and regional scale atmospheric circulation, solar radiation, temperature, hydrocarbon, and nitrogen oxide concentrations, among others (Dodge 1984; Lin et al. 1988; Davies et al. 1992, and references therein; Vukovich 1994). A great number of studies have shown that elevated O_3 concentrations in rural areas are caused by transport of O_3 and its precursors from urban and industrialized regions over distances of hundreds of kilometers. Episodes of high concentrations of surface O_3 over large parts of Europe and North America usually occur during the summer in stagnant air under dry, sunny weather conditions. Under these conditions O_3 precursors such as nitrogen oxides, carbon

Ecological Studies, Vol. 127
Sandermann et al. (eds) Forest Decline and Ozone

Table 1.1. Percentiles of hourly ozone concentrations for a rural and two "remote" sites in eastern Canada. (Fuentes and Dann 1994)

Site type	Site name	Location	Observation period	No. of hourly observations	Percentiles ($nl\ l^{-1}$)							($nl\ l^{-1}$)		
					5	10	25	50	75	90	95	Max	Mean	SD
Remote	Hawkeye Lake	Northern Ontario	1983–1991	43267	5	9	18	27	35	43	49	95	26.7	13.3
Rural	Simcoe	SW Ontario	1980–1991	96581	6	10	17	26	37	52	63	190	29.0	17.2
Remote	Kejimkujik	SW Nova Scotia	1984–1991	53948	10	13	20	27	33	40	45	150	27.0	11.2

monoxide, and volatile organic compounds tend to accumulate. The peak O_3 concentrations reach levels at which adverse environmental effects are expected for human health, crops, and natural vegetation. During most summers, preliminary critical O_3 levels proposed to protect the most sensitive plant species were exceeded in large parts of Europe (Beck and Greenfelt 1994).

Different units are commonly used in atmospheric chemistry to report ozone concentrations: ppb(v), ppm(v), $\mu g\ m^{-3}$, $nl\ l^{-1}$, and $\mu l\ l^{-1}$. Throughout this book ozone concentrations are reported as volume fractions: $nl\ l^{-1}$ and $\mu l\ l^{-1}$ which are equivalent to ppb(v), parts per billion by volume, and ppm (v), parts per million by volume, respectively. If ozone concentrations are reported in $\mu g\ m^{-3}$, the conditions of temperature and presssure must be given. At a temperature of 20 °C and pressure of 1013 hPa, an ozone concentration of 2 $\mu g\ m^{-3}$ is equivalent to 1 $nl\ l^{-1}$.

Exposure indices have been developed in an attempt to characterize O_3 exposure, and there has been considerable discussion in the literature concerning the selection of the optimum exposure indices for vegetation (Lefohn and Benedict 1982, 1985; Tingey 1984; EPA 1986, 1988; Krupa and Kickert 1987; Ashmore 1988; Hogsett et al. 1988; Lee et al. 1988; Lefohn et al. 1988, 1989; Parry and Day 1988; Runeckles 1988). An "exposure index" is not really a measure of exposure or dose, but rather it is supposed to be a "biologically meaningful" dose surrogate that is based on a chosen function of time and O_3 concentrations. For example, there is an urgent need to further investigate the complex micrometeorological transfer processes and the chemical processes above and within the forest canopy because plants respond most directly to O_3 fluxes, rather than ambient O_3 concentrations (Musselman et al. 1994).

The purpose of a dose surrogate is to provide a cumulative index that describes the highest exposure, while not ignoring the lower, biologically less important exposures. Possible approaches to different weighting functions for air pollutant concentrations have been discussed extensively. For example, Lefohn and Runeckles (1987) proposed an exposure index that uses the sigmoidal weighting of individual hourly average concentrations of O_3 and summing over time. The sigmoidal weighting function is multiplied by each of the hourly average concentrations. The lower, less biologically effective concentrations are included in the integrated exposure summation and are not eliminated because no threshold is used.

Exposure indices that have been used in the past to characterize hourly mean O_3 concentrations over monthly and seasonal periods are:

1. The sum of all hourly mean O_3 concentrations using no threshold concentration. This is commonly referred to as "total dose" (SUMO).
2. AOT30, 40, ... is the sum of l-h mean O_3 concentrations (whole day or daylight hours) above a threshold of 30, 40, ...nl l^{-1}, respectively, over a defined time period (e.g., 1 year, a number of months). The unit is nl l^{-1} h^{-1}; the threshold and the time period should also be indicated.
3. The sum of all hourly mean O_3 concentrations (W126) where each hourly concentration is weighted by a sigmoidal weighting function.
4. The sum of all hourly mean O_3 concentrations equal to or greater than 0.070 μl l^{-1} (SUM07).
5. The number of hourly mean O_3 concentrations equal to or greater than 0.070 μl l^{-1}.
6. The seasonal means of the average of the daily 7-h (9:00-15:59 h) concentrations.
7. The seasonal means of the average of the daily 12-h (7:00-18:59 h) concentration.

These results provide a variety of indices which can be used to compare O_3 exposures in control chambers with those experienced, for example, at relatively clean forested sites in rural areas (for an example, see Table 1.2; EPA 1988). More details on O_3 exposure and critical levels are given in Fuhrer and Achermann (1994), which addresses agricultural crops and forest trees as well as mapping and modeling. The AOT indices are widely used in Europe; for trees, AOT values based on 24-h time periods are typically used, and for agricultural crops such as wheat, AOT calculations based on daylight hours alone are preferred. Calculated AOT40 values are critically dependent on the completeness of the data record. Missing data will introduce large errors in the calculated indices unless missing data corrections are applied. Another overview on guideline values can be found in the work of Laurila and Lättilä (1994), which mainly focuses on observations from Finland.

Table 1.2. Ozone concentrations and exposure data for select rural locations. (EPA 1988)

Site	Elevation (m)	Year[a] 1980	1981	1982	1983	1984	1985	1986	1987
Kejimkukik, Nova Scotia	127							26	
								5.4	
								220	
Theodore Roosevelt	727					29	32	31	
National Park, Maryland						8.0	6.7	6.5	
						250	240	200	
Custer National Forest,	1006	44			35				
Montana		23			12				
		280			280				
Ochoco National Forest,	1364	38	31	34	34				
Oregon		20	5.8	9.1	8.0				
		300	240	260	250				
White River	1600			39	44				
Oil Shale Project, Utah				20	32				
				340	350				
Fortress Mountain,	2103							41	48
Alberta								25	84
								340	340
Apache National Forest,	2424	47	35	41	38				
Arizona		81	10	27	17				
		400	290	330	300				
Whiteface Mountain,	1483	41	36	33	36	39	39	40	42
New York		68	34	32	38	43	41	32	58
		330	260	270	260	310	320	300	340

[a] First Row: the annual average of the hourly mean concentrations; units in nl l^{-1}. Second Row: the value of the W126 Sigmoidal exposure parameter calculated over the annual period; units in $\mu l\ l^{-1}$ h. Third Row: the value of the SUMO exposure parameter calculated over the annual period.
units in $\mu l\ l^{-1}$ h.

1.2 The Chemistry of Tropospheric Ozone Formation

The atmosphere consists of several layers due to altitude-dependent temperature variations which determine the extent of mixing (Finlayson-Pitts and Pitts 1986; Warneck 1988). A layer is well mixed if the atmospheric temperature decreases with increasing altitude. The lowest layer is known as the troposphere and here the temperature decreases with height. The troposphere may be divided further into two sublayers. The boundary layer is the lowest sublayer and here emissions are rapidly mixed due to the turbulence caused by solar heating of the Earth's surface. Above the boundary layer is the free troposphere which is not as strongly affected by turbulence or surface deposition. In the free troposphere trace gases may be transported on a global scale. Between the troposphere and the stratosphere is a region known as the tropopause, a region with a relatively constant temperature and relatively neutral stability. Above the troposphere is the stratosphere which typically begins at an altitude between 10 and 15 km. The stratosphere is much more stable because the temperature increases with altitude. The ozone layer is found within the stratosphere.

Until the 1970s it was assumed that stratospheric O_3 was the main source of tropospheric O_3 (Junge 1963). It was believed that the O_3 was mainly transported within the troposphere until the O_3 was destroyed through deposition at the Earth's surface (Warneck 1988). Photochemical production of O_3 was considered to be important only in highly polluted urban regions. However, based on the work of Crutzen (1973), Fishman et al. (1979), and others, it became evident that photochemical production of O_3 from nitrogen oxides and volatile organic compounds is a major source of O_3 in the troposphere (Warneck 1988). As shown in Fig. 1.1, present-day O_3 concentrations are typically greater than 25 nl l^{-1} at remote locations due to tropospheric production (PORG 1993).

It is now recognized that O_3 sources in the continental boundary layer over Europe and North America include both transport from the stratosphere and in situ photochemical production. Recent studies have demonstrated that 50% of tropospheric O_3 is due to in situ production (Müller and Brasseur 1995). The export of O_3 and its precursors from source regions to more remote regions has a direct impact on global O_3 concentrations. For example, during the summer season O_3 concentrations can be below 5 nl l^{-1} over highly remote ocean regions such as the tropical Pacific, but over the North Atlantic the O_3 concentrations are typically much higher, ranging from 15 to 50 nl l^{-1} due to transport from North America (Müller and Brasseur 1995).

The photochemical production of O_3 in the troposphere was originally identified by Haagen-Smit (1952) as being responsible for air pollution in Los Angeles. The overall reactions in the "photochemical smog mechanism" are well known. However, many important aspects of the organic chemistry component remain unknown and are active research topics. In the troposphere the source of O_3 is the photolysis of nitrogen dioxide to produce ground state oxygen atoms, $O(^3P)$ [reaction (1.1); $h\nu$ represents photons of light and ultraviolet radiation]. The ground state oxygen atoms react with molecular oxygen to produce O_3 [reaction (1.2); M is either a nitrogen or oxygen molecule].

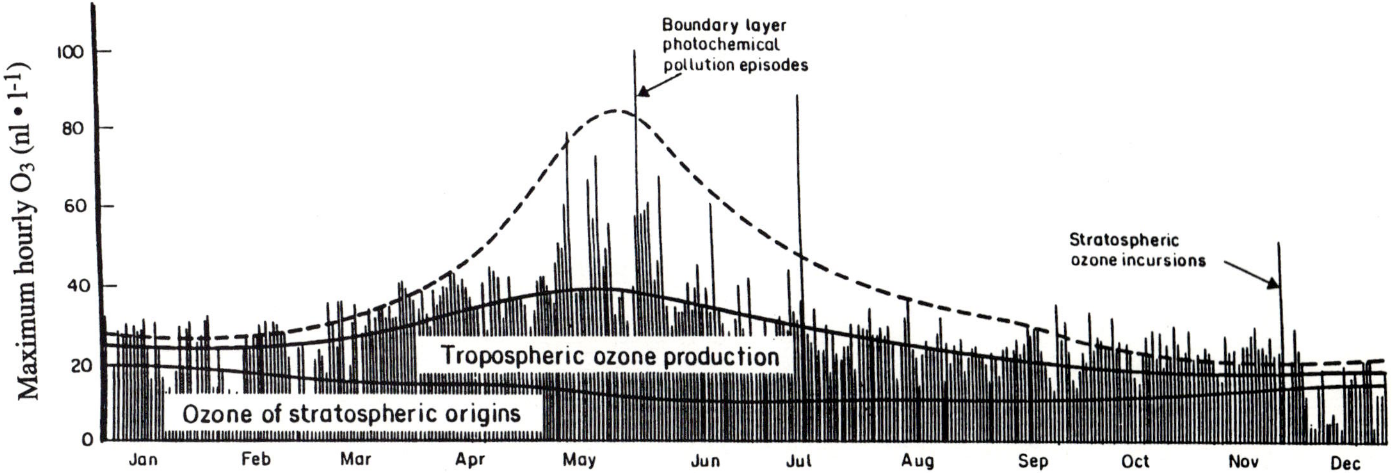

Fig. 1.1. Contribution from the major O_3 sources to the O_3 concentrations observed at ground level at a rural site (after Derwent and Kay 1988). The *bars* indicate daily maximum hourly means. As indicated by the *curves*, the contribution of photochemical O_3 production in the troposphere is superimposed on a natural background related to stratospheric influx. Peak concentrations in summer are largely due to enhanced O_3 production during pollution episodes. (PORG 1988)

$$NO_2 + h\nu \rightarrow NO + O(^3P) \ . \tag{1.1}$$

$$O(^3P) + O_2 + M \rightarrow O_3 + M \ . \tag{1.2}$$

When NO_x is present, O_3 reacts with NO to reproduce NO_2 (here, the modern definition is used: NO_x = NO + NO_2; in the early literature NO_x was the sum of NO, NO_2, HNO_3, NO_3, N_2O_5 ... which is nowadays denoted as NO_y).

$$O_3 + NO \rightarrow NO_2 + O_2 \ . \tag{1.3}$$

If the atmosphere contained only N_2, O_2, and nitrogen oxides, the O_3 concentration would be given by the "NO-photostationary state equation", Eqn (1.4).

$$[O3] = \frac{J_1[NO_2]}{k_3[NO]} , \tag{1.4}$$

where J_1 is the photolysis frequency of reaction (1.1), k_3 is the rate constant for reaction (1.3), $[NO_2]$ is the concentration of nitrogen dioxide, and [NO] is the concentration of nitric oxide. Equation (1.4) shows that the O_3 concentration is directly related to the NO_2 concentration and inversely related to the NO concentration. In the absence of CO or organic compounds, O_3 formation does not occur in the troposphere because reactions (1.1) through (1.3) only recycle O_3.

The photolysis of O_3 produces additional radicals which react with carbon monoxide or organic species to produce additional O_3 (Finlayson-Pitts and Pitts 1986; Seinfeld 1986). One of the O_3 photolysis reactions produces an excited oxygen atom, $O(^1D)$.

$$O_3 + h\nu \rightarrow O(^1D) + O_2 \ . \tag{1.5}$$

A fraction of the excited oxygen atoms reacts with water to produce HO• radicals.

$$O(^1D) + H_2O \rightarrow 2HO\bullet \ . \tag{1.6}$$

The HO• radicals react with CO and organic compounds (RH) to produce peroxy radicals (HO_2 or RO_2). The peroxy radicals react with NO to convert it back to NO_2 which photolyzes to produce additional O_3.

$$CO + HO\bullet(+O_2) \rightarrow CO_2 + HO_2\bullet \tag{1.7}$$
$$RH + HO\bullet \rightarrow R\bullet + H_2O \tag{1.8}$$
$$R\bullet + O_2 + M \rightarrow RO_2\bullet + M \tag{1.9}$$
$$RO_2\bullet + NO \rightarrow RO\bullet + NO_2 \tag{1.10}$$
$$RO\bullet + O_2 \rightarrow HO_2\bullet + R'CHO \tag{1.11}$$
$$HO_2\bullet + NO \rightarrow HO\bullet + NO_2 \tag{1.12}$$

The net sum of reactions (1.8) through (1.12) plus twice reactions (1.1) and (1.2) is:

$$RH + 4O_2 + 2h\nu \rightarrow R'CHO + H_2O + 2O_3 \ . \tag{1.13}$$

The initial reaction of an organic chemical and an HO• results in the production of two O_3 molecules and a carbonyl species, either an aldehyde (R'CHO) or a ketone (R′CR″O) if NO_x concentrations are great enough. The carbonyl compounds may react with HO• or photolyze to produce more peroxy radicals which further react to produce more O_3 molecules (Finlayson-Pitts and Pitts 1986; Seinfeld 1986).

The chemistry of O_3 formation is a highly nonlinear function of nitrogen oxides and organic chemistry (Dodge 1984; Liu et al. 1987, 1988). Both NO_x and organic compounds are consumed in the process of O_3 formation. Nitrogen oxides are slowly lost when HO• reacts with NO_2 to form nitric acid.

$$HO\bullet + NO2 \rightarrow HNO3 \ . \tag{1.14}$$

Reaction (1.14) tends to lower O_3 formation rates by reducing HO_x radical concentrations and by removing NO_x from the photooxidation cycle. In the lower troposphere there is little return of NO_x from HNO_3 because nitric acid reacts slowly in the lower troposphere and it is rapidly removed due to dry and wet deposition.

The production of O_3 may be limited by either the amount of available NO_x or organic species (Milford et al. 1989). In urban locations reactive alkenes and aromatic hydrocarbons (usually but not exclusively of anthropogenic origin) are among the most important organic precursors of O_3. In more rural environments, reactive biogenic compounds such as isoprene and terpenes are often the dominating organic compounds (Trainer et al. 1987; Chameides et al. 1988; Blake et al. 1993). The total O_3 production induced by an organic chemical is related to the number of NO to NO_2 conversions affected by the chemical and its decomposition products over its entire photooxidation cycle. The greater the number of NO to NO_2 conversions affected by the hydrocarbon, the greater the amount of O_3 produced. The catalytic production efficiency of NO_x can be defined as the ratio of the rate at which NO molecules are converted to NO_2 to the total rate of NO_x loss through conversion of NO_x to nitric acid, organic nitrates or loss through deposition (Liu et al. 1987; Lin et al. 1988; Hov 1989). Ozone production efficiency is inversely related to the NO_x concentration for most atmospheric conditions. In the continental boundary layer, NO_x concentrations vary over a range of three orders of magnitude which causes the O_3 production efficiency to vary also over a wide range of values.

Over 350 organic compounds have been found to be emitted from vegetation (Graedel 1979). Isoprene and terpenes are among the most reactive emissions from forest canopies. Isoprene is the most abundant organic species emitted from deciduous trees and it is an important emission of coniferous trees (Guenther et al. 1994). Isoprene has an estimated emission rate of 4.4 Mt year^{-1}, which is very significant when compared with anthropogenic emissions of 27 Mt year^{-1}. Terpenes are also emitted by coniferous trees in significant quantities. The structures of isoprene and a few typical biogenic compounds are shown in. 1.2 (Finlayson-Pitts and Pitts 1986).

These biogenic compounds react rapidly with HO• and O_3. For example, the lifetime of isoprene with respect to reaction with HO• during daylight hours has been estimated to be 6 h and the lifetime of *d*-limonene has been estimated to be 4 h (Finlayson-Pitts and Pitts 1986). However, terpenes contribute less to the formation of O_3 than is suggested by their reaction rates since these species and their decomposition products react rapidly with O_3. α-Pinene has a lifetime of 4 h with respect to reaction with O_3; *d*-limonene has a lifetime of about 35 min (Finlayson-Pitts and Pitts 1986). The reaction of O_3 with unsaturated hydrocarbons, such as isoprene and terpenes, may be especially important near forest canopies (Hov et al. 1983; Paulson and Seinfeld 1992). The presence of large amounts of natural biogenic emissions may lead to strong gradients in O_3 concentrations near forest canopies (Enders et al. 1989).

The photochemical oxidation of biogenic compounds yields a wide variety of organic compounds including peroxyacetyl nitrate (PAN), methyl vinyl ketone, methacrolein and 3-methylfuran, and organic aerosols, and it may produce additional O_3 if NO_x is present (Paulson and Seinfeld 1992). The organic products may have important effects on the vitality of vegetation. For example, recent studies indicate that PAN is even more toxic to vegetation than O_3. Sun and Huang (1995) have ob-

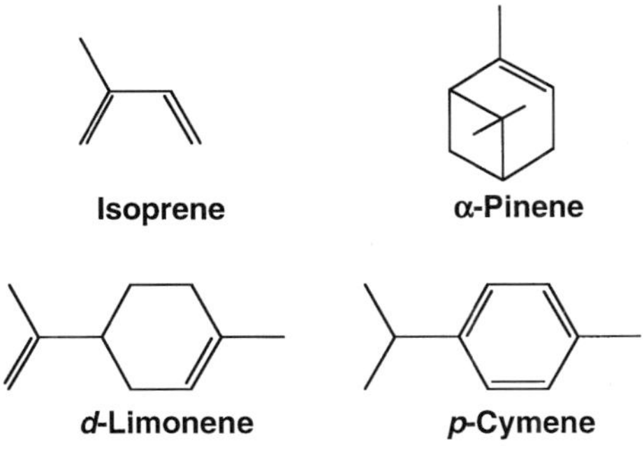

Fig. 1.2. The chemical structure of isoprene and typical biogenic organic compounds. (Finalayson-Pitts and Pitts 1986)

served that PAN exposure can produce toxic effects at concentrations as low as 4–5 nl l^{-1}.

1.3 Urban Ozone Formation and Transport to Mountain Sites

The complex interactions between meteorology, emissions, and photochemistry can be illustrated by examining an example of O_3 formation in an urban region. The highly populated Los Angeles urban region is surrounded by the Pacific Ocean and by several mountain ranges. This geographical location gives the region a set of meteorological conditions which promote the formation of O_3. Sea breezes cause air to flow from the ocean to inland areas during the early morning hours and back to the ocean during the afternoon and evening. The back and forth flow of the air is constrained by the surrounding mountains which allows the air to become highly enriched with pollutants. High levels of NO_X and organic compounds are contained within the large basin where they photochemically react to produce large quantities of O_3. In order to illustrate the formation of O_3 and its transport to the surrounding mountain ranges, a sequence of surface O_3 concentrations are shown for a typical summer day (Fig. 1.3, See Appendix). These O_3 distributions were calculated through the use of the California Institute of Technology (CIT) urban airshed model (McRae et al. 1992a,b,c).

The first plot in the sequence (Fig. 1.3, see Appendix) shows O_3 surface concentrations during the very early morning hours. Throughout the region the O_3 concentrations are extremely depleted due to surface deposition and nitric oxide emission. Later in the morning O_3 concentrations in the center of Los Angeles areas remain low due to nitric oxide emissions which react with O_3 [reaction (1.3)]. Photochemistry has produced O_3 in the more remote areas including the San Bernardino, the San Jacinto and the Santa Ana Mountain areas. In the mountain regions the O_3 concentrations are already near 100 nl l^{-1}.

During the mid-afternoon, the O_3 concentrations in the downtown areas are higher than during the early morning but not especially high. During the afternoon the O_3 concentrations are almost 100 nl l^{-1} in the San Gabriel mountains and the surrounding cites of San Bernardino and Riverside. The highest O_3 concentrations range from 150 to over 200 nl l^{-1} and occur in the San Bernadino mountains. In the late afternoon, the O_3 concentrations fall due to decreasing photochemical production rates, increasing nitric oxide emissions, and deposition. This type of O_3 be-

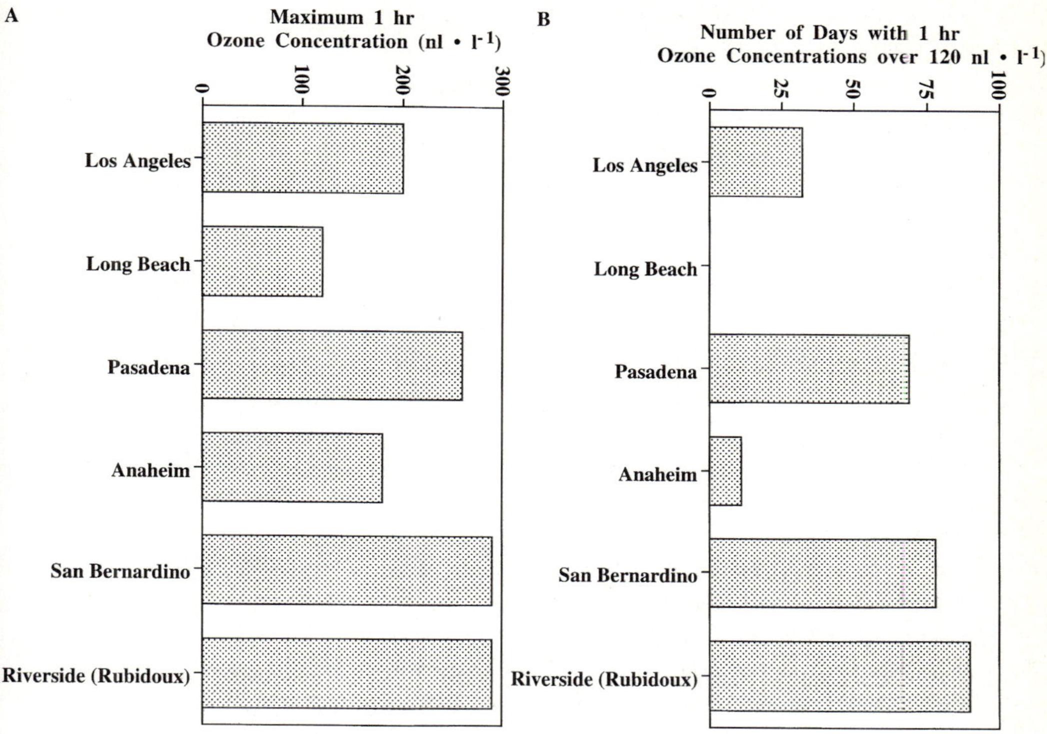

Fig. 1.4. **A** Maximum 1-h O_3 concentrations in the Los Angeles urban airshed. **B** The number of days where the 1-h O_3 concentrations were greater than 120 nl 1^{-1} for the year 1990. (SCAQMD 1990)

havior is not limited to the single episode given here. The maximum O_3 concentrations and the number of days when the O_3 concentrations were greater than 120 nl l^{-1} for the year 1990 (SCAQMD 1990) are shown in Fig. 1.4. The outlying regions of Pasadena, San Bernadino, and Riverside have generally higher O_3 concentrations than the central cities.

1.4 Meteorological Effects on Tropospheric Ozone Concentrations

Processes in the lower troposphere are the predominant cause of O_3 variability near the surface. For individual monitoring stations, long-term variations of tropospheric O_3 concentrations are partly due to site-specific characteristics such as the influence of regional meteorology on strato-sphere transport processes, local photochemistry, or the long-range transport of pollution from other tropospheric regions (Vukovich 1994). For example, regional scale meteorology strongly affects O_3 and precursor concentrations due to its impact on transport processes (Feister and Balzer 1991; Davies et al. 1992). It has been found that changes in regional O_3 concentrations may also be related to long-term temperature changes if higher temperatures reflect a greater frequency of clear weather conditions that are favorable for photochemical ozone production (Zurbenko et al. 1995).

During the daytime, O_3 is generally well mixed within the planetary boundary layer (PBL). The maximum amount of mixing is usually reached in the afternoon when the absorption of solar radiation has caused the Earth's surface temperature to reach a maximum. At night, lower thermal gradients reduce the rate of vertical mixing. During the night, the O_3 in air masses near the ground is titrated by nitric oxide from surface sources. Additional O_3 depletion may result from deposition to vegetation and surfaces. As shown above for the Los Angeles urban region, surface O_3 concentrations exhibit a distinct diurnal pattern with high mid-daytime and low night values. In contrast, mountains or elevated regions in more pristine areas undergo less of a nighttime loss of O_3 and have more flattened diurnal concentration profiles. An overview of the average diurnal variations (April–September, 1990–1992) observed at a large number of European sites was given by Hjellbrekke (1995).

Rapid changes in O_3 concentrations tend to be associated with changes in meteorological conditions on a time scale of synoptic processes (a few days). The concentrations of O_3 and its precursors can differ considerably as a function of altitude, depending on atmospheric conditions. Simultaneous short-term variations in trace gas concentrations and meteorological parameters are often very strong during the passage of meteorological fronts, when efficient vertical transport of pollutants from the boundary layer to the free troposphere is taking place. For example, at mountain sites in southern Germany, the occurrence of maximum O_3 concentrations during the summer was associated with the passage of meteorological fronts (Scheel et al. 1995).

On a local or regional scale, solar radiation has a strong impact on the photochemistry of O_3 production (Volz et al. 1989; Scheel et al. 1992). As an example, Fig. 1.5 shows the annual distribution of daily O_3 maxima classified according to the relative sunshine duration (RSD). The comparison is based on O_3 records from three neighboring Alpine sites located at ground level and nearly 1000 and 2200 m above ground level. Under cloudy sky conditions similar O_3 levels are observed during the year at the mountain stations, however, peak O_3 concentrations in the valley are lower by about 12 nl l^{-1}. In contrast, under conditions of intense solar radiation,

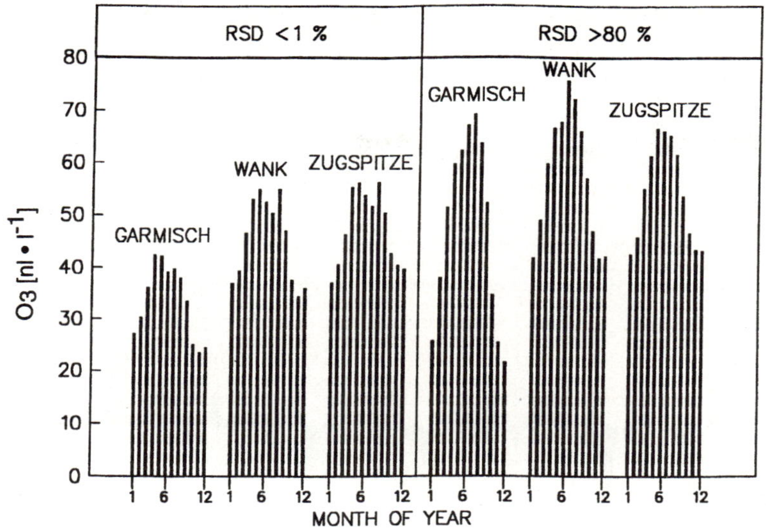

Fig. 1.5. Average annual distributions of daily O_3 maxima (1978-1991) as a function of sunshine, here represented by low and high values of the relative sunshine duration (*RSD*). The sites are located in the German Alps at 740 (*Garmisch*), 1780 (*Wank*) and 2962 (*Zugspitze*) m.a.s.l. (Scheel et al. 1992)

the peak of the distribution for the valley site even surpasses the peak for the site at the highest altitude due to a significant contribution from local photochemical O_3 production. Usually under conditions of intense solar radiation, the highest O_3 levels were observed at the altitude of about 1000 m above the ground. This observation is in agreement with model calculations (McKeen et al. 1989) which showed that optimum conditions for photochemical O_3 production should be encountered near this altitude. If intense photochemical episodes are relatively frequent, this maximum O_3 production near 1000 m above the ground may even be reflected in the monthly mean concentrations. Thus, for selected months during the summer, O_3 maxima in the vertical O_3 distribution were noticed within this altitude range at Alpine sites, such as in the vertical profiles constructed from measurements at locations in Austria (Puxbaum et al. 1991) and in southern Germany (Sladkovic et al. 1994b).

At the neighbouring Alpine sites, Wank and Zugspitze, the influence of both meteorological conditions and pollution was found to be stronger at an altitude of near 1800 m than at about 3000 m. Ozone concentration differences between these two elevations exceeding 20 nl l^{-1} over short time periods, ranging from hours to days, are frequently observed (Fig. 1.6). These differences are primarily caused by reduced vertical mixing, stratospheric intrusions of O_3, and also by loss processes and

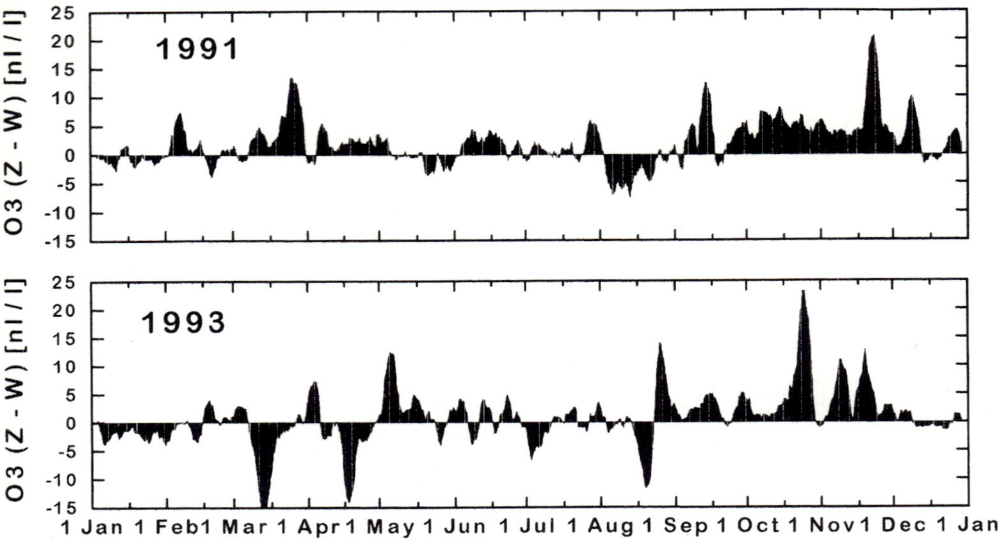

Fig. 1.6. Variations in the difference between the O_3 concentrations at 3000 m (Zugspitzez) and 1800 m (Wankw) during the years 1991 and 1993. The figure shows 5-day moving averages, which account for the typical duration of different weather patterns. The comparison of 2 years indicates significant interannual variability. (Scheel et al. 1995)

photochemical production which are usually greater at lower altitudes. Due to the processes involved, the difference in the O_3 concentrations between the altitudes of 1800 and 3000 m changes its sign depending on meteorological and air chemical conditions, including the stability of the atmosphere and levels of pollution at the respective altitudes. However, when observations were averaged over periods of months or years, there were only minor differences in concentration (Fig. 1.7).

1.5 Ozone Networks and the Analysis of Measurements

Relatively much is known about rural O_3 over the northern and the central part of Europe (Hakola et al. 1991; PORG (1993) and references therein; Beck and Grennfelt 1994; Laurila and Lättilä 1994; Pedersen and Lefohn 1994; Hjellbrekke 1995). A large data base of surface O_3 measurements has been accumulated within the network of the Cooperative ProgrammE for Monitoring and Evaluation of the Long Range Transport of Air Pollutants in Europe (EMEP). The EMEP network consists of about 70 monitoring stations in over 15 countries mainly located in central, eastern, and northern Europe (Hjellbrekke 1995 and references therein).

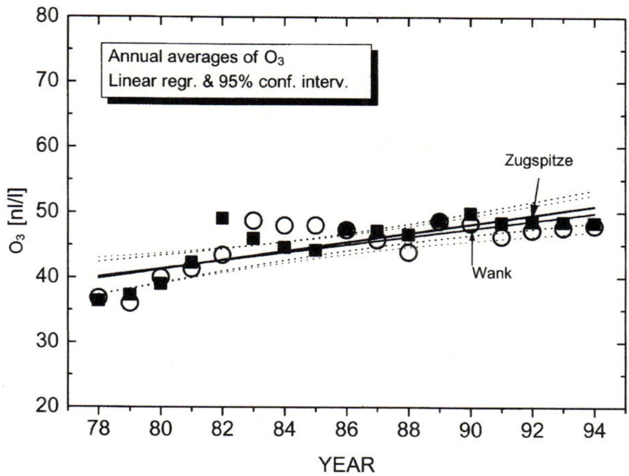

Fig. 1.7. Annually averaged concentrations of O_3 at the Alpine sites *Zugspitze* (2962 m.a.s.l.) and *Wank* (1780 m.a.s.l.) together with linear regression curves and 95% confidence intervals. The average rates of increase are around 0.8 nl l^{-1} $year^{-1}$. (Sladkovic et al. 1994a)

There are limits to the conclusions which may be drawn from the EMEP data because there has been no overall evaluation of the representativeness of the sites and no general intercalibration between them (Hjellbrekke 1995). In contrast, less detailed information is available from the regions south of the Alps, especially from Mediterranean countries, where photochemical production may be the dominant source for tropospheric O_3 (PORG 1993).

For North America, a network of surface measurement sites across the United States provides information on the seasonal, synoptic, and diurnal patterns of O_3. The United States Environmental Protection Agency's (EPA) principal statistical measure of O_3 concentrations is calculated from O_3 measurements made at 431 monitoring sites (EPA 1991a). The EPA commonly uses two kinds of measures to monitor year-to-year trends: concentration indicators and threshold indicators. An example of a concentration indicator is the measurement of the second highest 1-h maximum O_3 concentration per year. An example of a threshold indicator is the American "National Ambient Air Quality Standard" (NAAQS). The NAAQS is defined as 120 nl l^{-1} for the daily maximum 1-h average, and this limit should not be exceeded for more than 3 days in 3 years.

An awareness of the importance of the representativeness of observations has grown in recent years. Ozone concentrations observed within the boundary layer can be biased by a significant degree of local influence because the observed concentrations are the net result of a variety of different meteorological and chemical processes at an individual site. In particular the data sets from ground-level sites should be screened to remove local effects, such as NO titration or nighttime deposition. Local effects due to both meteorological and chemical processes are reflected by the amplitude and shape of the diurnal cycle of O_3 concentrations, for example. Frequency distributions of O_3 and O_X ($O_X = O_3 + NO_2$) yield valuable information on the representativeness of a measurement site. O_X is a conservative quantity over short time periods due to the rapid interconversion of O_3, NO, and NO_2. A comparison of the distributions of O_3 and O_X permits the estimation of the amount of O_3 removed due to NO titration (Kley et al. 1994). A large difference between the distributions of both species would be an indicator of local NO_X emissions.

Comparisons between different sites may yield misleading results unless there is detailed information available about each individual site including the statistical characteristics of the data. Only data which are representative of a well-mixed atmosphere should be used to meaningfully compare O_3 concentrations on a continental scale and to compare results derived from them, such as trend estimates. The representativeness of a site increases with altitude, which is reflected by decreasing amplitudes of the diurnal variations. However, under specific conditions, such as high wind speed, O_3 concentrations at ground level may be representative of the well-mixed boundary layer (e.g., Esser 1993).

Toupance and Aranda (1993) performed an analysis of the annual O_3 concentration profiles (seasonal variations) for a number of different sites on the European continent. They concluded that none of the O_3 measurements at these continental sites were fully representative of the background atmosphere. Sometimes even high mountain sites may be affected by air transported from the polluted boundary layer into the free troposphere. According to these authors ozone concentrations at the Zugspitze were affected by regional pollution about 30% of the time. An impression of the seasonal variations at different European sites can be obtained from the 3-year time series (1990–1992) given by Hjellbrekke (1995).

On a long-term timescale, the overall O_3 concentrations, as represented by monthly averages, are largely determined by the frequency of different meteorological situations. As a consequence, the long-term O_3 record reflects much of the year-to-year variability of the prevailing weather patterns. If the factors affecting long-term O_3 concentrations are to be determined, the data for analysis must be selected according to specific meteorological and/or air chemical conditions. A selection of "unpolluted" and "polluted" conditions by means of a known clean air sector (a selection made according to wind direction or according to the motion of air masses estimated through the calculation of backward trajectories) is usually appropriate for sites near the coast or in northern Europe (Hakola et al. 1991; Esser 1993).

However, in general, clean air sectors cannot be defined for locations inside the continent, for example in the central to southeastern part of Europe. Techniques making use of simultaneous measurements of trace species that are indicative of pollution, such as nitrogen oxides, SO_2, and CO, have been successfully employed for the classification of atmospheric conditions (Derwent et al. 1994; Laurila and Lättilä 1994). When ozone measurements from the Zugspitze and the Wank site were sorted into "polluted" and "unpolluted" categories, the result was a significantly different shape of the annual distribution (Sladkovic et al. 1994a). For measurements performed in clean air, the shape of the seasonal O_3 variations reflected hemispheric background conditions with a maximum in spring.

1.6 The Global Distribution of Ozone Concentrations and the Seasonal Cycle

On an annual basis, there is more O_3 in the northern hemisphere troposphere than in the southern hemisphere (Oltmans et al. 1989; Oltmans and Levy 1994). Approximately 80% of the NO_X emitted on a global basis is due to anthropogenic activity in the northern hemisphere (Müller and Brasseur 1995). The most important anthropogenic sources of nitrogen oxides include transportation, domestic heating, and electric power production, which are related to fossil fuel combustion (Warneck 1988). However, even without tropospheric production, O_3 concentrations are expected to be greater in the northern hemisphere because measurements suggest that the stratospheric source is stronger in the northern hemisphere (Oltmans et al. 1989). Model calculations excluding tropospheric O_3 production predict greater amounts of O_3 in the northern hemisphere originating from the stratospheric source (Levy et al. 1985).

Seasonal cycling is a prominent feature of the tropospheric O_3 distribution. Data obtained through the use of the Schönbein method show that O_3 concentrations reached a maximum during the spring at the end of the last century. The seasonality of surface O_3 was first reported by Junge (1962) from his interpretation of O_3 concentration data from Arosa (Switzerland). More recent analysis shows that for both the northern and southern Hemisphere surface, O_3 annual distributions exhibit maxima between April and July (Logan 1985; Scheel et al. 1992; Oltmans and Levy 1994). For the northern hemisphere, the maximum springtime O_3 can be at least partly interpreted in terms of the seasonality of stratosphere/troposphere exchange. Radioactive tracer observations have shown that the maximum stratospheric O_3 intrusion rate during spring is about three times the minimum rate during the autumn (Danielsen 1968; Reiter 1975).

Presently, the contribution of photochemical O_3 production in the free troposphere is assumed to increase in April due to the accumulation of O_3 precursors during the winter. About 15 nl l^{-1} of O_3 was estimated to be associated with springtime tropospheric photochemical O_3 production (Penkett and Brice 1986). However, over the continents of Europe and north America the major photochemical O_3 production takes place during the summer (Logan 1995). This summertime continental production causes a shift of the annual maximum towards June/August for sites that are not completely remote. When annual distributions from different European sites were compared, both a time shift of the maximum towards the summer and an increase in amplitude in the northwest to southeast direction were observed (Toupance and Aranda 1993). This is generally consistent with the spatial distribution of precursor (Volatile Organic Compounds, i.e., VOCs and NO_X) emissions.

Annual profiles of surface O_3 concentrations in the southern hemisphere likewise show a maximum in the middle of the year; however, this behavior is due to different processes. The maximum in the austral winter is mainly attributed to increased stratosphere/troposphere exchanges (Liu et al. 1980), while the minimum can be interpreted in terms of photochemical O_3 destruction in a low-NO_X environment (Fishman et al. 1979). This process peaks in the austral summer, thus leading to the O_3 minima observed around January.

1.7 Long-Term Trends in Tropospheric Ozone Concentrations

Near the end of the last century, O_3 measurements were made through the use of the Schönbein method at several locations in Europe. A careful reevaluation of these measurements has yielded insight into the O_3 concentrations prevailing in the boundary layer and the free troposphere over Europe a hundred years ago (Volz and Kley 1988; Anfossi et al. 1991; Marenco et al. 1994). The results indicate that boundary layer O_3 concentrations usually ranged between 5 and 15 nl l^{-1} at that time. The comparison of recent observations with those made at Montsouris between 1876 and 1910 (Volz and Kley 1988) suggests that the surface concentration of O_3 at mid- to high latitudes has more than doubled during the past 100 years.

Enhanced photochemical O_3 production in the lower troposphere appears to be responsible for the increase in O_3 observed during the last decades in the northern hemisphere (Logan 1985). This increase is believed to be mainly due to increased anthropogenic emissions of precursor substances in the industrialized regions of the northern hemisphere (Fishman et al. 1979; Logan 1985). Model calculations show that observed rate of increase in O_3 concentration of between 1 and 2% per year during the last two decades are consistent with precursor emissions (Isaksen and Hov 1987; Hough and Derwent 1990).

Figure 1.8 compares recent O_3 concentrations for September at different altitudes with measurements performed about six decades ago (Staehelin et al. 1994). The figure shows that there has been an overall increase at all altitudes between the surface and about 3000 m. The profile is a fairly good representation of an average vertical O_3 distribution although it was constructed with results from different locations. As the data points show, O_3 levels are now greater than 40 nl l^{-1} during the late summer and are typically around 50–60 nl l^{-1} in the lower free troposphere.

Ozone measurements made at several European mountain sites during different time periods show that O_3 concentrations in the lower free troposphere have increased (Fig. 1.9). Although the increase accelerated from about 1930, most of the increase occurred between the 1950s and 1970s (Staehelin and Schmid 1991; Marenco et al. 1994). More intensive measurements were performed at several locations during recent decades. In situ measurements at several mountain stations have established the existence of a long-term trend in O_3 concentrations in the lower troposphere (altitudes 1.5–3 km) over large parts of the northern hemisphere (e.g. Kley et al. 1994). All of the European measurements have shown a statistically significant increase in O_3 concentrations in the free troposphere. An annual increase of 1 to 1.5% was reported for Payerne, Switzerland, based on the period between 1969 and 1988 (Staehelin and Schmid 1991). An average increase of 2.1% per year was determined from O_3 sonde measurements at Hohenpeissenberg in southern Germany between 1967 and 1986 (Wege et al. 1989). Since the middle of the 1980s the Hohenpeissenberg record indicates that there has been a decrease in the rate of increase in O_3 concentration in the free troposphere (Logan 1994), and there is very recent evidence that O_3 concentrations are decreasing (Claude 1995).

The timing of the greatest changes in O_3 concentrations in the free troposphere differs between some of the stations. A large increase was observed at Payerne between 1982 and 1988 (3–4% annual increase above

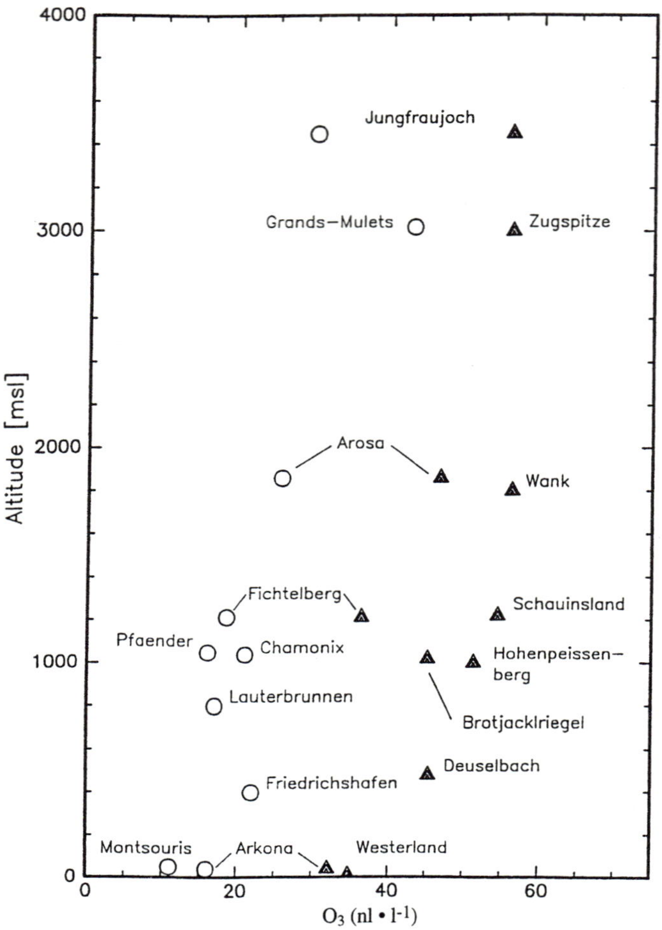

Fig. 1.8. Surface O_3 concentrations dating from the 1930s (*circles*) and recent O_3 levels from measurements during August to September (1988–1991; *triangles*) in Europe as a function of altitude. Note that the measurements are from different sites and that the historic data are based on measurements performed only on individual days of the respective months. (Staehelin et al. 1994)

2 km; Staehelin and Schmid 1991), while, for example, at the mountain sites Zugspitze and Wank (German Alps) the rate of O_3 increase was highest prior to 1982 (Fig. 1.7). In agreement with the observations at Hohenpeissenberg, the growth of O_3 concentrations at the Zugspitze has been close to zero in recent years (Sladkovic et al. 1994a). However, the entire time series for both the Wank and Zugspitze still indicate a statistically significant increase between 1978 and 1994, at an average rate of about 0.8 nl l^{-1} year^{-1}. The increase has led to annual averaged O_3 concentrations of around 48 nl l^{-1} at altitudes of 1800 and 3000 m above sea

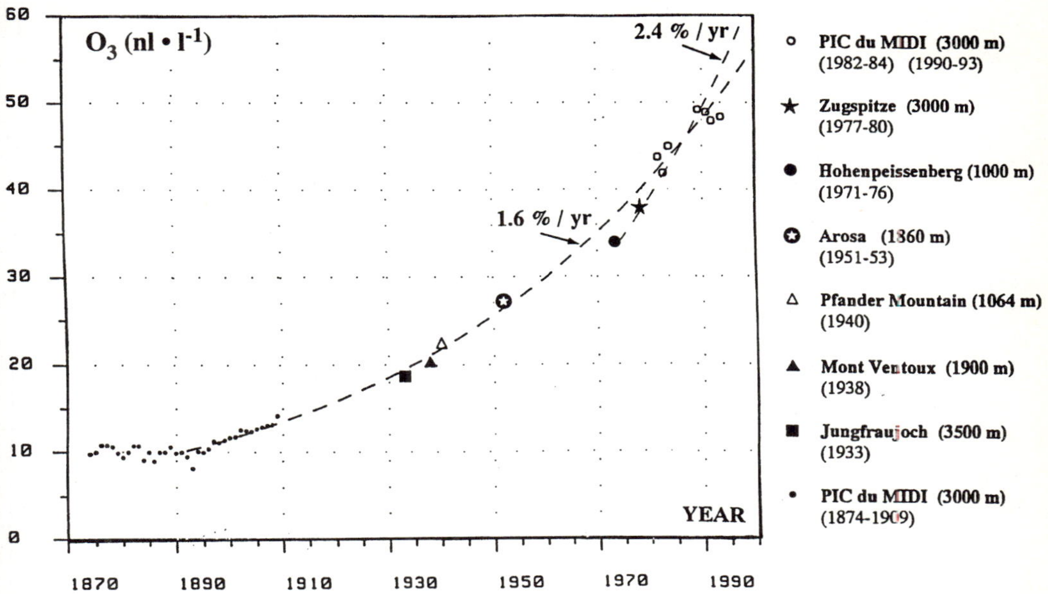

Fig. 1.9. Estimate of O_3 evolution in the free atmosphere over western Europe, from measurements at the Pic du Midi (southern France) and various European stations at high altitude. (Marenco et al. 1994)

level. In contrast to the free troposphere, recent trend estimates for individual sites in the boundary layer over Europe yield quite variable increase rates. There are even some cases where trends of opposite sign were observed at neighbouring sites (Low et al. 1992). At present, there is no known explanation to reconcile the differences between the free troposphere and boundary layer observations.

The Whiteface Mountain site, located in the Adirondack State Park of northern New York, provides one of the longest series of O_3 observations in the United States (Demerjian et al. 1992). These O_3 measurements show a statistically significant upward trend of 0.6% per year over the entire 20-year monitoring period of 1973–1993 (Fig. 1.10). A trend can be observed in these data only because of the long time span which is a full two decades. In contrast, the majority of sites with measurement histories of less than 10 years show no positive or negative trend according to the Mann-Kendall test (NAPAP 1989).

In general, ozone concentrations over the United States increased by 1% between 1979 and 1988 (Fig. 1.11; EPA 1991a). Ozone concentrations in 1989 were the lowest of the decade as a result of meteorological variations, thus reversing the positive trend established for the preceding decade. This demonstrates that American O_3 concentrations, like Eu-

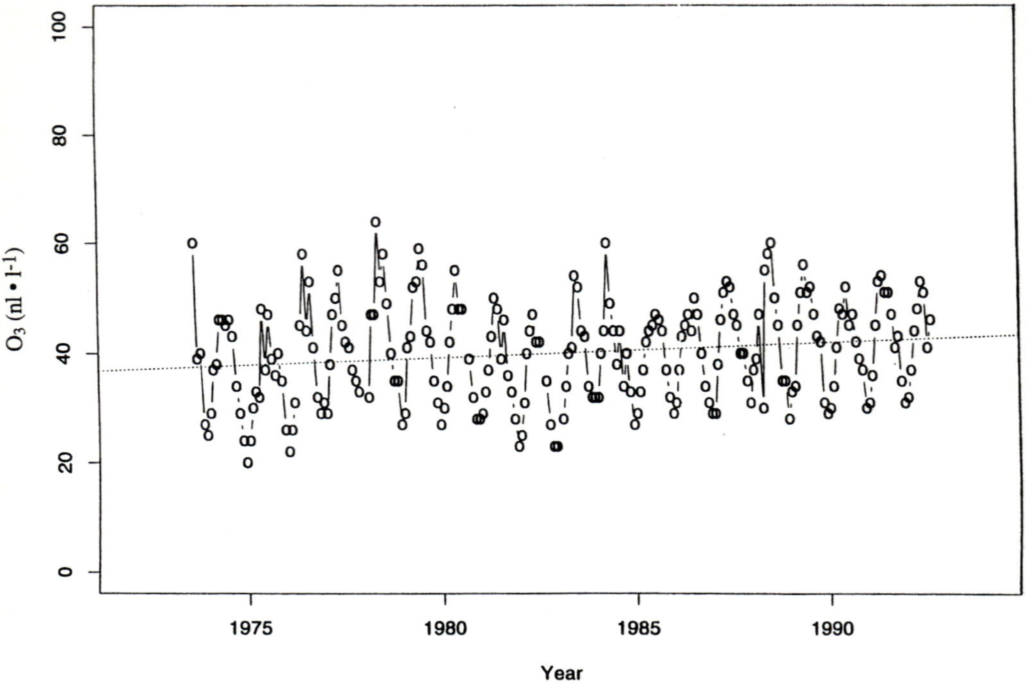

Fig. 1.10. Whiteface Mountain (New York State) monthly O_3 means. (Demerjian et al. 1992)

ropean concentrations, are highly dependent upon meteorological conditions. This variability, which is due to fluctuations in weather patterns, is not removed from the data by the EPA's principal statistical measures.

Zurbenko et al. (1995) have analyzed O_3 concentrations for much of the eastern United States. It is difficult to determine long-term trends in the spatial distribution because the EPA requires the states to monitor O_3 only during the "ozone season". The definition of the ozone season varies from place to place but it includes the spring and summer months. Zurbenko et al. (1995) show that surface level O_3 concentrations increased over the central eastern United States, especially over Ohio and Michigan, during the time period between 1980 and 1992. During this same period the O_3 concentrations mostly decreased over the east coast.

However, the apparent increases in O_3 over the eastern United States may be misleading because the summers of 1983 and 1988 were unusually hot. The temperature trends for this time period show that much of the temperature increase also occurred over central eastern United States. Because temperature is used an indication of those meteorological conditions which favor photochemical O_3 production (Zurbenko et al. 1995)

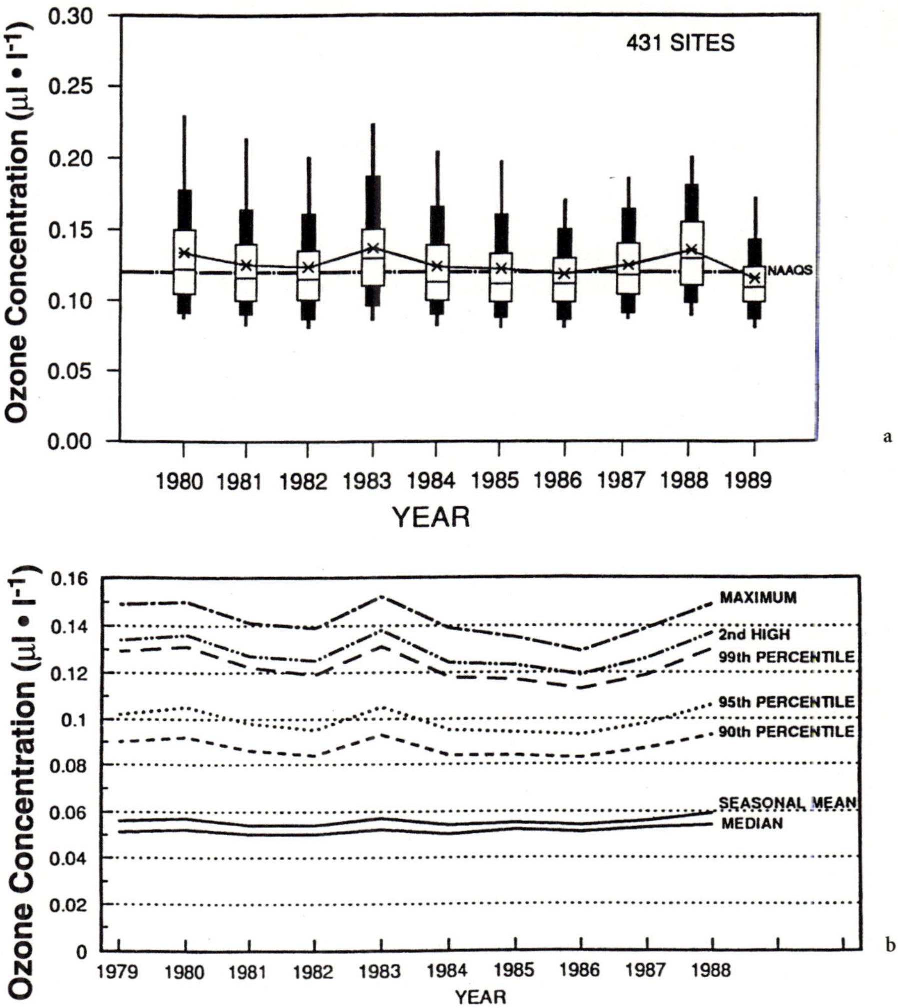

Fig. 1.11. a Box plot comparisons of trends in annual second highest daily maximum 1-h O_3 concentration of 431 US monitoring sites, from 1980 to 1989. The trends in ambient air quality, presented as *box plots*, display the 5th, 10th, 25th, 50th (median), 75th, 90th and 95th percentiles of the data; the *curve* indicates the composite average. For example, 90% of the sites would have concentrations equal to or lower than the 90th percentile. (EPA 1991a) **b** 10-year trends in various O_3 summary statistics. Ozone concentrations represent composite US averages for each summary statistic. (Chock 1988)

Rao and Zurbenko (1994, 1995) have developed methods for determining the concentration of "temperature-independent ozone". When the temperature-independent O_3 over the United States is considered, a very different picture emerges. Over the eastern United States there is either little change or a slight decrease in O_3 concentrations, except at a few mountain sites over the time period. Since 1989, regulations have required the use low volatility gasoline. Therefore, Zurbenko et al. (1995) examined the temperature-independent O_3 for the periods between 1980 and 1988, and 1988 and 1992 separately. They found that the O_3 concentrations increased between 1980 and 1988, especially along the central east coast and in western New York and eastern Pennsylvania. After 1988 they found a decrease in the temperature-independent O_3 concentrations.

For the southern hemisphere long-term O_3 records are available for only a few sites (Oltmans 1981; Scheel et al. 1992; Oltmans and Levy 1994). A significant increase in O_3 has not been observed at any of these sites (Schnell et al. 1991; Scheel et al. 1992), and it appears that in large parts of the southern hemisphere O_3 concentrations remain close to the historical natural background. Average O_3 concentrations over the ocean are only about 15–21 nl l^{-1} in the southern hemisphere (Winkler 1989; Scheel et al. 1992). However, extremely high O_3 concentrations have been observed over the tropical and subtropical continents during dry seasons. The high O_3 concentrations, which are observed from satellite data, are associated with biomass burning which causes substantial emission of O_3 precursor substances (Fishman et al. 1991; Levine 1991).

1.8 Present Ozone Concentrations and Ozone Exposure over Europe and North America

Ozone concentrations in the lower troposphere over Europe show a strong spatial dependence that is related to both the distribution of O_3 percursor emissions and the prevailing meteorological conditions (Davies et al. 1992; Beck and Grennfelt 1994). Episodes of high concentrations of O_3 were observed over most parts of Europe in summer (Cox et al. 1975; Guicherit and von Dop 1977; Schjoldager et al. 1981; Grennfelt and Schjoldager 1984). Ozone was found to reach levels above ambient air quality standards over large regions, especially during periods with anticyclonic conditions and stagnant air. During recent years the average daytime boundary layer O_3 concentrations in winter have been about 32 nl l^{-1} at the western border of the continent. They decrease to about 25 nl l^{-1} in the Scandinavian region and 21 nl l^{-1} in the region around the Alps. In summer, the respective average O_3 concentrations range from 35–40 nl l^{-1} near the Atlantic to 40 nl l^{-1} in Scandinavia and 50–55 nl l^{-1} in southern Germany and Austria (Beck and Grennfelt 1994; Scheel 1994). Over the North Atlantic, the average O_3 concentration is about 30 nl l^{-1} within the latitude belt 25–65 N (Winkler 1989).

Over Europe the relative contribution of photochemically produced O_3 to the observed annual distribution increases from west to east. In general, peak O_3 concentrations and the frequency of summertime O_3 episodes are higher in central than in northern Europe. Through an assessment of O_3 production and destruction rates over northwestern Europe, Beck and Grennfelt (1994) have estimated that the summertime mean internal production of O_3 is typically 10–15 nl l^{-1} in some regions. Their analysis also pointed to a net wintertime O_3 destruction which causes O_3 concentrations to be the lowest over the southeastern part of central Europe (Austria, Hungary) due to the prevailing gradient in the level of pollution from north-west to south-east. On an annual average, the geographic distribution of O_3 in the boundary layer does not seem to indicate an overall net O_3 production. In fact, the annual average suggests that the boundary layer might act as a net sink of O_3 in some regions of Europe and as a net source in others (Derwent et al. 1994; Scheel 1994).

The spatial distribution of exposure as indicated by AOT40 indices over Europe is shown in Fig. 1.12 (see Appendix). This figure shows a color map of AOT40 values determined from measurements at EMEP sites in central Europe and Scandinavia under daylight conditions during the period May–July, 1992 (Hjellbrekke 1995). The color scale indicates an AOT40 range of 2600 to 14400 nl l^{-1} h. The map shows that there is a pronounced gradient of O_3 exposure over Europe with lowest values in Scandinavia and the western part of the UK and over Ireland and highest values in the southern part of central Europe, including the region of the Alps. The indices, calculated on a 24-h basis, ranged roughly between 4000 and 28 000 nl l^{-1} h during the growing season (April–September, 1992) (Hjellbrekke 1995). Results for an earlier time interval (April–September, 1989) were given, for example, by Beck and Grennfelt (1994), where the exceedances of O_3 above a threshold of 30, 40 and 50 nl l^{-1} are listed. The authors state that the United Nations Economic Commission for Europe 1-h guide value of 75 nl l^{-1} was exceeded regularly at nearly all sites of the network.

The exposure indices increase with altitude because the O_3 concentrations in the boundary layer generally increase with altitude. Figure 1.13 shows average annual AOT40 indices calculated on a daytime basis for 1987–1993 time period for four neighbouring sites of different altitude in the German Alps, including the stations Wank and Zugspitze. For the ground-level site Garmisch, located in a valley, the AOT40 value is about 20 000 nl l^{-1} h, while annual AOT40 values are as high as about 45 000 and 50 000 nl l^{-1} h at the Wank and Zugspitze, respectively. As seen in Fig. 1.7, the average annual O_3 concentrations at these mountain sites were around 48 nl l^{-1} for recent years. In terms of SUMO, this corre-

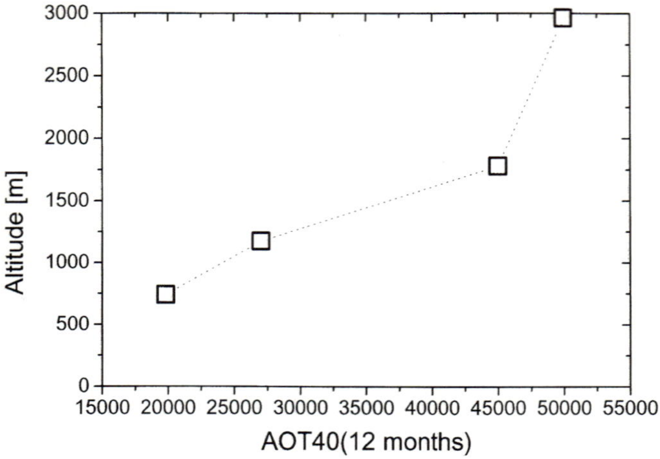

Fig. 1.13. Vertical distribution of average annual AOT40 indices (daytime, 1987–1993) determined from measurements at the Alpine sites Garmisch (740 m a.s.l.), Wank-Mittelstation (1175 m a.s.l.), Wank (summit, 1780 m a.s.l.), and Zugspitze (2962 m a.s.l.) in southern Germany

sponds to 420.5 $\mu l\ l^{-1}$ h. Based on an annual average O_3 concentration of 23.8 $nl\ l^{-1}$ (1989–1993) at ground level (Garmisch), the SUMO value amounts to 208.9 $\mu l\ l^{-1}$ h.

Large areas of forested land throughout the United States are exposed to O_3 levels above the natural background level (EPA 1991a; McKee 1994). Ambient air over vast portions of eastern North America maintains an average O_3 concentration of 50–70 $nl\ l^{-1}$ on clear summer days with frequent peak concentrations of 80–110 $nl\ l^{-1}$. At remote Canadian sites (Hawkeya Lake and Kejimkujik) and rural locations (Simcoe) readings show low mean concentrations of 27–29 $nl\ l^{-1}$ (Table 1.1) with values for the 95th percentile in the 45–63 $nl\ l^{-1}$ range associated with isolated long-range transport events, noticeable during late spring and summer (Fuentes and Dann 1994).

Altshuller and Lefohn (1996) have estimated from observations made from April through October that the current O_3 background for coastal United States locations is 30 ± 5 $nl\ l^{-1}$ (7-h averages, 9:00 to 15:59) and that for inland sites, the O_3 background is 35 ± 10 $nl\ l^{-1}$. Over western United States the observed maximum hourly ozone concentrations ranged from 50 to 98 $nl\ l^{-1}$ and over coastal locations the maximum hourly ozone concentrations ranged from 44 to 80 $nl\ l^{-1}$. However, over many parts of the United States the O_3 concentrations often exceed the NAAQS of 120 $nl\ l^{-1}$ (NAS 1991) in spite of an expensive control effort which began in 1970 through the passage of the Federal Clean Air Act. Over 100 American urban areas failed to meet the ambient air quality standard for O_3 in the late 1980s. Trainer et al. (1995) discuss evidence for O_3 production in urban plumes which may strongly affect much of the rural United States. Much of their recent studies were made as part of the "Southern Oxidant Study" (SOS) and the "Rural Oxidants in the Southern Environment" (ROSE) programs for the study of air pollution in south eastern United States. They find that there is enough NO_X emission to support a rate of photochemical O_3 production of 16 nl l^{-1} day^{-1} in the boundary layer over the entire state of Alabama.

1.9 Dry Deposition of Ozone

The loss of any atmospheric pollutant to a surface is called dry deposition. When exposed to low levels of air pollution, vegetation and soil function solely as sinks or receptors of O_3. Over forest canopies, micrometeorology and chemistry affect O_3 dry deposition, making the measurement of O_3 fluxes extremely difficult. Dry deposition rates generally depend upon the micrometeorological transfer properties of the atmospheric surface layer, the capability of vegetation and soil to take up matter, and the prevailing chemical and/or photochemical environment (Kramm 1989).

There are only two measuring techniques to directly measure dry deposition fluxes of O_3: (1) eddy correlation techniques of micrometeorology, and (2) enclosure techniques which can involve either static or dynamic chamber systems. Eddy correlation measurement techniques have the advantage of producing data representative of several hectares (Delany et al. 1986). However, fast-response physico-chemical sensors are required, which have been developed for only a few trace gases, such as water vapour, CO_2, and O_3 (Lenschow and Hicks 1989). In contrast, measurements of O_3 dry deposition by relatively simple enclosure techniques provide flux data which are representative for areas no larger than about a square meter and these techniques may alter the local environment (Baldocchi et al. 1988). Data representative of larger areas require the use of an array of chambers.

Measurements of O_3 concentrations and O_3 fluxes (F_c) together with the mean values of wind speed (U), temperature (Θ), relative humidity (RH), and global radiation (R_g; Table 1.3) were, for example, obtained over an east Tennessee oak-hickory forest (Hutchinson et al. 1986). The measurements were made by eddy correlation techniques in August 1986 during a period of prolonged drought (Meyers and Baldocchi 1988). The instruments were located at a height of 30 m which was 7 m above the

Table 1.3. Diurnal cycles of ozone flux (F_c) and deposition velocity (v_d), together with wind speed (U), air temperature Θ, relative humidity (RH) and global radiation (R_g) over a deciduous (oak-hickory) forest in August 1986. All data represent 30-min averages (with reference to Meyers and Baldocchi 1988). Note that here the upward ozone flux is taken to be positive. (Hutchinson et al. 1986)

Julian day	Time	U ($m\ s^{-1}$)	θ (°C)	RH (%)	R_g ($W\ m^{-2}$)	$[O_3]$ ($nl\ l^{-1}$)	F_c ($nl\ l^{-1}\ m^{-1}\ s^{-1}$)	v_d ($cm\ s^{-1}$)
234	12:30	1.4	27.0	74	519	51.7	−0.19	0.37
	13:00	1.7	27.2	72	431	51.8	−0.25	0.48
	13:30	1.5	28.2	67	737	57.3	−0.24	0.42
	14:00	1.5	28.8	65	866	59.3	−0.26	0.43
	14:30	2.1	28.8	62	747	62.5	−0.19	0.30
	16:00	2.0	28.9	61	577	60.4	−0.26	0.43
	16:30	1.8	29.1	61	467	62.7	−0.24	0.38
	17:00	2.1	29.1	61	353	63.0	−0.12	0.19
	17:30	2.2	28.9	61	257	60.2	−0.16	0.27
235	09:30	2.0	24.3	84	525	24.6	−0.20	0.83
	10:00	1.8	25.1	81	621	31.1	−0.15	0.47
	10:30	2.0	25.8	77	697	34.9	−0.24	0.69
	13:30	2.6	27.7	71	457	35.6	−0.11	0.31
	14:00	2.6	28.0	70	402	32.2	−0.08	0.26
	15:00	2.8	29.0	64	571	32.2	−0.10	0.30
	15:30	2.6	28.9	63	280	33.0	−0.14	0.42

mean canopy height. The O_3 deposition flux ranged from –0.26 to –0.08 nl $l^{-1}\ m^{-1}s^{-1}$ with a mean value of -0.18 ± 0.06 nl $l^{-1}m^{-1}s^{-1}$. These values lead to estimates of the O_3 deposition velocity, v_d, ranging from 0.19 cm s^{-1} to 0.83 cm s^{-1} with a mean value of 0.41 ± 0.16 cm s^{-1}. The direct O_3 flux measurements are difficult to explain using only micrometeorological theory. The best O_3 flux estimates are provided by a detailed multilayer canopy model that includes a water stress correction (Meyers and Paw 1987). However, large differences between the calculated values and measurements remain. The differences may well be due to chemical reactions and advection both of which can affect O_3 flux profiles.

Measurements performed over a Bavarian forest in June 1987 further illustrate the complexity involved in estimating the dry deposition of O_3 in forest canopies. The measurements were made at the Schachtenau site in the southern part of the "Grosse Ohe" watershed (Gietl and Rall 1986; Enders et al. 1989, 1992). The Schachtenau site is at an altitude of 807 m and is relatively level with uniform terrain in all directions up to a distance of 500 m (1500 m in the westerly main wind direction). The site is completely forested with 86% spruce [*Picea abies* (L.) Karst.] and 16% beech. The spruce trees were between 80 and 95 years old with an average height of 28 m and needles down to 16 m; the density of the stand was 712 trees ha^{-1}. Because the site is in a national park which stretched over an area of 170 km^2 there were no local sources of SO_2 and NO_x. Measurements were made at a 51-m-tall tower constructed at the center of the site. The measurements included wind direction, air pressure, precipitation, and soil heat flux; vertical profiles of wind speed, dry and wet bulb temperature (11 levels), concentrations of O_3, NO_x, and SO_2 (3 levels), short-wave and long-wave radiation (2 levels), and the eddy fluxes of momentum, sensible and latent heat, and O_3, NO_2, and SO_2 (Enders et al. 1989, 1992).

The daily patterns of O_3 at three levels above the canopy (51, 41, 31 m) for the period 27–30 June 1987 are shown in Fig. 1.14. The meteorological conditions consist of wind direction, air temperature, and global radiation. The O_3 concentrations at all three levels exhibit pronounced diurnal cycles with a maximum in the late afternoon and a minimum during the night. The O_3 maximum occurred slightly after the maximum of global radiation. Occasionally, the O_3 minimum was broken up by a short-term increase in concentration. This may have been due to the vertical transfer of O_3 when the short-term reenforcement of turbulence disturbed the nocturnal boundary layer. Such disturbances may be initiated by strong vertical wind shear. The station could have been affected by nocturnal drainage of cold air from the ridges to the north, especially during the high pressure situation prevailing at that time. During this period the wind speed at the 51-m height was generally low with a maximum of only 2.6 m s^{-1} and a mean value of 1.6 m s^{-1}.

When the air approached the canopy during the night the O_3 concentration was observed to decrease. But after sunrise the situation completely changed; "inverse" O_3 vertical gradients (i.e., concentrations decreasing with altitude) were observed (Enders et al. 1989). During the first two days this process was negligible, but it became more important with increasing solar radiation until on June 30, the day with maximum insolation, it resulted in differences of 5 nl l^{-1} 20 m^{-1} (30-min average). This "irregular" behavior of O_3 above the canopy layer was observed on many occasions. Based on the measurements taken from June to September 1987, "inverse" O_3 gradients above the canopy were observed during 80% of all days. The duration of such events ranged from only a few minutes to more than 10 h. Their occurrence was not limited to periods of high solar radiation although the events were more pronounced under these conditions.

However, days with high insolation do not necessarily cause inverse gradients. In spite of undisturbed solar radiation throughout the day, Enders et al. (1989) found days where only the "normal" increase in O_3 concentration with increasing height was observed. In a few cases inverse

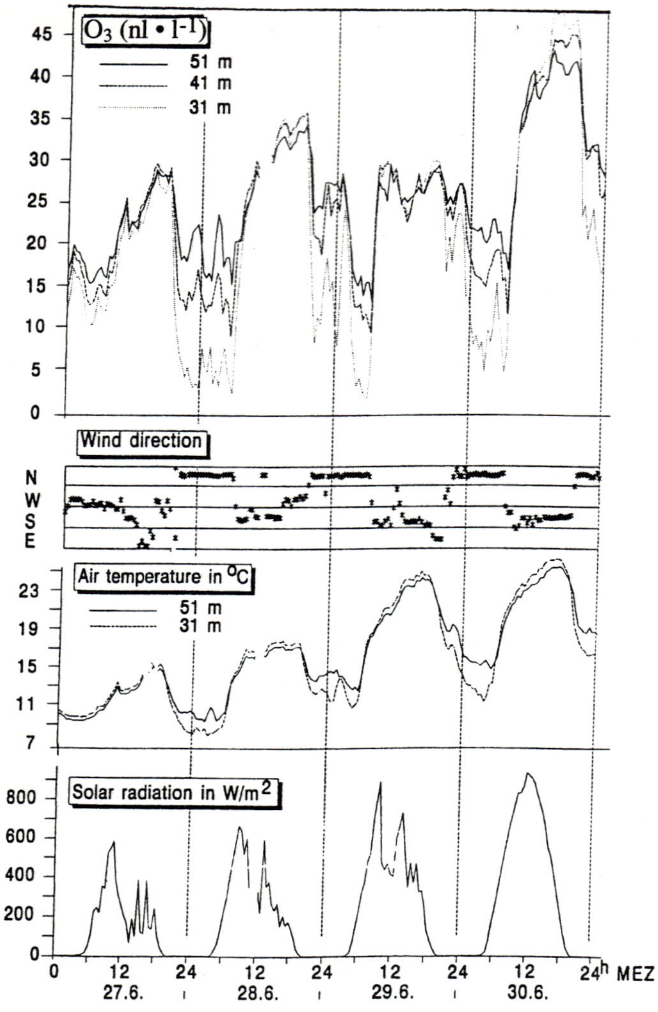

Fig. 1.14. Daily cycles of O_3, air temperature, wind direction, and solar radiation over the canopy at "Schachtenau" (southern Germany), June 27-June 30, 1987. All data represent 30-min averages. (Enders et al. 1989)

gradients were observed even at night! Increased destruction of O_3 far from the canopy or enhanced O_3 production either close to or within the canopy are two possible explanations.

As described above, ozone is produced through the photochemical conversion of NO to NO_2 by the reaction of NO with peroxy radicals, HO_2 and RO_2. These peroxy radical reactions shift the photochemical equilibrium towards increased O_3 concentrations. The concentrations of peroxy radicals can be increased through the reactions of biogenic compounds with HO• or O_3. This makes it possible for enhanced biogenic emissions to lead to a stronger increase in O_3 close to (or within) the canopy, thus causing the "irregular" O_3 behavior. The NO_x measurements showed no significant concentration differences with height. Unfortunately, it was not possible to measure NO and NO_2 separately during this study. Therefore, it is not yet possible to conclude that there is a causative relationship between the behavior of O_3 and biogenic compounds. In contrast, normal, downward O_3 fluxes were observed by eddy correlation techniques at the same site during a measuring period in September 1989 with only a few exceptions (Enders et al. 1992).

The deposition of ozone depends upon complex micrometeorological transfer processes such as wind direction, air temperature, and global radiation. Chemical processes above and within the forest canopy may either increase or decrease O_3 fluxes to the plants. As a result of the interaction between micrometeorological and chemical factors, "regular" and "inverse" vertical ozone gradients are observed. Because the dose of pollutants to plant surfaces depends upon the ozone flux, these results illustrate some of the problems in using ambient concentrations or exposure indices as dose surrogates. Further evaluation of the effect of ozone and other air pollutants on vegetation under field conditions requires investigations of micrometeorological transfer processes and the chemical processes above and within the forest canopy.

1.10 Conclusions

Tropospheric O_3 concentrations are affected through the influence of regional meteorology on stratosphere transport processes, local photochemistry, and the long-range transport of pollution from source regions to more remote regions. Although the stratospheric source is very important, recent studies show that nowadays over 50% of the O_3 observed in the boundary layer of the northern hemisphere is due to photochemical production. There is more O_3 in the northern hemisphere troposphere than in the southern hemisphere on an annual basis due the higher emission rate of O_3 precursors resulting from the high level of anthropogenic activity.

Analyses of historic O_3 measurements strongly suggests that the surface concentration of O_3 at mid to high latitudes has more than doubled during the past 100 years. More recent measurements have established that O_3 concentrations have increased in both the free troposphere and boundary layer over large parts of the northern hemisphere. The increases in O_3 concentration appear to be due to increases in the emissions of O_3 precursors, nitrogen oxides, and organic compounds. The observed rate of increase in O_3 concentration during the last two decades, between 1 and 2% per year, is consistent with precursor emission rates according to atmospheric chemistry models. Short-term rapid changes in O_3 concentrations tend to be associated with changes in meteorological conditions, especially the passage of meteorological fronts.

Indices such as SUMO and AOT40 are used to express O_3 exposure, but plants respond directly to O_3 fluxes rather than to ambient O_3 concentrations. Therefore, the complex micrometeorological transfer processes and the chemical processes above and within the forest canopy need to be investigated in order to adequately evaluate the effect of O_3 and other air pollutants, including PAN, on plants. Over Europe the relative contribution of photochemically produced O_3 to the observed annual distribution mainly increases from northwest to southeast. The AOT40 indices over Europe show that there is a strong gradient of O_3 exposure over Europe with lowest values in Scandinavia and the western part of the UK and over Ireland and highest values in the southern part of central Europe, including the region of the Alps. The exposure indices increase with altitude in mountainous terrain.

There remains an insufficient scientific basis for mandating O_3 control strategies. Studies by the American National Academy of Sciences (NAS 1991) and other research agencies have shown the need for a comprehensive program of tropospheric O_3 research with coordination between governments, universities, and private-sector interests. Particularly in Europe additional research is needed. European programs, such as EUROTRAC (European Experiment on Transport and Transformation of Environmentally Relevant Trace Constituents in the Troposphere over Europe) were established to meet this need for ozone research. Changes in conceptual models of the temporal and spatial variations in O_3, precursor concentrations, and emission rates (especially of biogenic organic compounds) are required for further progress (NAS 1991).

References

Altshuller AP, Lefohn AS (1996) Background ozone in the planetary boundary layer of the United States.J Air Waste Manage Assoc 46:134–141

Anfossi D, Sandroni S, Viarengo S (1991) Tropospheric ozone in the nineteenth century: the Moncalieri series. J. Geophy. Res. 96:17349–17352

Ashmore MR (1988) A comparison of indices that describe the relationship between exposure to ozone and reduction in the yield of agricultural crops. Atmos Environ 6:695–696

Baldocchi DD, Hicks BB, Meyers TP (1988) Measuring biosphere–atmosphere exchanges of biologically related gases with micrometeorological methods. Ecology 69:1331–1340

Beck J, Grennfelt P (1994) Estimate of ozone production and destruction over north-western Europe. Atmos Environ 28:129–140

Blake NJ, Penkett SA, Clemitshaw KC, Anwyl P, Lightman P, Marsh ARW (1993) Estimates of atmospheric hydroxyl radical concentrations from the observed decay of many reactive hydrocarbons in well-defined urban plumes. J Geophys Res 98:2851–2864

Chameides WL, Davis DD, Gregory GL, Sachse G, Torres AL (1988) Ozone precursors and ozone photochemistry over eastern North Pacific during the spring 1984 based on the NASA/CITE 1 airborne observations. J Geophys Res 94:9799–9808

Chock DP (1988) A need for a more robust ozone air quality standard. In: Proc 81st Annu Meet and Exhibition of the Air Pollution Control Association. Dallas, Texas, 19–24 June, Pap 88–121.5

Claude H (1995) Ozonbulletin des Deutschen Wetterdienstes Nr. 2, In: Claude H, Dehne K, Kohler U, Schubert U, Vandersee W, Vogel W (eds) Ozonbulletin des Deutschen Wetterdienstes, Gesamtausgabe Nr 1–12, Dez 1993–Dez 1994. Meteorologisches Observatorium, Hohenpeissenberg (in German)

Cox RA, Eggleton AJE, Derwent RG, Lovelock JE, Pack DH (1975) Long-range transport of photo-chemical ozone in north-western Europe. Nature 255:118–121

Crutzen PJ (1973) The role of NO and NO_2 in the chemistry of the troposphere and stratosphere. Annu Rev Earth Planet Sci 7:443–472

Danielsen EF (1968) Stratospheric-tropospheric exchange based on radioactivity, ozone and potential vorticity. J Atmos Sci 25:502–518

Davies TD, Kelly PM, Low PS, Pierce CE (1992) Surface ozone concentrations in Europe: links with the regional-scale atmospheric circulation. J Geophys Res 97:9819–9832

Delany AC, Fitzjarrald DR, Lenschow DH, Pearson R Jr, Wendel GJ, Woodruff B (1986) Direct measurements of nitrogen oxides and ozone fluxes over grassland. J Atmos Chem 4:429–444

Demerjian KL, Roychowdhury UK, Jin S, Lanni TR (1992) Measurements of trace chemical constituents at Whiteface Mountain. In: Trout M (ed) ASRC Report 1988–1991. Atmospheric Science Research Center, Albany, pp 25–34

Derwent RG, Kay PJA (1988) Factors influencing the ground level distribution of ozone in Europe. Environ Pollut 55:191–220

Derwent RG, Simmonds PG, Collins WJ (1994) Ozone and carbon monoxide measurements at a remote maritime location, Mace Head, Ireland, from 1990 to 1992. Atmos Environ 28:2623–2637

Dodge MC (1984) Combined effects of organic reactivity and NMHC/NO_x ratio on photochemical oxidant formation – a modeling study. Atmos Environ 18:1657–1665

Enders G, Teichmann U, Kramm G (1989) Profiles of ozone and surface-layer parameters over a mature spruce stand. In: Georgii H-W (ed) Mechanisms and effects of pollutant-transfer into forests. Kluwer Dordrecht, pp 21–35

Enders G, Dlugi R, Steinbrecher R, Clement B, Daiber R, v. Eijk J, Gäb S, Haziza M, Helas G, Herrmann U, Kessel M, Kesselmeier J, Kotzias D, Kourtidis K, Kurth H-H, McMillen RT, Roider G, Schürmann W, Teichmann U, Torres L (1992) Biosphere/atmosphere interactions: Integrated research in a European coniferous forest ecosystem. Atmos Environ 26A:171–189

EPA (1986) Air Quality Criteria for Ozone and Other Photochemical Oxidants, vol III EPA 600/8–84/020cF. US Environmental Protection Agency, US Environmental Criteria and Assessment Office, Research Triangle Park, North Carolina

EPA (1988) Summary of selected new information on effects of ozone on health and vegetation: Draft supplement to air quality criteria for ozone and other photochemical oxidants. EPA 600/8-88/105A. US Environmental Protection Agency, Office of Health and Environmental Assessment, Washington DC

EPA (1991a) National Air Quality and Emissions Trends Report, 1989. EPA/450/4-91-003. US Environmental Protection Agency, Office of Air Quality Planning and Standards, Research Triangle Park, North Carolina

EPA (1991b) Ozone and carbon monoxide areas designated nonattainment US Environmental Protection Agency, Office of Air Quality, Planning and Standards, Research Triangle Park, North Carolina

Esser PJ (1993) The effect of local and regional influences on ground level ozone concentrations under north European conditions. TNO-Rep, IMW-R93/098, Delft

Feister U, Balzer K (1991) Surface ozone and meteorological predictors on a subregional scale. Atmos Environ 25A:1781–1790

Finlayson-Pitts BJ, Pitts JN (1986) Atmospheric chemistry: fundamentals and experimental techniques. John Wiley, New York

Fishman J, Solomon S, Crutzen PJ (1979) Observational and theoretical evidence in support of a significant in-situ photochemical source of tropospheric ozone. Tellus 31:432–446

Fishman J, Fakhruzzaman K, Cros B, Nganga D (1991) Identification of widespread pollution in the southern hemisphere deduced from satellite analysis. Science 252:1693–1696

Fuentes JD, Dann TF (1994) Ground-level ozone in eastern Canada: seasonal variations, trends, and occurrences of high concentrations. J Air Waste Manage Assoc 44:1019–1026

Fuhrer J, Achermann B (eds) (1994) UN-ECE workshop report: critical levels for ozone. Federal Research Station for Agricultural Chemistry and Environmental Hygiene, Liebefeld-Bern

Gietl G, Rall A (1986) Bulk deposition into the catchment "Gross Ohe". Results of neighbouring sites in the open and under spruce at different altitudes. In: Georgii H-W (ed) Atmospheric pollutants in forest areas. Reidel, Dordrecht, pp 79–88

Graedel TE (1979) Terpenoids in the atmosphere. Rev Geophys Space Phys 17:937–948

Grennfelt P, Schjoldager J (1984) Photochemical oxidants in the troposphere: a mounting menace. Ambio 13: 61–67

Guenther AP, Zimmerman P, Wildermuth M (1994) Natural volatile organic compound emission rate estimates for U.S. woodland landscapes. Atmos Environ 28: 1197–1210

Guicherit R, van Dop H (1977) Photochemical production of ozone in western Europe (1971–1975) and its relation to meteorology. Atmos Environ 11:145–155

Haagen-Smit AJ (1952) Chemistry and physiology of Los Angeles smog. Ind Eng Chem 44:1342–1346

Haagen-Smit AJ, Fox MM (1956) Ozone formation in photochemical oxidation of organic substances. Ind Eng Chem 48:1484–1487

Haagen-Smit AJ, Darley EF, Zaitlin M, Hull H, Nobel WM (1952) Investigation of injury to plants from air pollution in the Los Angeles area. Plant Physiol 27:18–34

Haagen-Smit AJ, Bradley CE, Fox MM (1953) Ozone formation in photochemical oxidation of organic substances. Ind Eng Chem 45:2086–2089

Hakola H, Joffre S, Lättilä H, Taalas P (1991) Transport, transformation and sink processes behind surface ozone variability in north European conditions. Atmos Environ 25A:1437–1447

Hjellbrekke A-G (1995) Ozone measurements 1990–1992. EMEP/CCC 4/95, Norwegian Institute for Air Research N-2007 Kjeller

Hogsett WE, Tingey DT, Lee EH (1988) Exposure indices: concepts for development and evaluation of their use. In: Heck WW, Taylor OC, Lee DT (eds) Assessment of crop loss from air pollutants. Elsevier, London pp 107–138

Hough AM, Derwent RG (1990) Changes in the global concentration of tropospheric ozone due to human activities. Nature 344:645–648

Hov Ø (1989) Changes in tropospheric ozone: A simple model experiment. In: Bojkov RD, Fabian P (eds) Ozone in the atmosphere, Proc Quadrennial Ozone Symp 1988 and Tropospheric Ozone Worksh. Deepak, Hampton, Virginia, pp 557–560

Hov Ø, Schjoldager J, Wathne BM (1983) Measurement and modeling of the concentrations of terpenes in coniferous forest air. J Geophys Res 88:10679–10688

Hutchinson BA, Matt DR, McMillen RT, Gross LJ, Tajchman SJ, Norman JM (1986) The architecture of an east Tennessee deciduous forest canopy. J Ecol 74:635–646

Isaken ISA, Hov Ø (1987) Calculation of trends in the tropospheric concentration of O_3OH, CO, CH_4 and N_2O. Tellus 39B:271–285

Junge CE (1962) Global ozone budget and exchange between stratosphere and troposhere. Tellus 14:363–377

Junge CE (1963) Air chemistry and radioactivity. Academic Press, New York

Kley D, Geiss H, Mohnen VA (1994) Tropospheric ozone at elevated sites and precursor emissions in the United States and Europe. Atmos Environ 28A:149–158

Kramm G (1989) A numerical method for determining the dry deposition of atmospheric trace gases. Boundary-Layer Meteorol 48:157–176

Krupa SV, Kickert RN (1987) An analysis of numerical models of air pollutant exposure and vegetation response. Environ Pollut 44:127–158

Laurila T, Lättilä H (1994) Surface ozone exposures measured in Finland. Atmos Environ 28:103–114

Lee EH, Tingey DT, Hogsett WE (1988) Evaluation of ozone exposure indices in exposure-response modeling. Environ Pollut 53:43–62

Lefohn AS, Benedict HM (1982) Development of a mathematical index that describes ozone concentration, frequency, and duration. Atmos Environ 16:2529–2532

Lefohn AS, Benedict HM (1985) Exposure considerations associated with characterizing ozone ambient air quality monitoring data. In: Evaluation of the scientific basis for ozone/oxidants standards. Air Pollution Control Association, Pittsburgh, pp 17–31

Lefohn AS, Runeckles VC (1987) Establishing a standard to protect vegetation – ozone exposure/dose considerations. Atmos Environ 21:561–568

Lefohn AS, Laurence JA, Kohut RJ (1988) A comparison of indices that describe the relationship between exposure to ozone and reduction in the yield of agricultural crops. Atmos Environ 22:1229–1240

Lefohn AS, Runeckles VC, Krupa SV, Shadwick DS (1989) Important considerations for establishing a secondary ozone standard to protect vegetation. Air Control Assoc 39:1039–1045

Lenschow DH, Hicks BB (eds) (1989) Global tropospheric chemistry – chemical fluxes in the global atmosphere. National Center for Atmospheric Research (NCAR), Boulder

Levine JS (1991) Introduction. In: Levine JS (ed) Global biomass burning: atmospheric, climatic, and biospheric implications. MIT Press, Cambridge

Levy H II, Mahlman JD, Moxim WJ (1985) Tropospheric ozone: the role of transport. J Geophys Res 90:3753–3772

Lin X, Trainer M, Liu SC (1988) On the nonlinearity of the tropospheric ozone production. J Geophys Res 93:15879–15888

Liu SC, Kley D, McFarland M, Mahlmann JD, Levy H II (1980) On the origin of tropospheric ozone. J Geophys Res 85:7546–7552

Liu SC, Trainer M, Fehsenfeld FC, Parrish DD, Williams EJ, Fahey DW, Hübler G, Murphy PC (1987) Ozone production in the rural troposphere and the implications for regional and global ozone distributions. J Geophys Res 92:4191–4207

Logan JA (1985) Tropospheric ozone: Seasonal behavior, trends, and anthropogenic influence. J Geophys Res 90:10463–10882

Logan JA (1994) Trends in the vertical distribution of ozone: an analysis of ozone sonde data. J Geophys Res 99:25533–255585

Low PS, Kelly PM, Davies TD (1992) Variations in surface ozone over Europe. Geophys Res Lett 19:1117–1120

Marenco A, Gouget H, Nedelec P, Pages J-P (1994) Evidence of a long-term increase in tropospheric ozone from Pic Du Midi data series: consequences: positive radiative forcing. J Geophys Res 99 16617–16632

McKee DJ (ed) (1994) Tropospheric ozone: human health and agricultural impacts. CRC Press, Boca Raton

McKeen S, Kley D, Volz A (1989) The historical trend of tropospheric ozone over western Europe: A model perspective. In: Bojkov RD, Fabian P (eds) Ozone in the atmosphere. Proc Quadrennial Ozone Symp 1988 and Tropospheric Ozone Worksh. Deepak, Hampton, Virginia, pp 552–556

McRae GJ, Russell AG, Harley RA (1992a) CIT photochemical airshed model – systems manual. Report to the Coordinating Research Council under project SCAQS-8. Carnegie Mellon University, Pittsburgh, and California Institute of Technology, Pasadena

McRae GJ, Russel AG, Harley RA (1992b) CIT photochemical airshed model – data preparation manual Report to the Coordinating Research Council under Project SCAQS-8. Carnegie Mellon University, Pittsburgh, and California Institute of Technology, Pasadena

McRae GJ, Russell AG, Harley RA (1992c) CIT Photochemical airshed model – installation manual. Report to the Coordinating Research Council under project SCAQS-8. Carnegie Mellon University, Pittsburgh, and California Institute of Technology, Pasadena

Meyers TP, Baldocchi DD (1988) A comparison of models for deriving dry deposition fluxes of O_3 and SO_2 to a forest canopy. Tellus 40B:270–284

Meyers TP, Paw UKT (1987) Modelling the plant canopy micrometeorology with higher-order closure principles. Agric For Meteorol 41:143–163

Middleton JT, Kendrick JB Jr, Schwalm HW (1950) Injury to herbaceous plants by smog or air pollution. Plant Dis Rep 34:245–252

Milford JB, Russell AG, McRae GJ (1989) Spatial patterns in photochemical pollutant response to NO_x and ROG reductions. Environ Sci Technol 23:1290–1301

Müller JF, Brasseur G (1995) IMAGES: a three-dimensional chemical transport model of the global troposphere. J Geophys Res 100:16445–16490

Musselman RC, McCool PM, Lefohn AS (1994) Ozone descriptors for an air quality standard to protect vegetation. J Air Waste Manage Assoc 44: 1383–1390

NAPAP (1989) National acid precipitation assessment program vol III. Atmospheric processes. Government Printing Office, Washington DC

NAS (1991) (National Academy of Sciences) Rethinking the ozone problem in urban and regional air pollution. National Academy Press, Washington DC

Oltmans SJ (1981) Surface ozone measurements in clean air. J Geophys Res 86:1174–1180

Oltmans SJ, Levy H II (1994) Surface ozone measurements from a global network. Atmos Environ 28:9–24

Oltmans SJ, Komhyr WD, Franchois PR, Matthews WA (1989) Tropospheric ozone: variations from surface and ECC ozone sonde observations. In: Bojkov RD, Fabian P (eds) Ozone in the atmosphere, Proc Quadrennial Ozone Symp 1988 and Tropospheric Ozone worksh. Deepak, Hampton, Virginia, pp 539–543

Parry MAJ, Day W (1988) A comparison of indices that describe the relationship between exposure to ozone and reduction in the yield of agricultural crops. Atmos Environ 22:2057–2058

Paulson SE, Seinfeld JH (1992) Development and evaluation of a photochemical mechanism for isoprene. J Geophys Res 97:20703–20715

Pedersen U, Lefohn AS (1994) Characterizing surface ozone concentrations in Norway. Atmos Environ 28:89–101

Penkett SA, Brice KA (1986) The spring maximum in photo-oxidants in the north hemisphere troposphere. Nature 319:655–657

PORG (1993) Ozone in the United Kingdom 1993. 3rd Report of the United Kingdom Photochemical Oxidants Review Group, London

Puxbaum H, Gabler K, Smidt S, Glattes F (1991) A one-year record of ozone profiles in an alpine valley (Zillertal/Tyrol, Austria, 600–2000 m.a.s.l.). Atmos Environ 25A:1759–1765

Rao ST, Zurbenko IG (1994) Detecting and tracking changes in ozone air quality. J Air Waste Manage Assoc 44:1089–1092

Rao, ST, Zalewsky E, Zurbenko IG (1995) Determining temporal and spatial variations in ozone air quality. J Air Waste Manage Assoc 45:57–61

Reiter ER (1975) Stratospheric – tropospheric exchange processes. Rev Geophys Space Phys 13:459–474

Runeckles VA (1988) A comparison of indices that describe the relationship between exposure to ozone and reduction in the yield of agricultural crops. Atmos Environ 22:1241–1242

SCAQMD (1990) (South Coast Air Quality Management District) 1990 Air quality data. South Coast Air Quality Management District, Diamond Bar, California

Scheel HE (1994) Interpretation of TOR ozone measurements. In: Borrell PM, Borrell P, Cvitas T, Seiler W (eds) Proc EUROTRAC Symp '94. SPB The Hague, pp 92–98

Scheel HE, Sladkovic R, Brunke E-G, Seiler W (1992) Measurements of lower tropospheric ozone at mid-latitudes of the northern and southern hemisphere. In: Hudson RD (ed) Ozone in the troposphere and stratosphere, part 1. NASA Conference Publication, Greenbelt. Maryland. pp 11–14

Scheel HE, Sladkovic R, Seiler W (1995) Ground-based measurements of ozone related precursors at 47°N, 11°E. In: Final Report EUROTRAC/TOR, EUROTRAC Int Sci Secretariat, Garmisch-Partenkirchen (in press)

Schjoldager J, Dovland H, Grennfelt P, Saltbones J (1981) Photochemical oxidants in north-western Europe 1976–1979. A pilot project. Norwegian Institute for Air Research (NILU OR 19/81), Lillerstrøm, Norway

Schnell RC, Liu SC, Oltmans SJ, Stone RS, Hofmann DJ, Dutton EG, Deshler T, Sturges WT, Harder JW, Sewell SD, Trainer M, Harris JM (1991) Decrease of summer tropospheric ozone concentrations in Antarctica. Nature 351:726–729

Seinfeld JH (1986) Atmospheric chemistry and physics of air pollution. Wiley, New York

Sladkovic R, Scheel HE, Seiler W (1994a) Ozone climatology at the mountain sites Wank and Zugspitze. In: Borrell PM, Borrell P, Cvitas T, Seiler W (eds) Proc EUROTRAC Symp '94. SPB The Hague, pp 253–258

Sladkovic R, Scheel HE, Junkermann W (1994b) Operation of a measurement station for forest research in the Kalkalpen Projekt Rep, GSF Forschungszentrum für Umwelt und Gesundheit, Munich (in German)

Staehelin J, Schmid W (1991) Trend analysis of tropospheric ozone concentrations utilizing the 20-year data set of balloon soundings over Payerne (Switzerland). Atmos Environ 25A:1739–1749

Staehelin J, Thudium J, Buehler R, Volz-Thomas A, Graber W (1994) Trends in surface ozone concentrations at Arosa (Switzerland). Atmos Environ 28:75–88

Sun EJ, Huang MH (1995) Detection of peroxyacetyl nitrate at phytotoxic level and its effects on vegetation in Taiwan. Atmos Environment 29:2899–2904

Tingey DT (1984) The effects of ozone on plants in the United States. In: Grennfelt P (ed) Proc Int Worksh on the Evaluation of the effects of photochemical oxidants on human health. Agricultural Crops, Forestry, Materials and Visibility. Swedish Environmental Research Institute, Goeteborg

Toupance G, Aranda F (1993) Spatial evolution of annual ozone profiles over France and western Europe. In: Borrell PM, Borrell P, Cvitas T, Seiler W (eds) Proc EUROTRAC Symp '92, SPB The Hague, pp 133–139

Trainer M, Williams ET, Parrish DD, Buhr MP, Allwine EJ, Westberg HH, Fehsenfeld FC, Liu SC (1987) Models and observations of the impact of natural hydrocarbons in rural ozone. Nature 329:705–707

Trainer M, Ridley BA, Buhr MP, Kok G, Walega J, Hüber G, Parrish DD, Fehsenfeld FC (1995) Regional ozone and urban plumes in the southeastern United States: Birmingham, a case study. J Geophys Res Geophys Res 100:18823–18834

Volz A, Kley D (1988) Evaluation of the Montsouris series of ozone measurements made in the nineteenth century. Nature 332:240–242

Volz A, Geiss H, McKeen S, Kley D (1989) Correlation of ozone and solar radiation at Monsouris and Hohenpeissenberg: indications for photochemical influence. In: Bojkov RD, Fabian P (eds) Ozone in the atmosphere. Proc Quadrennial Ozone Symp 1988 and Tropospheric Ozone Workshop. Deepak, Hampton, Virginia, pp 447–450

Vukovich FM (1994) Boundary layer ozone variations in the eastern United States and their association with meteorological variations: long-term variations. J Geophys Res 99:16839–16850

Warneck P (1988) The chemistry of the natural atmosphere. Academic Press, New York

Wege K, Claude H, Hartmannsgruber R (1989) Ozone observation at Hohenpeissenberg. In: Bojkov RD, Fabian P (eds) Ozone in the atmosphere. Proc Quadrennial Ozone Symp 1988 and Tropospheric Ozone Workshop. Deepak, Hampton, Virginia, pp 109–112

Winkler P (1989) Meridional distribution of surface ozone over the Atlantic (83°N-76°S). In: Bojkov RD, Fabian P (eds) Ozone in the atmosphere. Proc Quadrennial Ozone Symp 1988 and Tropospheric Ozone Worksh. Deepak, Hampton, Virginia, pp 423–424

Zurbenko IG, Rao ST, Henry RF (1995) Mapping ozone in the eastern United States. Environ Manager 1: February, 24–30

2 Ozone and Its Known and Potential Effects on Forests in Western United States

P.R. Miller, M.J. Arbaugh and P.J. Temple

2.1 Introduction

California is the only state in the western United States where ozone damage to montane conifer forests is a significant problem (USEPA 1986). The combination of climate and high rates of precursor emissions from large urban areas in California results in levels of photochemical oxidant air pollution which are consistently higher than other urbanized areas of the United States (Ludwig and Shelar 1980). The transport of polluted air masses can be detected at distances as far as 300–350 km downwind from urban source areas (Carroll and Baskett 1977; Hoffer et al. 1982). Mixed conifer forests on distant mountain ranges are affected by chronic ozone exposure.

The species comprising the Californian mixed conifer forest type include ponderosa pine (*Pinus ponderosa* Dougl. ex Laws.), Jeffrey pine (*P. jeffreyi* Grev. & Balf.), sugar pine (*P. lambertiana* Dougl.), white fir [*Abies concolor* (Gord. & Glend. Lindl. ex Hildebr.)], incense cedar (*Calocedrus decurrens* Torr.), and black oak (*Quercus kelloggii* Newb.). These species occur in different mixtures over the elevational range of 1225–3050 m in the mountains of southern California. The order of highest to lowest sensitivity to ozone includes ponderosa and Jeffrey pines (about the same), white fir, black oak, incense cedar, and sugar pine. The least sensitive tree species are the best competitors in stands with long-term ozone exposure. In southern California, big-cone Douglas fir [*Pseudotsuga macrocarpa* (Vasey) Mayr] is distributed in the 600–2150 m range and mixes with some of the above species. A recent study (Peterson et al. 1995) shows that ozone effects can be identified along a gradient of high to low ozone exposure in the San Bernardino mountains. Its sensitivity appears to be similar to white fir and incense cedar. Coulter pine (*Pinus coulteri* D. Don) is also found at the lower elevation and drier limits of the mixed conifer type and its ozone sensitivity is similar to sugar pine. Because ponderosa and Jeffrey pines have a similar ozone sensitivity and

Ecological Studies, Vol. 127
Sandermann et al. (eds) Forest Decline and Ozone

because they typically occupy the 1500–2750 m ozone in both the Sierra Nevada and southern Californian mountains they have been used as bioindicators of chronic ozone injury (Miller 1992; Stolte et al. 1992).

Californian mixed conifer forests provide the best-documented example of the cumulative effects of long-term ozone exposure on a conifer forest ecosystem (USEPA 1986). A host of other abiotic and biotic stressors that act simultaneously or sequentially with ozone to cause injury to foliage, diminish carbon fixation and reduce the future growth of foliage, root, and stem tissue. The purpose of this chapter is to describe the long-term effects of ozone on mostly ponderosa and Jeffrey pines in California at several levels of biological organization, ranging from leaf tissue to whole stands of trees comprised of different size classes. Evidence from controlled studies and repeated measurements of forest plots will be discussed in relation to the influences of other biotic and abiotic stressors.

2.2 Spatial Distribution and Temporal Trends of Ozone in Relation to Geographic Setting and Summer Climate

Both landform and prevailing meteorological conditions are important features controlling the spatial distribution of ozone as well as the shape of the curve describing the diurnal changes in ozone concentration. California is divided into air basins defined as those valleys or regions where geographic features restrict horizontal transport and result in relatively distinct air masses (Fig. 2.1). The National Forests (and National Parks) of the Sierra Nevada are the recipients of polluted air from the Sacramento and San Joaqin Valley air basins. The South Coast air basin, including the Los Angeles metropolitan complex is the source area for the transverse mountain ranges occupied by the Angeles and San Bernardino National Forests (San Gabriel and San Bernardino Mountains, respectively). The polluted air is swept through the inland valleys and reaches the forested slopes and ridges of adjacent mountains by a combination of mass flow of marine air and upslope flow on heated mountain slopes (Fig. 2.2, see Appendix; Edinger et al. 1972). During the 1977–1985 period, the maximum 1-h ozone concentrations (averaged over 8 years) for the air basins contributing most to forest damage were 183 nl l^{-1} for the San Francisco Bay Area, 195 nl l^{-1} for the Sacramento Valley, 185 nl^{-1} l^{-1} for the San Joaquin Valley, and 413 nl l^{-1} for the South Coast Air Basin. The maximum 1-h ozone concentration of Lake Tahoe Basin was 91 nl l^{-1} (California Air Resources Board 1987).

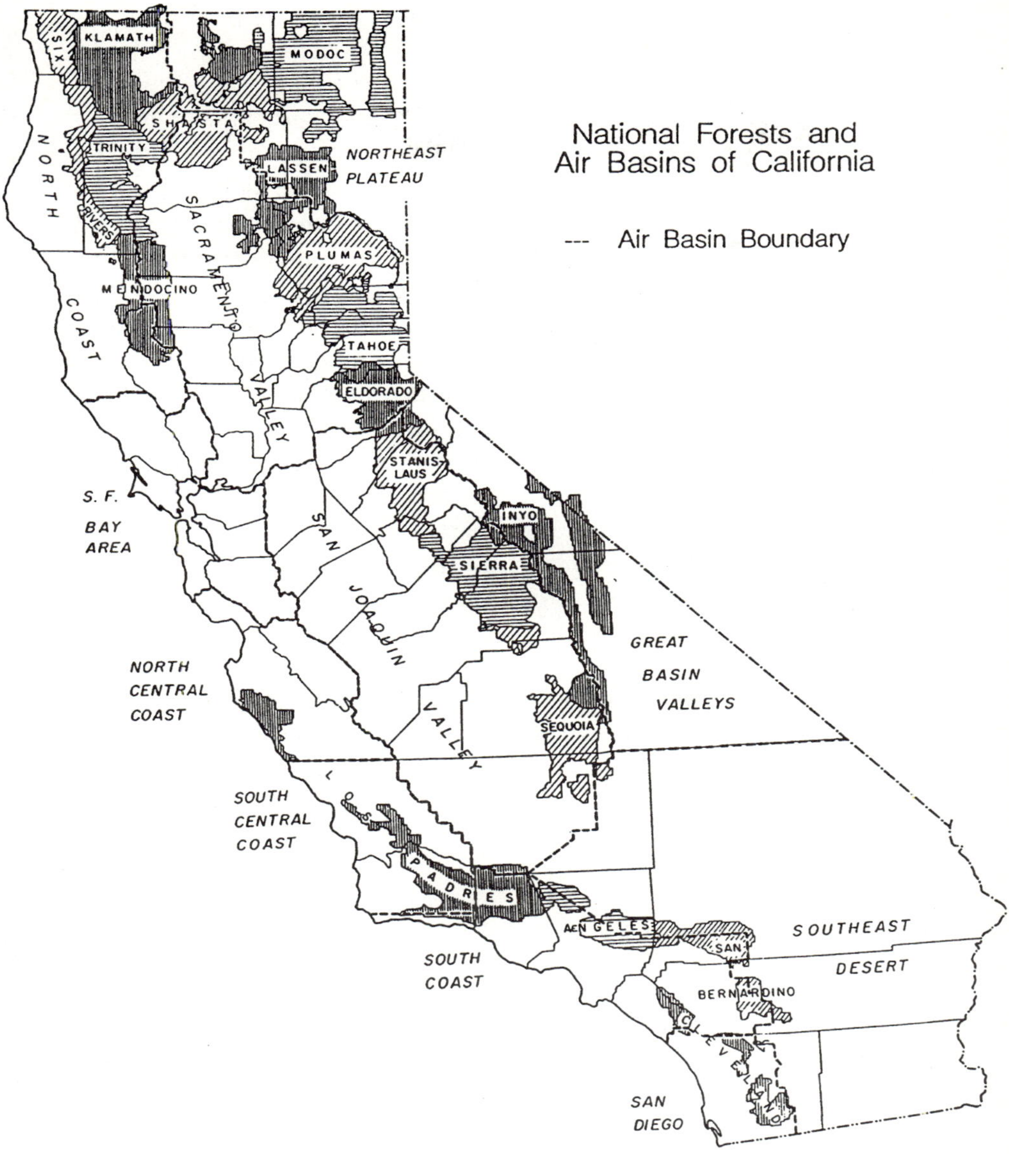

Fig. 2.1. California air basins in relation to national forests. The San Francisco Bay Area and San Joaquin Valley Air Basin are the sources of pollution influencing the western portions of the Eldorado, Stanislaus, Sierra and Sequoia National Forests. The South Coast Air Basin influences the Los Padres, Angeles, and San Bernardino National Forests

2.2.1 Daily Patterns of Ozone Concentrations in Rural Mountain Locations

The proximity to urban source areas and transport mechanisms result in different diurnal patterns in hourly ozone concentrations as defined by Bohm (1992). Van Ooy and Carroll (1995) studied the relationships between local climatology and patterns of ozone concentrations for six sites in the Sierra Nevada for the June through September period of 1992. Three sites did not exhibit a night time decrease in ozone, and afternoon peaks were not very pronounced. The wind directional frequencies at these sites were more broadly distributed, including strong north–south components. Three other sites had a distinct afternoon peak. The wind directional frequencies at these sites exhibited mainly flow from the west (the source area) during the day and from the east during the night. The source of the night time flow may be cleaner air from higher elevations or from cleaner continental areas to the east.

Elevation and local topographic features were also considered as controlling features. For example, higher elevation ridges may be exposed to high-elevation air masses with high ozone concentrations at night. Diurnal patterns of ozone concentration could influence ozone uptake by trees. Ozone flux to foliage is highest during daylight hours when stomatal conductance is highest and negligible at night (see Sect. 2.4).

2.2.2 Spatial Variation of Ozone in the Sierra Nevada and San Bernardino Mountains

The results of extended ozone monitoring in the Sierra Nevada (Van Ooy and Carroll 1955) show that local climate and terrain have an important influence on summer ozone concentrations. A uniform gradient of increasing ozone concentrations over the complex terrain of the western Sierra Nevada is not to be expected, although crop injury data from the San Joaqin Valley show increasing injury to sensitive crops from north to south. In the San Bernardino Mountains there is a definite gradient of decreasing ozone concentration from west to east (Miller et al. 1986).

2.2.3 Temporal Trends of Ozone and the Outlook for the Future

In the context of long life spans of trees, ozone air pollution is a recent phenomenon. Reliable records of ozone concentrations in California have been available only since the 1960s, and most data are available from monitoring stations located in or near urban areas. Between 1976 and

1991 the running mean of ozone data (which tend to minimize the year-to-year influence of weather) for the May through October "smog season" in the South Coast air basin shows that the number of days exceeding the Federal Standard (> 120 nl l^{-1}, 1-h average) has declined at an average annual rate of 2.27 days/year (Davidson 1993). The number of days with stage I episodes (> 200 nl l^{-1}, 1-h average) has declined at an average annual rate of 4.70 days $year^{-1}$ over the same time period. The total number of days/year exceeding the Federal Standard was as high as 159 in 1978 and the lowest number was 105 days in 1990. The highest number of stage I episode days was 108 in 1979 and the lowest was 33 days in 1990. The 1974 to 1988 trends of the May through October hourly average and the average of monthly maximum ozone concentrations for Lake Gregory, a forested area in the western section of the San Bernardino Mountains, have also shown a decline (Miller and McBride 1989).

2.3 History of Ozone Damage to Californian Forests

The establishment of ozone as a cause of chlorotic mottle symptoms on needles of pines was determined by independent studies (Miller et al. 1963; Richards et al. 1968). Ozone injury to sensitive species located on the western slopes of the Sierra Nevada was first described in 1970 (Miller and Millecan 1971). Repeated measurements of the long-term effects of ozone on ponderosa and Jeffrey pines in permanent plots began from about 1968 to 1974 – less than 30 years ago in most cases for the San Bernardino mountains (Miller 1992) and the Sierra Nevada (Pronos and Vogler 1981).

2.3.1 Field Identification of Ozone Injury

Appearance of Injured Trees and Stands. The transport of ozone into a mixed conifer forest and the appearance of injured trees, branches, and needles of ponderosa pine are shown in Fig. 2.2 (see Appendix). Controlled exposures and field observations of ozone effects on western conifer species have shown that chlorotic mottle on all needle surfaces begins with the degradation of the contents of mesophyll cells below the epidermis, causing the degradation of chlorophyll within the cell (Evans and Miller 1972a,b; Rice et al. 1983). Microscopically this appears as amorphous staining of cellular contents, plasmolysis of cell contents, and cell death. The degradation of chlorophyll beneath the epidermis is manifested on the needle surface as amorphous chlorotic blotches with diffuse borders that occur in irregular patterns, giving a yellow "mottled" appearance; hence the terminology chlorotic "mottle" (Fig. 2.2, see Appendix). This foliar injury symptom is visibly distinct from foliar symptoms induced by other air pollutants.

Chlorotic mottle first appears in the one-third of the needle surface nearest the tip on 1-year-old or older needles, and progresses basipetally until the entire needle is involved (Miller et al. 1963; Richards et al. 1968). The current year needles occasionally exhibit small amounts of chlorotic mottle if sufficiently elevated ozone exposures have occurred during the summer growing season. Tip necrosis or necrotic bands can result from acute ozone exposures in fumigation experiments, but chronic field exposures typically induce only chlorotic mottling. In the most severe cases of injury at the highest ozone exposure levels, mainly the current year's needles remain; however, they are shorter and are over 30% mottled.

2.3.2 Role of Other Biotic and Abiotic Stresses

Insects and Fungi. Foliar injuries resulting from biotic causal agents may appear on needle surfaces, but symptoms are sufficiently distinct to prevent misidentification as ozone injury. Chlorotic and necrotic spots or blotches caused by sucking insects such as aphids and pine needle scale may sometimes resemble ozone chlorotic mottle. The most important biotic agents are fungi (Scharpf 1993), and chewing (needle weevil) or sucking (scale or aphids) insects (Miller et al. 1977; Pronos et al. 1978). Chlorotic mottle can be differentiated from injuries brought about by fungi and insects by observation of the color and pattern of the symptoms on the needles.

Abiotic Factors. Foliar injuries from abiotic factors, other than ozone, may also be present on pine needles. Winter fleck of pine needles is apparently caused by snow and cold temperatures, and is a symptom that superficially resembles chlorotic mottle (Fig. 2.2, see Appendix). The lesions, when large, can be irregular in size and shape like chlorotic mottle, but are tan or brown in color and present only on the adaxial surface of the needle, suggesting the role of sunlight in their formation (Miller and Evans 1974). Also, smaller weather fleck lesions tend to be more round in shape, with distinct margins. Histological examination has shown that winter fleck results from a uniform plasmolysis of all mesophyll cells lying under the surface lesion and can cause collapse of the epidermis. In contrast, ozone injury affects only mesophyll cells under the epidermis randomly usually without structural loss to the epidermis. Symptoms of chlorotic mottle and winter fleck can be differentiated even when they are found together in significant amounts on the same needles (Fig. 2.2, see Appendix).

2.3.3 Crown Changes Associated with Chronic Ozone Injury

In many conifer species, foliage longevity can be measured by counting nodes on branches back from the branch tip, with each node separating annual whorls of needles or needle fascicles (dwarf shoots) corresponding to one year of growth. A direct relationship between the incidence of chlorotic mottle and accelerated abscission of needles has been shown for ponderosa pines in southern California and for ponderosa and Jeffrey pines in the Sierra Nevada of California (Miller and Van Doren 1982; Duriscoe and Stolte 1989). Reduction in needle longevity is recognized as an indicator of air pollution stress for these species when other factors leading to accelerated abscission of needles are taken into account (Fig. 2.2, see Appendix). In unpolluted areas of the west coast, ponderosa pines may be expected to retain foliage for three to four years, with an average of four annual whorls retained at any given time (Sudworth 1908). Foliage longevity is also related to other factors, most particularly the elevation at which a tree grows (Ewers and Schmid 1982), which is an indication of the length of the growing season at a particular site. Trees at higher elevations consistently retain their foliage longer (up to 40 years for bristlecone pine, *P. longaeva* D. Bailey). Persistent infections of needle cast fungi can lead to tree crowns that are extensively defoliated. Interpretation of data on foliage longevity must take into account these confounding factors.

2.3.4 Spatial Distribution and Temporal Changes of Ozone Damage

2.3.4.1 Southern California

In southern California a chain of mountain ranges extends from the north to the east of Los Angeles and includes the San Rafael, San Gabriel, San Bernardino, and San Jacinto mountains; the Los Padres, Angeles, and San Bernardino National Forest boundaries define the forested areas on these ranges (Fig. 2.1). For the most part the high points of these mountains form the boundary between the South Coast and the Southeast Desert Air Basin. Conifer forests on those mountain ranges most distant from Los Angeles, namely, the San Rafael and San Jacinto ranges, have generally very low levels of ozone injury, at the limit of detection. Forests of the San Gabriel range have a wide range of injury. This heterogeneity is partially caused by the influence of the extremely steep terrain that limits the number of favorable sites for forest establishment and also tends to channel the flow of polluted air as it emerges from the basin.

The best documented description of chronic ozone damage in southern California pertains to the San Bernardino Mountains. Table 2.1 summarizes the response of ponderosa and Jeffrey pines in plots located along a west to east gradient of decreasing ozone exposure (Miller et al. 1991). Crown injury measurements (ozone injury score) correspond to the ozone gradient; however, plot elevation, precipitation, and soil moisture holding capacity are quite variable among plots. The ozone injury score shows a decrease in foliar ozone injury between 1974 and 1988 (Miller 1992).

Table 2.1. Properties of 13 forest plots in relation to ozone injury score of ponderosa and Jeffrey pines in order of increasing injury level. (Miller et al. 1991)

Plot name	Elevation (m)	Estimated ozone dose 24-h average May-Oct 1988 (nl l^{-1})	Average precipitation (cm)	Available soil water in top 90 cm (%)	Average ozone injury score[a] Ponderosa/Jeffrey	Injury description
CP	1600	110	78	18	13.0/No JP	Severe
BP	1525	90	105	17	15.2/No JP	Moderate
SF	1705	100	109	19	17.2/No JP	Moderate
SCR	1830	80	70	8	18.2/No JP	Moderate
DWA	1725	100	90	15	19.8/No JP	Moderate
UCC	1610	100	90	13	21.9/No JP	Slight
CA	1830	90	114	NA	22.5/No JP	Slight
GVC	1950	70	99	10	24.8/26.9	Slight
BF	1900	70	61	21	26.1/24.0	Slight
CAO	2135	60	82	15	31.6/33.6	Very slight
BL	2320	70	90	12	No PP/35.9	Very slight
HB	2160	50	56	10	No PP/43.6	No injury
HV	2225	50	52	9	No PP/51.3	No injury

JP, Jeffrey pine; PP, ponderosa pine; CP, Camp Paivika; BP, Breezy point; SF, Sky Forest; SCR, Schneider Creek; DWA, Dogwood; UCC, U.C. Center; CA, Camp Angelus; GVC, Green Valley Creek; BF, Barton Flats; CAD, Camp Osceola; BL, Buff Lake; HB, Heart Bar; HV, Holcomb Valley; NA, not available.

[a]The ozone injury score is primarily a visual estimate of needle and crown condition by which very severe injury is 0–8, severe is 9–14, moderate is 15–25, slight is 26–35, and 36 or higher is very slight to no visible injury. This method has been replaced by the ozone injury index derived from close-up inspection of five branches pruned from the lower part of the tree.

2.3.4.2 Sierra Nevada

Peterson et al. (1991) sampled the crown condition of ponderosa pines at sites in seven Federal administrative units located from north to south in the Sierra Nevada, including Tahoe National Forest, Eldorado National Forest, Stanislaus National Forest, Yosemite National Park, and Sierra National Forest, Sequoia-Kings Canyon National Park, Sequoia National Forest. In July-August 1987, four symptomatic and four asymptomatic sites were visited in each unit. The symptomatic plots generally indicated increasing levels of ozone injury (reduction in the numbers of annual needle whorls retained and chlorotic mottle symptoms on progressively younger age classes of needles, e.g., as young as 2 years) from north to south.

Both permanent plots and cruise surveys have been employed in Sequoia, Kings Canyon (SEKI) and Yosemite (YOSE) National Parks to determine the spatial distribution and temporal changes of injury to ponderosa and Jeffrey pine within the Parks (Duriscoe and Stolte 1989). Comparisons of the same trees at 28 plots between 1980–1982 and 1984–1985 in SEKI showed increases of ozone injury to many trees and increases in the total number of trees with ozone injury. Ozone injury decreased with increasing elevation of plots. The hourly average ozone concentrations associated with the highest levels of tree injury in the Marble Fork drainage of the Kaweah River at approximately 1800 m elevation were frequently found to peak at 80–100 nl l^{-1}, but seldom exceeded 120 nl l^{-1}.

During a cruise survey in 1986 (Duriscoe and Stolte 1989) more than one-third of 3120 ponderosa or Jeffrey pines evaluated for ozone injury in SEKI and YOSE were found to have some level of chlorotic mottle. At SEKI, symptomatic trees comprised 39% of the sample (574 out of 1470) and at YOSE they comprised 29% (479 out of 1650). Ponderosa pines were generally more severely injured than Jeffrey pines.

The Lake Tahoe Basin is located in the central Sierra Nevada (near the Eldorado National Forest; Peterson et al. 1991). It is an isolated air basin with ozone precursor sources and air quality problems distinct from other Sierra Nevada sites. In 1987, a survey of 24 randomly selected plots in the basin included a total of 360 trees of which 105 (29.2%) had generally slight amounts of foliar injury (Pedersen 1989). Of 190 trees in 16 cruise plots that extended observations to the east outside the basin, 21.6% had very slight injury – less than in the basin.

2.4 Experimental Exposures of Western Conifers to Ozone

Experimental exposures of conifers to ozone were first used to establish that ozone was the causal agent of the "chlorotic decline" syndrome of ponderosa pine widespread in the mountains of southern California (Miller et al. 1963). Chamber studies have also been used to determine the relative susceptibilities of western conifers to ozone (Miller et al. 1983) and other air pollutants (Hogsett et al. 1989). Recent experimental work has emphasized physiological responses of western conifers to ozone (e.g., Beyers et al. 1992; Grulke and Miller 1994) in an effort to elucidate the mechanisms of ozone toxicity and repair in trees. This section will examine the extent to which the results of experimental chamber exposures to ozone can be extrapolated to the responses of trees observed in the field.

2.4.1 Foliar Injury Symptomatology

The relationship between the development of chlorotic mottle injury and subsequent needle senescence and abscission has been explored in experimental exposures and in the field. In an open-top chamber experiment conducted as part of the Responses of Plants to Interacting Stresses (ROPIS) program (Goldstein and Ferson 1994), foliage of ponderosa pine seedings exposed to ozone which exhibited >30% of leaf surface per annual whorl with chlorotic mottle in September had abscised by the end of the next growing season (Temple and Miller 1994). A similar pattern was observed with Jeffrey pine growing in Sequoia and Kings Canyon National Parks (SEKI), where needles typically abscised after 30% of the needle surface had become chlorotic (Patterson and Rundel 1995).

Ozone injury development on the foliage of giant sequoia [*Sequoiadendron giganteum* (Lindl.) Buch.] provides a good example of synergism between field observations and experimental results. Experimental exposures of giant sequoia branches to ozone, either in branch chambers (Grulke and Miller 1994) or as young trees (Temple 1988), confirmed that mature foliage is highly resistant to ozone. However, exposure of current-year seedlings to elevated ozone in open-top chambers in the field showed that some young seedlings developed injury symptoms absent from those in charcoal-filtered control chambers. Furthermore, symptom development appeared to be greatest on seedlings partially shaded from full sunlight (Miller et al. 1994). Examination of current-year giant sequoia seedlings naturally regenerating in Sequoia National Park showed similar foliar injury symptoms on some of the seedlings (Miller et al. 1994). Clearly, current levels of ozone appear to have little effect on giant sequoia saplings or mature trees; however, ozone may be a selective agent, influencing the natural regeneration of giant sequoia seedlings (Grulke and Miller 1994).

Observations of foliar ozone injury symptoms across ozone gradients in the mountains of southern California (Miller et al. 1989) and from south to north along the western slopes of the Sierras (Peterson and Arbaugh 1992) suggest that the "threshold" for development of significant foliar injury and needle abscission on ponderosa and Jeffrey pines is a seasonal (May to October) mean ozone level of 50–60 nl l^{-1}. Experimental exposures of western conifers to ozone have confirmed these observations.

In the ROPIS study, 74% of well-watered (WW) and 83% of drought-stressed (DS) ponderosa pine seedlings exposed for 3 years to a growing season ambient mean ozone level of 54 nl l^{-1} (SUM0 = 488 nl l^{-1} h) had no visible injury symptoms. Seven percent of WW seedlings developed moderate injury, but injured DS seedlings developed only slight amounts of injury (Temple, unpubl. data). This level of injury was less than that observed with ponderosa pines in the surrounding mixed conifer forest, where ambient ozone levels averaged at 63 nl l^{-1} (SUM0 = 542 nl l^{-1} h) for the same period (Duriscoe and Stolte 1989). Seedlings exposed to a seasonal mean of 87 nl l^{-1} (SUM0 = 795 nl l^{-1} h^{-1}) developed an average of 17% chlorotic mottle and over 30% of the seedlings had moderate to severe injury symptoms (Temple et al. 1992). This level of injury is similar to that observed on ponderosa pines in heavily polluted sites in the mountains of southern California (Miller et al. 1989).

The ROPIS open-top chamber study also demonstrated the significant effects of plant genotype and co-occurring stress, particularly drought, on the development of ozone injury (Temple et al. 1992). In some half-sib families of ponderosa pine nearly all the seedlings had moderate to severe symptoms of ozone injury in response to a seasonal mean of 87 nl l^{-1}, while seedlings in other half-sib families showed little or no ozone injury response. Seedlings watered weekly developed three to four times greater amounts of ozone injury than those subjected to periodic drought stress (Temple et al. 1992). These experimental results highlight the complex interrelationships among ambient ozone concentrations, plant genotypes, and environmental conditions that make it difficult to predict tree responses to ozone in the field.

2.4.2 Physiological Responses: Gas Exchange and Photosynthesis

Because ozone injury to leaf tissues can occur only if the molecules pass through the stomata into substomatal cavities, the role of gas exchange in controlling plant responses to ambient ozone has received considerable attention. This is particularly true for forests in California, where the Mediterranean climate and highly variable amounts of winter precipitation can induce periods of moderate to severe summer drought lasting from one to several years.

In the ROPIS study, drought-stressed ponderosa pine seedlings exhibited significantly less ozone injury and less of a reduction in photosynthetic rates than well-watered trees, as discussed above, and the DS trees also had on average a significantly lower stomatal conductance than the non-stressed seedlings (Beyers et al. 1992; Temple, unpubl.). This observation has been confirmed by multi year evaluations of ozone injury symptoms on ponderosa and Jeffrey pines in the mountains of southern and central California. Less foliar ozone injury and greater needle retention have been observed in years with low precipitation the previous winter than in years with normal or above normal winter precipitation (Miller et al., in prep.). Conversely, in years with low ambient ozone and high winter precipitation, trees growing on sites with low moisture holding capacity recovered less from ozone injury than trees growing on favorable sites (Miller et al. 1989). These observations suggest that combinations of ozone and drought stresses may be protective if both stressors are present simultaneously, but sequential exposures to one stress followed by the other may not permit the trees to recover fully from the adverse effects of either stress.

Whether trees with higher intrinsic rates of stomatal conductance are more susceptible to ozone injury because of their greater rates of gas exchange has not yet been determined. Patterson and Rundel (1989) reported that severely ozone-injured Jeffrey pine growing in SEKI had higher rates of stomatal conductance than ozone-resistant trees growing nearby. The susceptible trees also showed less stomatal response to vapor pressure deficits (VPD), i.e., stomata remained open longer at higher VPD in comparison with resistant Jeffrey pines (Patterson and Rundel 1989). Greater rates of stomatal conductance were also observed on 1-year-old needles of mature ponderosa pine trees with severe ozone injury in the San Bernardino Mountains relative to trees with lower amounts of injury (Coyne and Bingham 1982). However, Helms et al. (1994), in a study that exposed three half-sib families of ponderosa pines to ozone using both branch and open-top chambers, found no relation between intrinsic rates of gas exchange and susceptibility to ozone in these trees.

In addition, ozone injury itself can influence stomatal conductance. Ozone-injured older age cohorts of needles have significantly lower rates of conductance and photosynthesis than noninjured needles, as observed in experimental exposures (Beyers et al. 1992) and in the field (Coyne and Bingham 1981). These reductions in rates of conductance and photosynthesis become apparent when the chlorotic mottle covers more than 30% of the leaf (Beyers et al. 1992; Patterson and Rundel 1995). However, in the ROPIS study in which ponderosa pine seedlings were exposed to ozone and other stresses for three growing seasons, current-year foliage of severely injured seedlings had higher rates of conductance and photosynthesis than current-year needles of control trees (Table 2.2; Beyers et al. 1992).

Miller (1965) reported the first evidence of increased rates of CO_2 fixation in current-year foliage of severely ozone-injured ponderosa pines. This suggests that pines that have lost older age classes of needles because of ozone-induced senescence and premature needle abscission can compensate for the loss by increased rates of gas exchange and carbon fixation in the remaining foliage. This compensatory increase in gas exchange in the remaining age classes of needles may explain, in part, the persistence of severely ozone-injured ponderosa pines that retain a maximum of only two age classes of needles.

Table 2.2. Net photosynthesis at light saturation (A^{max}, $\mu mol\ CO_2\ m^{-2}\ s^{-1}$) ± SE of three age classes of ponderosa pine needles measured during late September, 1990. Values are means of three fascicles on each of four trees, except where indicated

Watering regime	Age class	Ozone treatment		
		Carbon-filtered air	Nonfiltered air	150% of nonfiltered ozone
Well-watered	2-year-old	2.94 ± 0.12	2.73 ± 0.19	3.76 ± 0.22[a]
	1-year-old	3.84 ± 0.21	4.11 ± 0.11	3.12 ± 0.27[b]
	Current	3.57 ± 0.20	3.64 ± 0.14	5.01 ± 0.42
Drought-stressed	2-year-old	2.65 ± 0.12	2.93 ± 0.19	2.82 ± 0.18[c]
	1-year-old	3.87 ± 0.15	3.69 ± 0.14	4.04 ± 0.11[c]
	Current	4.22 ± 0.14	3.71 ± 0.15	4.33 ± 0.20

[a]Mean of three fascicles on one tree and one fascicle on another.
[b]Mean of three fascicles on each of two trees.
[c]Mean of three fascicles on each of three trees.

2.4.3 Growth Responses of Seedlings and Saplings

Because of the inherent variability of growth patterns in natural populations of trees, growth as a response variable has been difficult to study in experimental ozone exposures. Naturally regenerating ponderosa pine saplings, approximately 20–25 years old, were grown in charcoal-filtered and nonfiltered greenhouse enclosures from May to September from 1968 to 1973 in the San Bernardino Mountains (Miller and Elderman 1977). The growth of these saplings and of a similar cohort of sapling exposed to ambient air was measured for that period. After 5 years, during which summer ozone concentrations were typically high (Miller 1973), the weights of needles retained per age class and both radial and height growths of saplings in the filtered-air house were significantly greater than those of nonfiltered or ambient trees. This increased growth of the filtered-air trees became statistically significant after 4 years of growth.

The mass of needles retained in each age class for mature ponderosa and Jeffrey pines varies as a function of the effect of soil moisture availability, number of needles, and length of needles per whorl. Needle whorls were harvested and weighed for every year present from a sample of five branches from each of 131 trees in 1991. Current year (1991), 1-year-old (1990), 2-year-old (1989), and 3-year-old whorls were present on all 131 trees at Barton Flats in the San Bernardino mountains where moderate ozone injury is present (Fig. 2.3). The progressively lighter weight from 1988 to 1990 reflects a severe drought period. Needle weight recovered in 1991 when adequate precipitation returned. Fewer trees retained a complete needle whorl, of a particular year of origin, as whorl age became progressively older, namely, 1987 to 1983. This reduction reflects a combination of normal aging and the accelerated needle abscission caused by ozone injury.

This example illustrates how drought stress drastically reduces the quantity of photosynthetically active needle mass (1989 and 1990 whorls). Ozone exposure remains fairly constant each year so recovery from the drought was hampered by ozone injury to the 1989 and 1990 needle whorls in 1991 and later years. This illustrates how ozone and drought stresses are coupled over time, mainly as sequential events. The frequency of past severe drought periods can be characterised by the absence of measureable tree rings (Fig. 2.5).

Experimental ozone exposures of seedlings or saplings lasting 2 years or less have yielded no significant radial or height growth responses of ponderosa pine (Bytnerowicz and Temple 1993; Helms et al. 1994) or other species (Bytnerowicz 1992) to ozone. In the ROPIS study, in which ponderosa pine seedlings were exposed to three levels of ozone for three

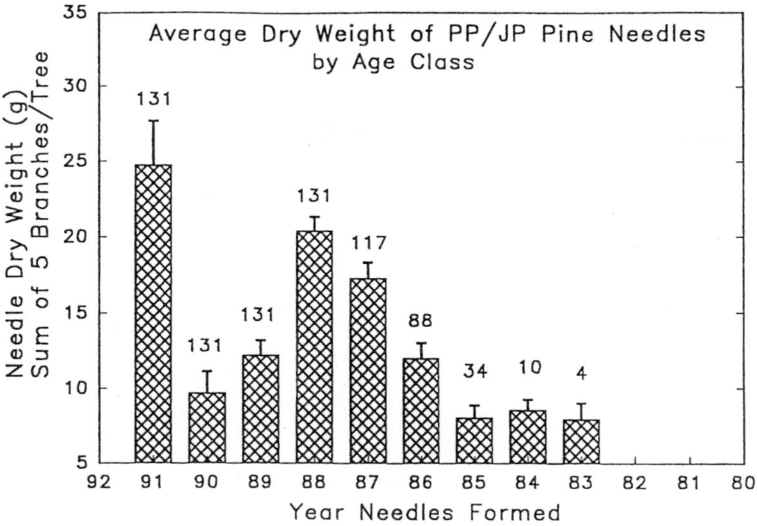

Fig. 2.3. Needle dry weights for individual needle whorls formed each year from 1983 to 1991 illustrate the effect of prolonged drought on reducing needle mass (1989 and 1990). The number above each bar indicates the number of trees (total = 131) retaining a particular age of whorl during the 1983-1991 period, for a stand with moderate ozone injury. *PP* Ponderosa pine; *JP* Jeffrey pine

growing seasons, radial growth of well-watered trees was significantly reduced by the high ozone treatment (Temple et al. 1993). Reductions in growth were proportional to the amount of ozone injury, but significant decreases in radial growth became apparent only when seedlings had lost two or more years of needles due to ozone-induced premature needle abscission (Fig. 2.4; Temple and Miller 1994). However, drought-stressed trees exposed to ozone had significant amounts of ozone injury but no significant reductions in radial growth (Temple et al. 1993). These results suggest that dendroecological studies of incremental growth of ponderosa pine in response to ozone would show a positive correlation between ozone exposure and tree growth only for trees with severe amounts of ozone injury, i.e., only one or two annual whorls retained (see Sect. 2.5).

2.5 Effects of Chronic Ozone Exposure on Radial Growth in the Sierra Nevada and San Bernardino Mountains

Mature tree productivity differs substantially from seedlings. Mature trees have larger carbohydrate reserves, rooting area, and different stemwood

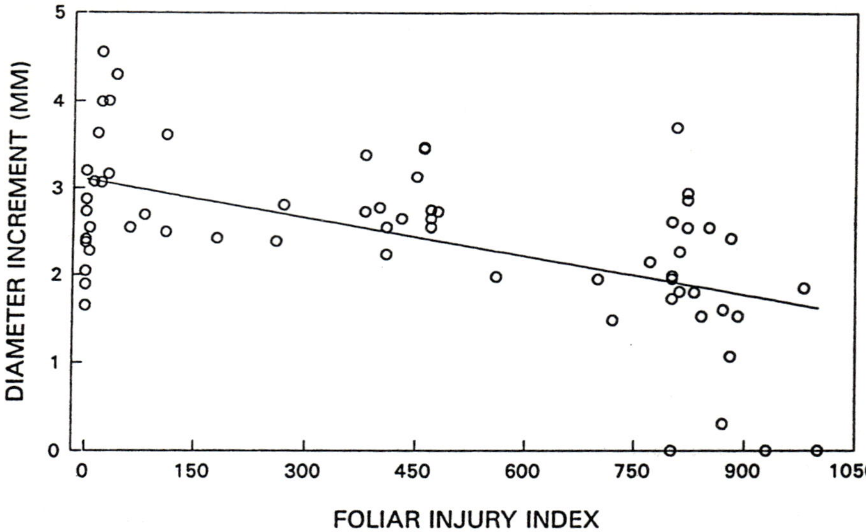

Fig. 2.4. Seedling diameter growth in relation to an index of foliar injury to ponderosa pine seedlings exposed for three growing seasons (June-October) to a mean of 87 nl l^{-1} ozone (Temple and Miller 1994)

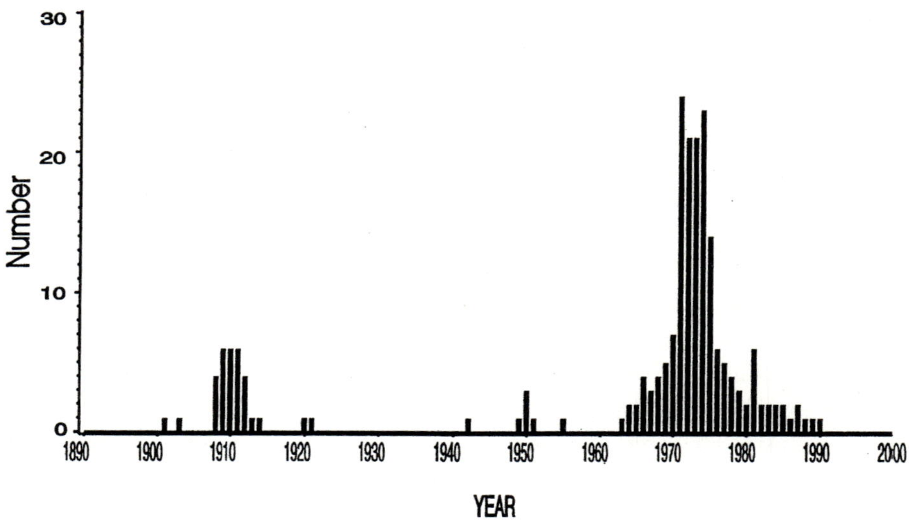

Fig. 2.5. Numbers of missing rings occuring at 11 sites located in high ambient ozone locations in the San Bernardino Mountains. The missing rings around 1910 are probably due to a prolonged drought occuring during 1895–1907

accumulation patterns than seedlings. Seedlings survive if they can rapidly establish rooting areas to exploit deeper soil water, and rapidly increase height growth to compete with understory vegetation and other seedlings for light (van der Valk 1992). This makes it difficult to directly extrapolate from seedling growth to mature tree growth.

2.5.1 Seasonal Radial Growth Responses

Increasing ozone exposure reduces carbohydrate production due to reduced needle retention and function. Carbohydrate partitioning and allocation to woody tissue and roots is also reduced (Cooley and Manning 1988). Radial and volumetric growth may then be reduced during the growing season. Loss of seasonal ring patterns may also occur. Several studies of ponderosa pine (Ohmart and Williams 1979; Gemmill et al. 1982; Peterson et al. 1995) in the San Bernardino Mountains reported difficulties in cross-dating due to low growth rates and loss of ring boundaries at sites with high levels of ozone exposure. This was also observed in a recent study of ponderosa pine growth at high ambient ozone locations (Arbaugh and Miller, unpubl.). This may indicate that seasonal phenology of severely affected trees has been altered, since latewood formation is reduced or eliminated.

2.5.2 Long-Term Radial Growth Responses

The long-term effects of ozone pollution on mature trees are difficult to separate from those of other stress factors affecting tree growth. Competition, soil water stress, weather changes, and pathogens may cause growth changes similar to those caused by ozone. Interactions between various stressors may also synergistically increase or reduce the effect of single stressors, making it difficult to identify a single cause for the observed growth change (Arbaugh and Peterson 1992).

2.5.2.1 Direct Effects of Ozone on Radial Growth

Severe ozone may cause partial or complete reduction of radial growth, or alteration of the pattern of tree-ring formation for one or more decades (Gemmill et al. 1982; Ohmart and Williams 1979; Peterson et al. 1995). Several studies have observed sustained radial growth decreases (ring width or basal area increment) in areas with high ambient ozone (Peterson et al. 1987; Miller 1992). In the Sierra Nevada, Jeffrey pine had decreased residual ring widths during the pollution period 1960–1988 relative to prepollution growth and growth at protected sites. In the San Bernardino Mountains, trees at high ambient ozone sites have had large reductions in radial growth since the 1950s, 1–10 years of missing rings, and altered ring formation. Missing and altered rings were also found in a study of big-cone Douglas fir in areas exposed to high pollution (Peterson et al. 1995) and in a study of 11 ponderosa pine sites in a high ambient ozone location (Fig. 2.5). Trees located further in the interior of the mountain range (0.5–1 km) had less evidence of growth decline relative to trees on the first ridge overlooking the San Bernardino valley where ozone flux is highest (Fig. 2.6).

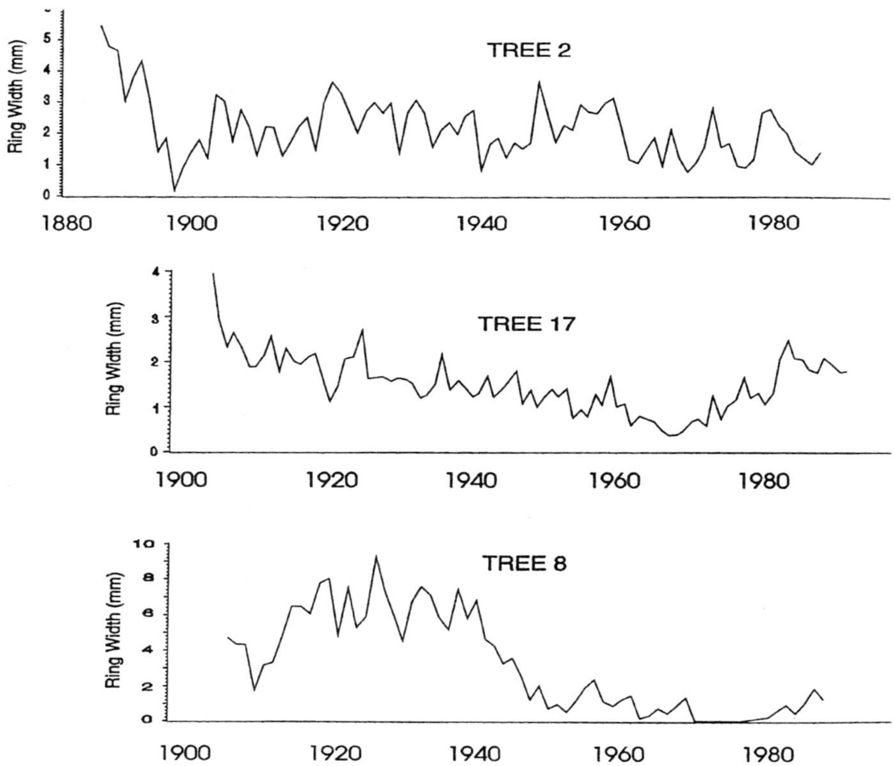

Fig. 2.6. Three individual ring width patterns from a high ambient ozone location in the San Bernardino Mountains. *Tree 2* is largely unaffected by pollution, while *Tree 17* has a temporary growth suppression starting during the 1960s and ending in the 1970s. *Tree 8* has suppressed growth starting during the 1940s, and missing rings in the 1970s. The initial growth decline in the 1940s is partially due to a prolonged drought starting at that time

2.5.2.2 Effects of Ozone Together with Climate on Tree Growth

The amount of precipitation and its pattern of change from year to year is the single most important factor influencing tree growth and establishment of new seedlings of the Californian mixed conifer type. Ring width growth of Jeffrey pine in the Sierra Nevada and big-cone Douglas-fir in the San Bernardino Mountains was found to be greater at sites with high water availability and severe crown damage than at sites with low availability of water and low crown damage. Ponderosa pine in the Sierra Nevada did not have significantly different radial growth at sites with severe crown damage relative to sites with low crown damage (Peterson et al. 1987, 1989) when water availability was comparable. Differences between ponderosa pine and Jeffrey pine growth in the Sierra Nevada was attributed to shallower, undeveloped soils that typify Jeffrey pine sites and deeper, more developed soils found at ponderosa pine sites. Peterson et al. (1995) also observed that historically slower growing trees, possibly those at a competitive disadvantage, had longer periods of reduced growth following droughts.

2.5.3 Relationship Between Needle Retention and Radial Growth Reductions

The relationship between foliar damage and radial growth changes has proved difficult to quantify due to the strong dependence of radial growth on water availability. Studies in the San Bernardino Mountains (Ohmart and Williams 1979; Miller 1992) found growth declines only for trees with the highest crown damage levels. In a study by Peterson et al. (1995) lower needle retention did not affect average basal areas increment for big-cone Douglas-fir in the San Bernardino Mountains, but there were significantly more trees with large growth declines at sites with high ozone levels. Similarly, Miller (1992) found that the most poorly growing trees had the worst foliar damage, but total relative growth between sites did not vary. This pattern was also observed in a ponderosa pine site with growth decline in the Sierra Nevada. The median basal area increment increased while the slowest growing 5% decreased growth during the pollution period, but not before (Arbaugh et al. 1988).

2.5.4 Development and Composition of Ozone-Exposed Forests Under Euro-American Settlement Influences

Many factors influence forest development in the Sierra and San Bernardino Mountains, most of which are either directly or indirectly related to Euro-American settlement. The first recorded timber harvests in the San Bernardino mountains began in 1852 when settlers from the nearby San Bernardino valley built a crude road to access the western portion of the mountains. At that time the climax forest consisted primarily of large overstory pines, including mainly ponderosa pine with smaller numbers of sugar pine, white fir, incense cedar, and black oak. This forest structure was considered a fire-determined succession because it was perpetuated by frequent wildfires, which tended to thin out competing species occupying the understory and those mature individuals of species that did not have a sufficiently thick bark to protect them from fire damage. Harvesting of these ancient forest stands progressed eastwards, culminating in about 1912 when most of the easily available timber supply had been exhausted. Fire exclusion started shortly after that time, and continues through to the present. As a result, today's stands are second or third growth, and contain larger numbers of fire-sensitive (and ozone-tolerant) species than the ancient forest.

High ambient ozone levels probably accelerate forest succession due to fire exclusion. A group of 18 plots established for the multidisciplinary study in 1978 (Miller and Elderman 1977; Taylor 1980) were remeasured at 4-5-year intervals from 1978 (Fig. 2.7). In areas with low ozone concentrations, ponderosa pine and Jeffrey pine increased their basal area, even in drier years. At sites with high concentrations of ozone, the total basal area either remained the same or declined. However, it was less than that of competing species over the period of the study.

2.5.4.1 Chronic Ozone Injury and Mortality of Pines Due to Bark Beetle

Ozone injured trees are more vulnerable to immediate mortality from insect/disease complexes. Bark beetle (Scolytidae, Coleoptera) attack is almost always a direct cause of mortality (Stark et al. 1968). Annual counts of trees killed by the bark beetle are available for the Lake Arrowhead area since 1921 (Taylor 1973). This record shows the expected peaks associated with dry years and troughs during moist years until 1951 when counts of killed trees remained high every year until the end of the record in 1971. The sustained high level of mortality is coincident with the beginning and continuation of tree injury due to ozone. In remeasured

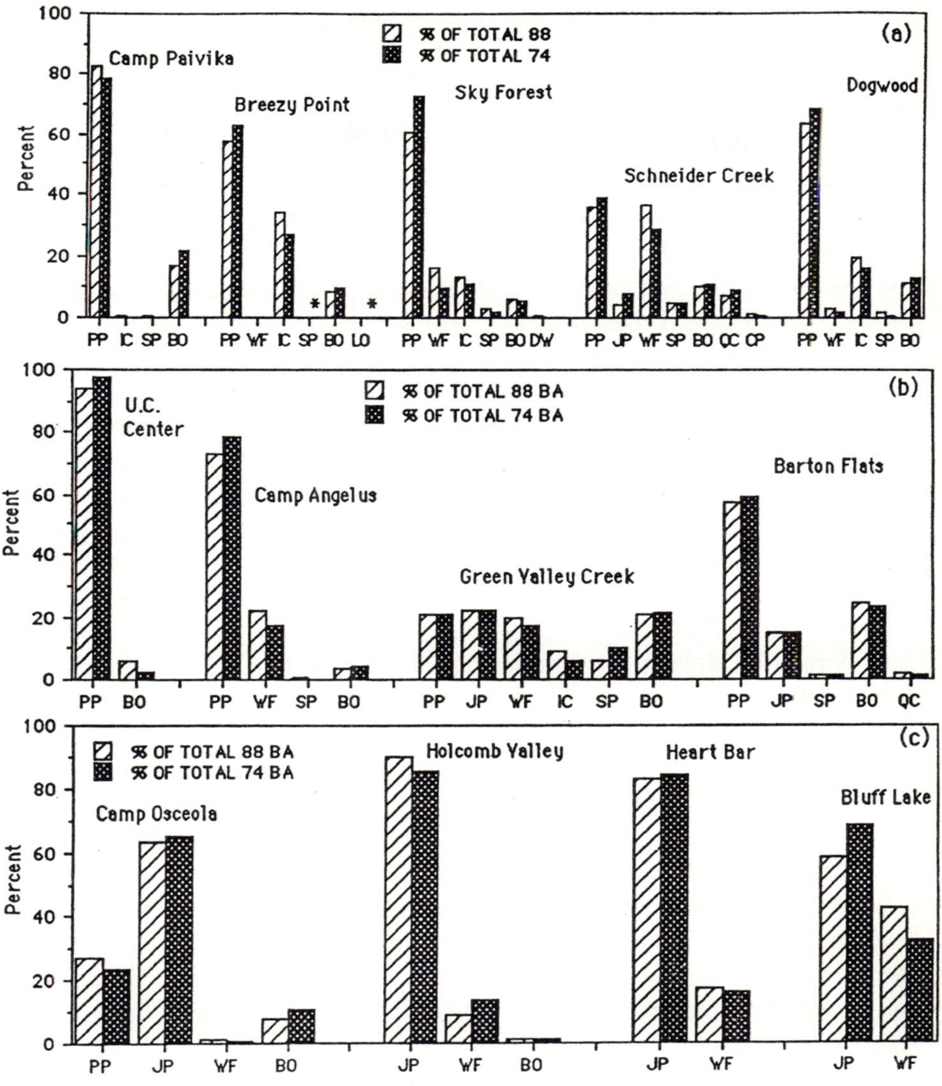

Fig. 2.7. Total basal area for each species as percent of the total basal area for all species in 1974 and 1988 on **a** plots with severe to moderate injury; **b** plots with slight injury; **c** plots with very slight injury or no visible symptoms. *PP* ponderosa pine; *WF* white fir; *IC* incense cedar; *SP* sugar pine; *CP* coulter pine; *JP* Jeffrey pine; *BO* black oak; *DW* dogwood; *QC quercus crysolepis*

plots from a high ozone area of the San Bernadino Mountains, 25% of the ponderosa pines died between 1973 and 1992, while the highest mortality for other species was 15%. Mortality in severe beetle mortality plots in the San Jacinto Mountains was also much higher for ponderosa, Jeffrey, and Coulter pines than for other species (Savage 1994).

2.5.4.2 Ozone Influences on Recruitment and Mortality of Ponderosa Pine Seedlings

Regeneration in present-day montane forests is higher for tree species which are less susceptible to ozone (Miller 1973; Savage 1994). Fire exclusion is probably the primary cause of changes in ozone susceptibility. Ozone may accelerate the development of the difference in regeneration rates via a reduction in seed production, and poorer germination conditions due to higher rates of needle abscission that lead to thicker litter layers. The average recruitment and average mortality of ponderosa pines (≤ 1 m tall) was determined for 1974–1977 (higher ozone, lower moisture), 1978–1983 (lower ozone, higher moisture), and 1984–1988 (lower ozone, lower moisture) at eleven plots (Miller and McBride 1989). Recruitment was higher than mortality only during the 1978–1983 period, and lowest during 1984–1988. Mortality was highest during the 1974–1977 period of higher ozone and lower precipitation, and lowest during the moist 1978–1983 period.

2.5.5 Status of Models for Simulating Ozone Effects on Tree Growth and Stand Development

Several models have been developed to characterize ozone effects on tree and forest growth. These include modification of PROGNOSIS (Droessler 1989), a growth-and-yield model, and the JABOWA-based Gap model AIRSILVA (Miller and Ladd 1989).

Most models presently being used in ozone studies are physiological models. These models are able to track element flows through the system, allowing more information about ozone uptake, carbohydrate reduction, and water and nutrient usage to be directly incorporated into the model structure. Presently, TREGRO (Weinstein et al. 1994), plant growth simulation model (PGSM; Chen et al. 1994), and the nutrient cycling model (NUCM; Johnson and Poth, pers. comm.; Fenn et al. 1966) are being examined for use in ozone-stressed forests. The parameterization and testing of these models are in the initial stages, and they are likely to be modified before final versions are released.

In addition to single model efforts, an attempt is being made to link models, (Weinstein et al. 1994). In this effort a physiological model, TREGRO, is being used to develop individual tree response surfaces that are used in a JABOWA-based gap model, ZELIG, to predict changes in stand development. As with the physiological models, this model is also in the initial stages of development. In most models, ozone is assumed to reduce actual photosynthesis, generally as a simple functional relationship between ambient ozone concentrations and reduced maximum photosynthesis. Various ozone exposure indices are utilized in testing models, for example Sum0, Sum06, W126, or simply 7- and 12-h daylight averages.

2.6 Summary

Chronic ozone damage to forests of the western part of the United States is confined to the western slope of the central and southern Sierra Nevada, and to the sections of the transverse mountain ranges in southern California nearest the source of ozone pollution. California's summer climate, topography, and densely urbanized air basins combine to create ideal conditions for the accumulation of pollution and its long-range transport to adjacent forest areas. The species comprising the Californian mixed conifer forest type, particularly ponderosa pine and Jeffrey pine, are the most affected by chronic ozone exposure. Air pollution control measures have resulted in a gradual decrease of ozone concentrations in both urban and rural locations. An evaluation of whether the current situation should be considered in the context of "forest decline" must be based on an agglomeration of information ranging from mechanisms of ozone injury at the leaf and single tree level to the response of competing species in the mixed conifer forest. The role of interacting stressors, particularly periodic droughts, must also be evaluated. There has been no compelling incentive to examine changes in leaf antioxidant systems or to identify biochemical markers of early ozone injury to foliage since the appearance of visible symptoms (chlorotic mottle) serves for detection.

Physiological studies with ponderosa and Jeffrey pine show that stomatal conductance must be high enough to permit the entrance of ozone into the leaf during the times when ozone polluted air is transported to forests. The concept that there is an "effective dose" required to cause leaf injury is very difficult to measure and interpret. Therefore, we cannot clearly describe the ozone exposure–response relationship for ponderosa and Jeffrey pine. The results of experiments with seedlings exposed to ozone in open-top chambers have provided valuable information on exposure response, but these results can not be extrapolated to larger trees in forest stands where other stressors cannot be easily evaluated.

Drought stress and its relationship to chronic ozone injury has been investigated the most by inducing drought in seedling experiments and by interpreting tree ring patterns over an extended time. These approaches suggest that simultaneous drought and ozone stress leads to a significant reduction of ozone injury in controlled experiments; in forest stands, the various sequential combinations of drought and ozone stress result in "episodes" of tree decline due first to extended drought followed by a more "effective" dose of ozone during recovery periods.

Since there is a range of sensitivity to ozone in a species population of ponderosa and Jeffrey pine, the most sensitive individuals will suffer greater competition from companion trees of the same or other species. White fir and incense cedar in all age classes and especially newly established seedlings are the most successful competitors in stands with the most ozone exposure. Tree-ring patterns, including consecutive years of missing rings, and records of higher rates of mortality suggest that "forest decline" is episodic, depending on the frequency of consecutive drought years and the levels of ozone exposure. Mature forest experimental manipulations need to be conducted to test this hypothesis and to more precisely determine the connections between single stressor seedling and sapling studies and mature forest observational studies. Underlying the ozone effect is the simultaneous accumulation of nitrogen in leaf litter and soils. The greater abundance of nitrogen in forest ecosystems that evolved under lower levels of nitrogen is an additional variable the interaction of which with long-term ozone injury are yet to be determined (Fenn et al. 1996). Early indications of lower soil pH and lower base saturation have been reported at heavily polluted sites in the San Bernardino mountains.

Seedling experiments have shown a remarkable capacity of individual trees to compensate for ozone injury by increasing the photosynthetic efficiency of the youngest needles. The induction of this behaviour in seedlings may take 2–3 years of exposure. Similarly, chronically injured saplings took 3–4 years before the needle and stem growth of trees maintained in a filtered-air greenhouse was significantly greater than those exposed to ambient, ozone-polluted air inside and outside of greenhouses. One of the most important consequences of these results i.e., the tree response capability, is that controlled experiments and remeasurement studies in the forest require at least 3–4 years before significant changes are detectable.

The outlook for ozone exposure and forest health in California depends on how existing and proposed control measures will be able to cope with larger numbers of ozone precursor sources as well as the geographic spread of these sources. The climate, including the pattern of drought years and a warming trend (if confirmed), will be the second important variable regulating tree response. Worst-case results from remeasurement plots in the San Bernardino mountains show the beginning of a shift from ponderosa and Jeffrey pine dominance (a product of frequent fires predating Euro-American settlement) toward an ascendancy of several ozone-tolerant, fire-sensitive species. The interpretation of possible outcomes may be aided by both tree growth and stand succession models which are now in the testing stages of development.

References

Arbaugh MJ, Peterson DL (1992) Stemwood production patterns in ponderosa pine: effects of stand dynamics and other factors. Res Pap PSW-RP-217, USDA Forest Service, PSW Experiment Station, Albany, California

Arbaugh MJ, Peterson DL, Strauss DJ (1988) Growth trends in the mixed conifer forest of the Sierra Nevada: An extended analysis. Final Report to Western Conifers Research Cooperative, Corvallis, Oregon

Beyers JL, Riechers GR, Temple PJ (1992) Effects of long-term ozone exposure and drought on the photosynthetic capacity of ponderosa pine (*Pinus ponderosa* Laws.). New Phytol 122:81–90

Bohm M (1992) Air quality and deposition. In: Olson RK, Binkley D, Bohm M (eds) The responses of western forests to air pollution. Ecological studies, vol 97. Springer Berlin Heidelberg, New York, pp 63–152

Bytnerowicz A (1992) Determination of the responses of big cone Douglas-fir to photochemical smog and mineral nutrition. Final report to Southern California Edison, Contract No C2061920, USDA Forest Service, PSW Experiment Station, Riverside, California

Bytnerowicz A, Temple PJ (1993) Growth, physiological, and biochemical responses of ponderosa pine (*Pinus ponderosa*) to ozone. Final report to California Air Resources Board, Contract No. A733–137, A833–083, A033–056, Statewide Air Pollution Research Center, Univ Calif Riverside, California

California Air Resources Board (1987) Effect of ozone on vegetation and possible alternative ambient air quality standards. ARB, Research Div Sacramento, California

Carroll JJ, Baskett RL (1977) Dependence of air quality in a remote location on local and mesoscale transports: a case study. J Appl Meteor 18:474–486

Chen CW, Tsai WT, Gomez LE (1994) Modeling responses of ponderosa pine to interacting stresses of ozone and drought. For Sci 40:267–288

Cooley RD, Manning WJ (1987) The impact of ozone on assimilate partitioning in plants: a review. Environ Pollut 47:95–113

Coyne PI, Bingham GE (1981) Comparative ozone dose response of gas exchange in a ponderosa pine stand exposed to long-term fumigations. J Air Pollut Control Assoc 31:38–41

Coyne PI, Bingham GE (1982) Variation in photosynthesis and stomatal conductance in an ozone-stressed ponderosa pine stand: light response. For Sci 28:257–273

Davidson A (1993) Update on ozone trends in California's south coast air basin. J Air Waste Manage Assoc 43:226–227

Droessler TD (1989) Methods for testing the variance of PROGNOSIS: Sensitivity to simulated ozone. In: Multiresource management of ponderosa pine forests. Gen Tech Rep RM-185, USDA, Rocky Mountain Forest and Range Experiment Station, Fort Collins, Colorado

Duriscoe DM (1990) Cruise survey of oxidant air pollution injury to *Pinus ponderosa* and *Pinus jeffreyi* in Saguaro National Monument, Yosemite National Park, and Sequoia and Kings Canyon National Parks. NPS/AQD-90/003. Available from: National Park Service, Air Quality Division, Denver

Duriscoe DM, Stolte KW (1989) Photochemical oxidant injury to ponderosa (*Pinus ponderosa* Dougl. ex Laws) and Jeffrey pine (*Pinus jeffreyi* Grev. and Balf.) in the national parks of the Sierra Nevada of California. In: Olson RK, Lefohn AS (eds) Effects of air pollution on western forests. Transactions Ser No 16. Air and Waste Management Association, Pittsburgh, pp 261–278

Edinger JJ, McCutchan MH, Miller PR, Ryan BC, Shroeder MJ, Behar JV (1972) Penetration and duration of oxidant air pollution in the South Coast Air Basin of California. J Air Pollut Control Assoc 22:882–886

Evans LS, Miller PR (1972a) Ozone damage to ponderosa pine – a histological and histochemical appraisal. Am J Bot 59:297–304

Evans LS, Miller PR (1972b) Comparative needle anatomy and relative ozone sensitivity of four pine species. Can J Bot 50:1067–1071

Ewers FW, Schmid R (1981) Longevity of needle fascicles of *Pinus longaeva* (bristlecone pine) and other North American pines. Oecologia 51:107–115

Fenn ME, Poth MA, Johnson DW (1996) Evidence for nitrogen saturation in the San Bernardino mountains of southern California. For Ecol Manage 81 (in press)

Gemmill B, McBride JR, Laven RD (1982) Development of tree-ring chronologies in an ozone air pollution-stressed forest in southern California. Tree-Ring Bull 42:23–31

Goldstein R, Ferson S (1994) Responses of plants to interacting stresses (ROPIS): program rationale, design and implications. J Environ Qual 23:407–411

Grulke NE, Miller PR (1994) Changes in gas exchange characteristics during the life span of giant sequoia: implications for response to current and future concentrations of atmospheric ozone. Tree Physiol 14:659–668

Helms JA, Anderson PD, Houpis JL (1994) Gas exchange by *Pinus ponderosa* in relation to atmospheric pollutants. Final report to California Air Resources Board, Contract no A132-101, Dept of Environ Sci Policy Manage, Univ Calif, Berkeley

Hoffer TE, Farber RJ, Ellis EC (1982) Background continental ozone levels in the rural US Southwestern desert. Science Total Environ 23:17–30

Hogsett WE, Tingey DT, Hendricks C, Rossi D (1989) Sensitivity of western conifers to SO_2 and seasonal interaction of acid fog and ozone. In: Olson RK, Lefohn AS (eds) Effects of air pollution on western forests, APCA Transactions Ser No 16, AWMA, Pittsburgh, pp 469–491

Ludwig FL, Shelar E Jr (1980) Empirical relationships between observed ozone concentrations and geographic areas with concentrations likely to be above 120 ppb. J Air Pollut Control Assoc 30:894–897

Miller J, Ladd L (1989) AIRSILVA: a model of mixed conifer forest response to multiple stress. In: Multiresource management of ponderosa pine forests. Gen Tech Rep RM-185, USDA Forest Service, Fort Collins, Colorado

Miller PR (1965) The relationship of ozone to the suppression of photosynthesis and to the cause of the chlorotic decline of ponderosa pine. PhD Thesis, Univ Calif, Berkeley, 129 pp

Miller PR (1973) Oxidant-induced community change in a mixed conifer forest. In: Air pollution damage to vegetation. Adv Chem Ser 122:101–117

Miller PR (1983) Ozone effects in the San Bernardino National Forest. In: Proc Worksh on Air Pollution and the productivity of the forest. Izaak Walton League of America, Arlington, pp 161–197

Miller PR (1992) Mixed conifer forests of the San Bernardino mountains, California. In: Olson RK, Binkley D, Bohm M (eds) The response of western forests to air pollution. Springer Berlin Heidelberg, New York, pp 461–497

Miller PR, Elderman MH (eds) (1977) Photochemical oxidant air pollution effects on a mixed conifer forest ecosystem. US EPA Rep no EPA 600/3-77-104. SAPRC, Univ Calif, Riverside

Miller PR, Evans LS (1974) Histopathology of oxidant injury and winter fleck injury on needles of western pine. Phytopathology 64:801–806

Miller PR, McBride JR (1989) Trends of ozone damage to conifer forests between 1974 and 1988 in the San Bernardino mountains of southern California. In: Bucher JB, Bucher-Wallin I (eds) Proc 14th Int Meet Specialists in air pollution effects on forest ecosystems. Interlaken, pp 61–68

Miller PR, Millecan AA (1971) Extent of oxidant air pollution damage to some pines and other conifers in California. Plant Dis Rep 55:555–559

Miller PR, Van Doren RE (1982) Ponderosa and Jeffrey pine foliage retention indicates ozone dose response. USDA, Forest Service, Pacific Southwest Research Station, Albeng, California, Gen Tech Rep PSW-58, p 621

Miller PR, Parmeter JR, Taylor OC, Cardiff EA (1963) Ozone injury to the foliage of *Pinus ponderosa*. Phytopathology 53:1072–1076

Miller PR, Kickert RN, Taylor OC, Arkeley RJ, Cobb FW Jr, Dahlsten DL, Gersper PJ, Luck RF, McBride JR, Parmeter JR Jr, Wenz JM, White M, Wilcox WW (1977) Photochemical oxidant air pollution effects on a mixed conifer forest ecosystem – A progress report, 1975–1976. Statewide Air Pollution Research Center, University of California, Riverside, 338 pp

Miller PR, Longbotham GJ, Longbotham CR (1983) Sensitivity of selected western conifers to ozone. Plant Dis 67:1113–1115

Miller PR, Taylor OC, Poe MP (1986) Spatial variation of summer ozone concentrations in the San Bernardino mountains. Pap 86–39.2, Annu Meet Air Pollution Control Association, Pittsbursh, 14 pp

Miller PR, McBride JR, Schilling SL, Gomez AP (1989) Trend of ozone damage to conifer forests between 1974 and 1988 in the San Bernardino Mountains of southern California. In: Olson RK, Lefohn AS (eds) Effects of air pollution on western forests. Transactions Ser No 16, Air and Waste Management Association, Pittsburgh, pp 309–324

Miller PR, McBride JR, Schilling SL (1991) Chronic ozone injury and associated stresses affect relative competitive capacity of species comprising the California mixed conifer forest type. In: Memorias del primar simposial nacional, Agricultura sostenible: Una opcion para el desarrollo sin deterioro ambiental. Available from: Comision de Estudios Ambientales, Colegio de Postgraduados, Montecillo, Edo. Mexico, pp 161–172

Miller PR, Grulke NE, Stolte KW (1994) Effects of air pollution on giant sequoia ecosystems. USDA, Forest Service, Pacific Southwest Research Station, Albany, California, Gen Tech Rep PSW-151, pp 90–98

Ohmart CP, Williams CB (1979) The effects of photochemical oxidants on radial growth increment for five species of conifers in the San Bernardino National Forest. Plant Dis Rep 63:1038–1042

Patterson MT, Rundel PW (1989) Seasonal physiological responses of ozone stressed Jeffrey pine in Sequoia National Park, California. In: Olson RK, Lefohn AS (eds) Effects of air pollution on western forests. APCA Transactions Seri No 16, AWMA, Pittsburgh, pp 419–427

Patterson MT, Rundel PW (1995) Stand characteristics of ozone-stressed populations of *Pinus jeffreyi* – extent, development, and physiological consequences of visible injury. Am J Bot 82:150–158

Pedersen BS (1989) Ozone injury to Jeffrey and ponderosa pines surrounding Lake Tahoe, California and Nevada. In: Olson, RK, Lefohn AS (eds) Effects of air pollution on western forests. Transactions Ser No 16, Air and Waste Management Association, Pittsburgh, pp 279–292

Peterson DL, Arbaugh MJ (1988) An evaluation of the effects of ozone injury on radial growth of ponderosa pine (*Pinus ponderosa*) in the southern Sierra Nevada. Air Pollut Control Assoc 38:921–927

Peterson DL, Arbaugh MJ (1992) Mixed conifer forest of the Sierra Nevada. In: Olson RK, Binkely D Bohm M (eds) The response of western forests to air pollution. Ecological studies 97. Springer Berlin Heidelberg, New York, pp 433–460

Peterson DL, Arbaugh MJ, Wakefield VA, Miller PR (1987) Evidence of growth reduction in ozone-injured Jeffrey pine (*Pinus jeffreyi* Grev. and Balf.) in Sequoia and Kings Canyon National Parks. Air Pollut Control Assoc 37:906–912

Peterson DL, Arbaugh MJ, Robinson LJ (1989) Ozone injury and growth trends of ponderosa pine in the Sierra Nevada. In: Olson RK, Lefohn AS (eds) Effects of air pollution on western forests. Transactions Ser No 16, Air and Waste Management Association, Pittsburgh, pp 293–308

Peterson DL, Arbaugh MJ, Robinson LJ (1991) Regional growth changes in ozone-stressed ponderosa pine (*Pinus ponderosa*) in the Sierra Nevada, California, USA. Holocene 1:50–61

Peterson DL, Silsbee DG, Poth M, Arbaugh MJ, Biles FE (1995) Growth response of big cone Douglas-fir (*Pseudotsuga macrocarpa*) to long term ozone exposure in southern California. J Air Waste Management Assoc 43:36–45

Pronos J, Vogler DR (1981) Assessment of ozone injury to pines in the southern Sierra Nevada, 1979/1980, USDA Forest Service, Pacific SW Region, San Francisco, Forest Pest Manage Rep 81–20

Pronos J, Vogler DR, Smith RS (1978) An evaluation of ozone injury to pines in the southern Sierra Nevada. USDA Forest Service Pacific Southwest Region Forest Pest Management Rep, 78–1. Available from: USDA Forest Service, San Francisco

Rice PM, Boldi RA, Carlson CE, Torangeau PC, Gordon CC (1983) Sensitivity of *Pinus ponderosa* foliage to airborne phytotoxins: Use in biomonitoring. Can J For Res 13:1083–1091

Richards BL Sr, Taylor OC, Edmunds GF Jr (1968) Ozone needle mottle of pine in southern California. J Air Pollut Control Assoc 18:73–77

Savage M (1994) Anthropogenic and natural disturbance and patterns of mortality in a mixed conifer forest in California. Can J For Res 24:1149–1159

Scharpf RF (1993) Diseases of Pacific coast conifers. USDA Forest Service, Agric Handbook 521, 199 pp

Stark RW, Cobb FW Jr (1969) Smog injury, root diseases and bark beetle damage in ponderosa pine. Calif Agric 23:13–15

Stark RW, Miller PR, Cobb FW Jr, Wood DL, Parmeter JR Jr (1968) Photochemical oxidant injury and bark beetle (Coleoptera:Scolytidae) infestation of ponderosa pine. I. Incidence of beetle infestation in injured trees. Hilgardia 39:121–126

Stolte KW, Duriscoe DM, Cook ER, Cline SP (1992) Methods of assessing responses of trees, stands and ecosystems to air pollution. In: Olson R K, Binkley D, Bohm M (eds) The responses of western forests to air pollution. Ecological studies 97. Springer Berlin Heidelberg, New York, pp 259–330

Sudworth GB (1908) Forest trees of the Pacific slope. USDA, GPO Washington, DC

Taylor OC (1973) Oxidant air pollution effects on a western coniferous forest ecosystem: Task B, Historical background and proposed systems study of the San Bernardino mountain area. Statewide Air Pollution Research Center, Univ Calif, Riverside

Taylor OC (1980) Photochemical oxidant air pollution effects on a mixed conifer forest ecosystem. US Environmental Protection Agency, EPA-600/3-80-002, Corrallis Environmental Research Laboratory, 196 pp

Temple PJ (1988) Injury and growth of Jeffrey pine and giant sequoia in response to ozone and acidic mist. Environ Exp Bot 28:323–333

Temple PJ, Miller PR (1994) Foliar ozone injury and radial growth of ponderosa pine. Can J For Res 24:1877–1882

Temple PJ, Riechers GH, Miller PR (1992) Foliar injury responses of ponderosa pine seedlings to ozone, wet and dry acidic deposition, and drought. Environ Exp Bot 32:101–113

Temple PJ, Riechers GR, Miller PR, Lennox RW (1993) Growth responses of ponderosa pine to long-term exposure to ozone, wet and dry acidic deposition, and drought. Can J For Res 23:59–66

US Environmental Protection Agency (USEPA) (1986) Air quality criteria for ozone and other photochemical oxidants. EPA-600/8-84-020aF-ec. Available from NTIS, Springfield, Virginia, PB87-142949

Van der Valk AG (1992) Establishment, colonization and persistence. In: Glenn-Lewin DC, Peet RK, Veblen TT (eds) Plant succession theory and practice. Chapman & Hall, New York

Van Ooy DJ, Carroll JJ (1995) Spatial variation of ozone climatology on the western slope of the Sierra Nevada. Atmos Environ 29:1319–1330

Weinstein DA, Arthur MA, Laurence JA, Gollands B, Taylor G (1994) Predicting the effects of tropospheric ozone on regional productivity of ponderosa pine and white fir in the Pacific northwest. Report to the Environmental Protection Agecny, Ozone Program, Corvallis, Oregon

3 Ozone and Its Known and Potential Effects on Forests in Eastern United States

J.M. Skelly, A.H. Chappelka, J.A. Laurence and T.S. Fredericksen

3.1 Introduction

Tropospheric ozone is considered by many as the most important anthropogenic air pollutant confronting the forests of eastern United States. From the late 1970s through to the early 1990s, hundreds of investigations were initiated to investigate the potential role of acidic depositions as a direct cause of several notable declines of forest tree species indigenous to eastern United States (McLaughlin 1985). However, the review of the literature, regarding the existence of many previously described decline-type diseases, and the results of numerous investigations in the 1980s clearly demonstrated that many forest insects, pathogens, and abiotic stresses were parts of the causal complexes of the reported declines (Manion and Lachance 1992; Skelly and Innes 1994). Three of the six major conclusions of the National Acid Precipitation Assessment Program (NAPAP) "State of Science Report No. 16" addressing forest health and productivity in the US and Canadian forests are pertinent to this chapter. These three conclusions succinctly stated that:

1. Ozone is the pollutant of greatest current concern with respect to regional scale in North American forests.
2. Compared to ozone and many nonpollutant stress factors, acidic deposition appears to be a relatively minor factor affecting the health and productivity of most forests in the United States and Canada.
3. The vast majority of forests in the United States and Canada are *not* affected by decline. (Barnard et al. 1989)

The simple fact that such conclusions emerged from "acid rain" research, reported by the many cooperative programs of the NAPAP-Forest Response Program effort, is of interest. For the most part, these revelations concerning the importance of tropospheric ozone as an air pollutant for the health and productivity of North American forests are not new. Rather, these are historically correct statements regarding our previous under-

Ecological Studies, Vol. 127
Sandermann et al. (eds) Forest Decline and Ozone

standing of ozone as a mesoscale air pollutant with consequences for the forests of eastern United States (USEPA 1986). Such conclusions are well founded, based on the pioneering works of many air pollution and forest protection scientists of the late 1950s through to the 1970s. References to the known impacts of other "non-pollutant pests" on our eastern forest tree species numbered in the thousands in 1970 (Hepting 1971; Millers et al. 1989). Since there is no way to recognize each of the earlier contributors individually, we simply do so collectively.

As a scientific community we are now turning our attention to tropospheric ozone air pollution as a potential regional scale stressor of the long-term health and productivity of our eastern forests. A major difference in our approach in attempting to discern the potential role of ozone in altering forest health and productivity (compared to the approach of those driven by the "acid rain" funding) is the fact that the most current studies on ozone effects are not focussed on a specific decline concept. Rather, the more subtle effects which may be induced by ambient ozone exposures are being carefully pursued.

The current exposure response data regarding ozone-induced effects on eastern US forest tree species would suggest that sensitive species' responses range, for the most part, between chronic foliar injuries (manifested as adaxial stippling) and "hidden" injury. Hidden injury was defined by Stoklasa (1923) as productivity losses without the presence of visible foliar injuries. Keller (1977) later described such effects as "latent" injury.

Thus, within most current ozone and forest effects research programs, (with the exception of the San Bernardino Mountain studies; see Miller et al. this Vol.), we are not pursuing the dramatic decline of forest tree species as characterized through death of populations, or even acute symptoms of foliar necrosis following long-term exposures to regional scale ambient ozone. Rather, we are investigating far more subtle ozone induced effects on forest tree species in order to gain a better understanding of the previsual, physiological disturbances which may lead to growth and productivity effects. Such "hidden", "latent", or "subtle" effects may be of considerable importance to the health and productivity of the forests of eastern United States. As we begin to deal with complicated issues of forest management and environment quality, we realize that actions taken now may have consequences far into the future: at the beginning of the century, who could have foreseen the impact of automobiles and traffic patterns on the San Bernardino forest ecosystem in California (Miller et al. 1982)?

The purpose of this chapter is to review what is known about ozone-induced effects on several forest tree species within various regions of eastern United States (Fig. 3.1., see Appendix).

3.2 Recent Findings from Northeastern Forests

Acadia National Park, located on the Atlantic coast of Maine, experiences high concentrations of ozone due to transport from the heavily urbanized areas to the south (Eckert et al. 1994). A number of cases of injury to vegetation due to ozone have been reported (Treshow 1984; Sanchini 1986). Over the past 5 years, studies have been conducted to define the response, in terms of visible injury, changes in photosynthesis, and changes in growth habit, of a number of species native to the ecosystems of the northeastern part of the US (Kohut et al. 1992; Eckert et al. 1994).

Studies were of two types:

1. Exposures of seedlings (or small, clonally propagated material) to controlled concentrations of ozone (about 0.5, 1, 1.5, and 2 times the level of ambient ozone, 8 h/day from May until late September) in open-top chambers.
2. Field surveys designed to assess the prevalence of ozone-type foliar injury on park vegetation.

In the case of open-top chamber studies, a number of very sensitive species (or clones of species), based on symptoms or changes in growth (such as leaf production and/or abscission), were identified. In particular, large-leaf aster (*Aster macrophyllus*, L.), dogbane (*Apocynum cannabinum*, L.), black cherry (*Prunus serotina*, Ehrh.), and some clones of aspen (*Populus tremuloides*, *Michx.*) were found to be sensitive to ozone at ambient or near-ambient concentrations, (ca 0.035–0.040 $\mu l\ l^{-1}$ O_3, 12-h daytime growing season mean) compared to plants grown in charcoal-filtered air. Field surveys identified plants growing in the wild that exhibited the same symptoms as those attributed to ozone under conditions of controlled ozone concentration. Considerable variability in response, presumably related to the genotype of the plants, was found to exist in the population.

White pine (*Pinus strobus*, L.), reported to be sensitive to ozone and injured by the pollutant at Acadia, was not affected by ozone under open-top chamber conditions. Clones made from trees located in the park and classified as "symptomatic" or "asymptomatic" (Skelly et al. 1988) with respect to visible foliar necrosis of unknown cause (but believed by some to be ozone-related) did not respond to controlled ozone exposures in any consistent fashion. Ozone did not cause visible symptoms, nor did it alter the rates of photosynthesis and respiration in exposed trees. Parent trees located in the park did display symptoms, however. Needle cast fungi have been isolated from needles in the past (Dreisbach 1989), but were not isolated from trees in open-top chambers. Needles of parent trees were cultured for pathogens, and needle cast fungi were isolated (Merrill. 1994). According to Merrill (1994), the symptoms match those of semi-mature-tissue needle blight (Linzon. 1967a, b), but seem to be due to needle infections by an as yet unidentified fungus; therefore, the exact cause (s) of the needle symptoms of eastern white pine remain unknown.

Relating the studies at Acadia to what might occur in the ecosystem is problematic, as is usually the case for air pollution studies. By necessity, due to limitations of technology, exposures were conducted with plants of small stature. In some cases, such as dogbane and large-leaf aster, this is entirely appropriate. In the case of large tree species however, the effects of longer-term ambient exposures provide an unsatisfying basis for predicting possible effects of pollutants on organisms that live for hundreds of years. The presence of ozone-induced symptoms on sensitive vegetation in the field indicates that the Acadia ecosystems (as well as those surrounding Acadia) are exposed to concentrations of ozone known to be harmful to plants. The long-term impact of ozone on forested ecosystems is unknown; however, the potential effects, not only on ecosystem composition, but also on genetic diversity in the population (see Karnosky 1991) and on the aesthetics of the area, may be significant.

From 1985 to 1993, considerable research was conducted in order to characterize the response of red spruce and sugar maple to ozone exposures in many areas of northeast North America. Air pollutant studies included exposure under carefully controlled conditions in growth chambers, in open-top chambers, and under field conditions. In addition to investigating the direct effects of ozone on the physiology of the two species, many of the studies were aimed at identifying potential interactions between ozone and acidic precipitation, or the indirect effects of one or more pollutants, perhaps through altering the ability of the trees to withstand low winter temperatures or drought stresses.

In studies with seedlings and grafted red spruce, no strong experimental evidence was found to support a role for ozone or acidic precipitation in the destruction of red spruce trees or stands of the northeast (Taylor et al. 1986; Alscher et al. 1989; Fincher et al. 1989; Laurence et al. 1989; Amundson et al. 1990, 1991; Kohut et al. 1990; Lee et al. 1990; Patton et al. 1991; Rebbeck et al. 1992; Thornton et al. 1992). In a 4-year study of 10–15-year-old red spruce saplings, trees were exposed in open-top chambers from bud-break until frost to ozone levels ranging from a total of 167 (0.5 times the ambient level) to 667 (twice the ambient level) $\mu l\ l^{-1}\ h^{-1}$, over four seasons. There were some indications that cold tolerance might be affected by ozone or acidic precipitation (Laurence et al. 1993) and that the dynamics of carbon storage might be altered (Woodbury et al. 1991), possibly leading to weakened trees after many years of exposure.

However, there were no changes in growth, photosynthesis, or respiration, attributable to pollutant exposure. It is possible that long-term stress due to pollutants could deplete carbon reserves necessary to withstand unfavorable periods, thus contributing to a decline that might involve other natural stresses (Laurence et al. 1994). However, since the experiments were conducted for a relatively short time, it is impossible to speculate on the life-long effects of ozone exposure.

Growth of sugar maple seedlings was not affected by exposure to ozone at concentrations of up to twice the ambient concentration, for three growing seasons in open-top chambers (Laurence et al. 1993). In these studies, the rates of photosynthesis, respiration, and growth were not altered by exposure to ozone at a total level (over three seasons) of 148 (0.5 times the ambient level) to 591 (twice the ambient level) $\mu l\ l^{-1}h$. Periodic whole-plant harvests showed that the allocation of carbon to various plant parts, as measured by dry weight, was not altered by either ozone or acidic precipitation. Furthermore, the storage carbon was also not affected by either pollutant. Foliar injury was never observed, nor was there a premature casting of leaves. It is important to note, however, that the exposures were conducted during a small part of the total life span of sugar maple, and that the response of seedlings may differ from that of mature trees (Edwards et al. 1994).

3.3 Recent Findings from Northcentral Pennsylvania

The leaves of black cherry, a valuable timber species in eastern USA, appear to be sensitive to ambient ozone levels which are reached across much of the Allegheny Plateau region of northcentral Pennsylvania (Davis and Skelly 1992; Simini et al. 1992). The metropolitan and highly industrialized areas of the Midwest have been identified as the source region (Comrie 1992). A major concern with determining the impact of ozone on forest-grown black cherry, however, is that most studies of air pollution effects on trees have been restricted to seedlings grown in field chambers or greenhouses.

Seedlings may differ from larger trees in many ways, including exposure to ozone, leaf and canopy morphology, gas exchange, and transport of water and carbohydrates (Reich 1987; Pye 1988; Cregg et al. 1989). Differences in physiological characteristics between seedlings and mature canopy black cherry trees, as well as differences in environmental conditions within the canopies of these trees may give rise to differences in response to ozone.

The objectives of recent studies have been to provide an assessment of seedling responses and to compare these responses with those of larger trees growing under ambient ozone and natural forest conditions (Fredericksen et al. 1995, 1996a, b). For three growing seasons (1993–1995) ambient ozone concentrations and leaf gas exchange were monitored, and ozone-induced foliar stipple was observed on seedling, sapling, and mature canopy black cherry in order to investigate the presence and magnitude of size-related differences in ozone uptake and foliar injury. The crowns of the larger trees were accessed by the use of scaffold towers. In addition, predawn and midday leaf water potentials were measured at weekly intervals throughout the growing season; soil moisture was also measured weekly with the use of soil moisture cells.

Black cherry seedlings growing near the forest floor had the lowest ozone exposure of any tree size class–light environment combinations. Ozone exposure was similar for foliage of trees growing at other locations within the forest canopy and in forest openings. Previous studies have suggested that the forest floor may be a significant source of nitric oxide which leads to the conversion of ozone to diatomic oxygen (Enders et al. 1989; Fuentes et al. 1992). In addition, uptake of ozone by foliage within the canopy may result in a significant filtration of ozone from the air by the time it reaches the forest floor.

Unlike the minor differences in ozone exposure, large differences in stomatal conductance, due to tree size and light environment, were observed. Stomatal conductance appeared to be the primary determinant of ozone uptake. Open-grown seedlings had the highest rates of stomatal conductance while forest understory seedlings and saplings had the lowest rates. In general, stomatal conductance increased with decreasing tree size and increased by 40 % from seedlings to mature canopy trees. It was also greater in high light compared to low light environments by 25 to 40 % depending on tree size. Average predawn leaf water potentials of seedlings were approximately 0.2 MPa higher than that of canopy trees, which can be attributed largely to differences in tree height. An increased xylem tension of 0.1 MPa should occur for each 10 m increase in height because of the difficulty in moving water against the force of gravity (Kramer 1983). Since study trees averaged 20 m and 1.5 m for canopy trees and seedlings, respectively, differences in water potentials are almost totally accounted for by differences in height.

Sensitivity to ozone within this study was evaluated visually and was defined as the percentage of foliage area displaying oxidant stipple and the percentage of injured leaves that also showed symptoms of early leaf senescence, including premature yellowing or early abscission. In general, higher ozone uptake (higher dose), should lead to a greater sensitivity to ozone. However, there are factors other than uptake that can be important in determining symptom expression. These include leaf defense and repair capabilities (Tingey and Taylor 1982), leaf morphological traits that might increase sensitivity to injury (Bennett et al. 1992), and phenological characteristics of leaf development that might allow for compensation of ozone injury (Pell et al. 1994).

As a result of factors not related to uptake, sensitivity to foliar ozone injury may be expressed in various ways: injury per unit ozone exposure to individual leaves; injury accumulated over whole-crown ozone exposure, and injury per unit ozone uptake.

In recent studies black cherry seedlings were found to be three times more sensitive than canopy trees per unit leaf exposure because of their higher rates of stomatal conductance (Fredericksen et al. 1996 a,b). However, compensation for this type of ozone sensitivity occurred for seedlings because of the juvenile trait possessed by many tree species whereby new leaves are produced throughout the growing season. This trait is referred to as "free growth" (Kramer and Kozlowski 1979) compared to "fixed growth" where leaf expansion and leaf production per shoot is completed early in the season followed by the formation of terminal buds.

In this seedling study, continual production of new uninjured leaves allowed open-grown plants to carry on physiological processes at normal rates on a whole-tree level despite an increasing amount of injury to older leaves due to high ozone uptake rates. Older, ozone-injured leaves often displayed early senescence and abscission, perhaps preceded by transfer of some of the resources in these leaves back to the plant for reallocation to more viable parts of the tree. As seedlings age, this juvenile trait becomes less apparent; hence, the youngest seedlings will have the greatest ability to compensate for ozone injury. By the time the trees reached sapling height, free growth was restricted to terminal shoots in the upper canopy.

Canopy trees in this study rarely displayed any free growth on any part of the tree, and most shoots had set bud by the end of June. Factors other than age, such as light environment, obviously also control this trait, since understory seedlings and shoots in the lower, more shaded crown of seedlings and saplings did not display free growth. Thus despite their tendency towards greater sensitivity per unit exposure on an individual leaf basis, seedlings possessed an ability to compensate for foliage injury. Seedlings were less sensitive than canopy trees when foliar injury is measured on a whole-crown basis, where many seedling leaves will not have been exposed to ozone for the entire season.

Finally, differences in sensitivity may occur due to factors operating within the leaf at the biochemical level, or perhaps due to internal leaf morphology. This sensitivity can be evaluated by measuring visible injury per unit of ozone taken up into the leaf, thus removing the effects of differences in higher stomatal conductance and leaf area development. In this study, when foliar injury was measured per unit ozone uptake, canopy trees again appeared to be more sensitive than seedlings. It is unclear what mechanisms account for greater sensitivity at this level, but some possible factors include differences in internal leaf structural properties or in respiration activities and the overall leaf carbon balance that might alter antioxidant defense and repair processes (Rennenberg and Polle 1989; Bennett et al. 1992; Winner 1994).

Similar mechanisms may account for the counter intuitive observation that although leaves in the upper crowns of mature trees and saplings in this study had higher rates of stomatal conductance, and hence ozone uptake, throughout the growing season than leaves in the lower crowns, yet, foliar ozone symptoms were more prevalent in the lower crown than in the upper crown. This apparent contradiction may be explained by leaf morphological attributes, such as blade thickness and specific leaf area which may cause increased ozone exposure per unit cell mesophyll volume and result in greater contact of ozone with cellular membranes (Bennett et al. 1992). Photosynthetic rates were also greater for leaves in the upper crown of mature trees and saplings than for those in the lower crown. Higher photosynthetic rates for foliage in the upper compared with the lower crown may have allowed for a greater carbon balance in this foliage and a greater ability to replace damaged tissue or prevent damage by the production of antioxidant chemicals (Volin et al. 1993). Average uptake per net photosynthesis differed by a factor of 1.5–4 depending on tree size for shaded, lower crown leaves and the sunlit upper crown leaves (Fredericksen et al. 1996b).

3.4 Recent Findings from the Shenandoah National Park, Virginia

Long-range transport of ozone into the Shenandoah National Park (SHEN), Virginia, from the midwestern urban and industrial centers of the United States has been repeatedly demonstrated (eg. Hayes and Skelly 1977; Wolff et al. 1977, 1982; Galloway and Skelly 1978; Aneja and Li 1992; Comrie 1992). Graphic summaries of the seasonal patterns of ozone distributions demonstrate that the SHEN often has the highest summer season ozone exposures in eastern United States (Fig. 3.1, see Appendix;) USDA-Forest Service 1992). The SHEN is comprised of mountainous terrain giving rise to well defined gradients of increasing ozone exposures with increasing elevation (Lefohn and Pinkerton 1988; Gilliam et al. 1989; Winner et al. 1989; Poulida et al. 1991; Lefohn 1992). Therefore on a seasonal basis, forest species situated at different elevations within this high ozone impact region receive quite different exposures. In addition, ozone exposures for the entire SHEN have been found to vary on an annual basis (Joseph and Flores 1993).

Chappelka et al. (1992), Hildebrand (1994), Skelly et al. (1996) and Hildebrand et al. (1996) have reported findings derived from trend plots established within the SHEN. Trend Plots were established permanent plots with permanently marked trees in order to provide a foundation for a long-term study whereby trends and responses can be analyzed on the basis of several years of observational data descriptive of responses of individual trees.

The initial study of ozone-induced foliar injuries was conducted in the late summers of 1991, 1992, and 1993. Ozone-sensitive tree species induced black cherry, yellow poplar (*Liriodendron tulipifera*, L.), and white ash (*Fraxinus americana*, L.; Fig. 3.5, see Appendix). Thirty trees of each species were selected at each of three trend plots established within an elevation range of 152 m and a radius of 4.4 km from the air quality monitoring stations (inclusive of ozone monitors) situated at Dickey Ridge (617 m a.s.l.), Big Meadows (1067 m a.s.l.), and Sawmill Run (457 m a.s.l.). Six branchlets holding at least 30 leaves each were sampled during each of the 3 years of observations. The presence of an ozone monitor at each site offered the opportunity of establishing statistical relationships between observed foliar injuries and ambient ozone exposures; the elevational gradients of ozone exposures were also confirmed for the interpretation of biological effects.

The details of the findings of this 3-year study may be found in Hildebrand (1994) and Hildebrand et al. (1996). Only two of the many findings will be reviewed here in support of the importance of our understanding of ozone exposure and associated controlling environmental factors (Figs. 3.2–3.4; Fig. 3.5, see Appendix).

1. Of the three species sampled, black cherry proved to be the most consistent in response to the varying ambient ozone exposures along the ozone elevational gradient (Figs. 3.2, 3.4; Fig. 3.5A, see Appendix). Significantly more trees were symptomatic with adaxial stipple at Big Meadows than at Dickey Ridge; similarly, more trees were symptomatic at the latter site than at Sawmill Run. Although not evident from the ozone data presented here, the annual variation in symptom occurrence closely followed the pattern of seasonal ozone exposures with 1991 > 1993 > 1992.

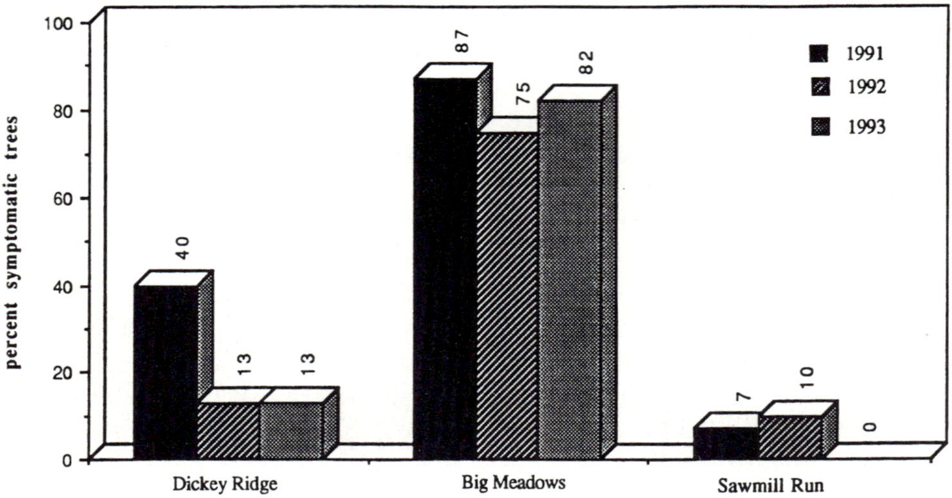

Fig. 3.2. Percentage of black cherry trees symptomatic for ozone damage within the three Trend Plots observed within the Shenandoah National Park, Virginia. Note the consistently high percentage of trees with symptoms at the high elevation, high ozone exposure site of Big Meadows. (Hildebrand et al. 1996)

2. The foliar symptom response of yellow poplar to ambient ozone exposures was obviously influenced by quite differing soil moisture conditions within the natural sites of the trend plots. While receiving the lowest ozone exposures for each of the three years of study, a consistent 87 % of the yellow poplar at the low elevation Sawmill Run site were symptomatic (Figs. 3.3, 3.4; Fig. 3.5B, see Appendix). All trees at the Sawmill Run site were growing along a stream bank with favorable soil moisture favoring pollutant uptake on a season-long basis. However, only eight of the 30 yellow poplar found for the 1992 and 1993 seasons at the Big Meadows trend plot were growing on moist soils while the remaining 22 trees were on a very dry west facing slope at this high elevation site. Symptoms of adaxial stipple were consistently observed amongst the eight trees growing on the moist site and never on the 22 "dry site" trees (Hildebrand et al. 1996).

Ozone exposure indices that gave more weight to higher ozone concentrations performed better than those that did not weight cumulative exposures. These observations point to the need for careful interpretations of ozone-induced exposure/response relationships on a site by site basis. As is evident from the studies of the physiological responses of black cherry in Pennsylvania (Fredericksen et al. 1995, 1996a,b) variations in mature tree canopy response occur within crowns and forest sites and

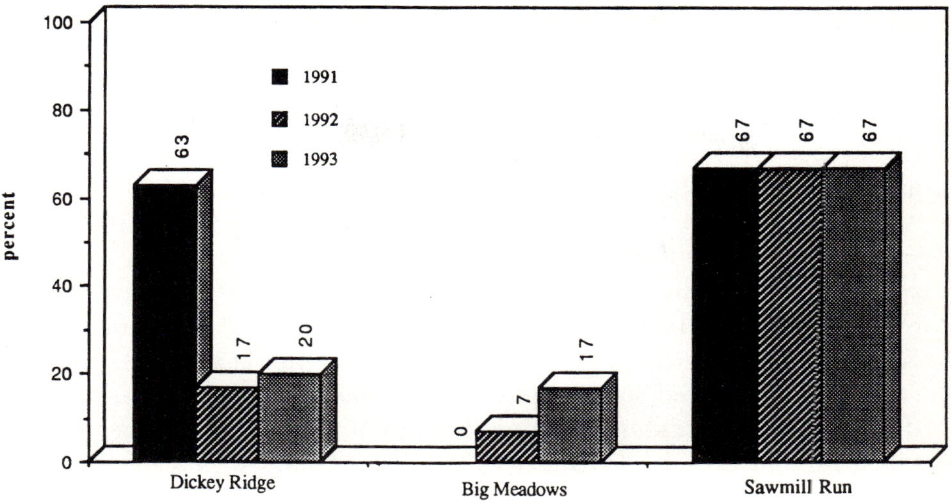

Fig. 3.3. Percentage of yellow poplar symptomatic for ozone damage within the three Trend Plots observed within the Shenandoah National park, Virginia. Note the consistently high percentage of trees with symptoms at the lowest elevation, low ozone exposure site of Sawmill Run. (Hildebrand et al. 1996)

as such broad-brush responses as a function of exposure should not be taken as reliable estimates of forest (species) responses using regional scale ozone exposures, ie., across the range of a given species.

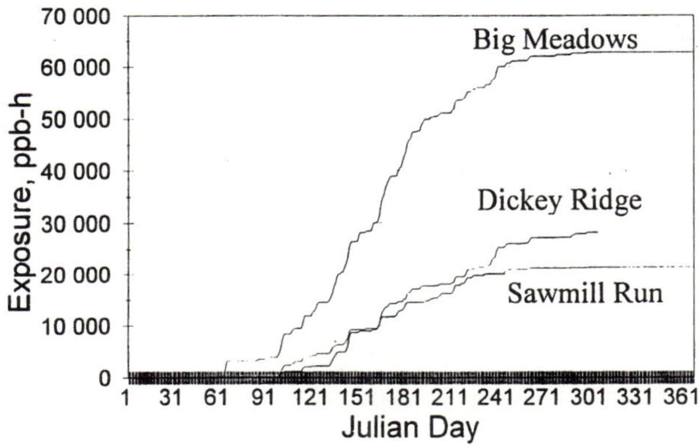

Fig. 3.4. Representative yearly exposure of cumulative sum60 ozone monitored at the three Trend Plots established within the Shenandoah National Park, Virginia. Measurements for 1992 are shown. Note the earlier season increase and the dramatic differences in the high elevation Big Meadows site over the two lower elevations. (Hildebrand et al. 1996; synopsis provided by the Monitoring and Data Analysis Branch, Air Quality Division, National Park Service, Lakewood, Colorado)

3.5 Recent Findings from Southern United States

Rural regions of southern United States occasionally experience ozone episodes above 0.100 $\mu l\ l^{-1}$ with typical summer daytime average levels (0900–1600 h) >0.050 $\mu l\ l^{-1}$ (Pinkerton and Lefohn 1987). In the south there are 26 areas currently classified as not attaining the standard of the National Ambient Air Quality Standard (NAAQS; National Research Council 1991), with approximately 8 and 12 % of forested land and harvested crops, respectively, located in these areas (Lefohn 1992). Ozone effects have been observed, however, in many other areas that often attain the standard of the NAAQS (Duchelle et al. 1982; Shaver et al. 1994; Taylor et al. 1994).

In southern United States, a recent (since 1960) unexplained decline in the radial growth of southern pines in natural stands has been reported (Sheffield and Cost 1987; Zahner et al. 1989). Although no cause–effect relationships were determined, air pollutants, primarily acidic deposition and/or ozone were implicated as potential causes or contributing factors to such declines. Findings from these studies and reports of declines in other parts of the US and Europe led to intensive research from the late 1980s to the early 1990s on the effects of air pollutants on southern forests, primarily southern pines (McLaughlin 1985; Flagler 1992; Kelly et al. 1993).

The majority of the early research on ozone effects on southern tree species has been conducted in greenhouse experiments (Chappelka and Chevone 1992). However, results from such studies need to be interpreted with caution since greenhouse plants differ both morphologically and physiologically from those growing naturally under ambient field conditions.

In a multispecies study, Kress and Skelly (1982) exposed 2–4-week-old seedlings of ten eastern forest species to ozone. In eight species the highest concentrations of ozone (0.15 $\mu l\ l^{-1}$) resulted in a significant reduction in height growth. Loblolly pine (*Pinus taeda* L.) and American sycamore (*Platanus occidentalis* L.) exhibited the greatest depressions in growth but showed no appreciable foliar injury; of the eight species tested, only sweetgum (*Liquidambar styraciflua* L.) developed symptoms manifested as leaf reddening.

In another study, Kress et al. (1982) subjected seedlings of 18 full-sibling families of loblolly pine to increasing ozone concentrations. They found a wide variation in foliar sensitivity, manifested as a chlorotic mottle of the adaxial needle surfaces; needle tipburn was evident only on trees exposed to the highest ozone treatments (i.e., 0.15 and 0.25 $\mu l\ l^{-1}$). The results suggested that some genetic dependence in response to ozone exposures was involved. These results were confirmed in a series of studies using 100 open-pollinated half-sibling families of loblolly pine grown throughout the southern pine region (McLaughlin et al. 1988; Reinert et al. 1988; Wiselogel et al. 1991). Although different types of exposure systems were used, ozone exposures >0.160 μl. 1^{-1} produced visible injury in all families. Reductions in growth, however, were not directly correlated with visible injury, with certain families exhibiting greater suppressions in growth than others.

Several studies have reported decreased biomass production of loblolly pine with increased ozone exposure (McLaughlin et al. 1988; Reinert et al. 1988; Horton et al. 1990; Wiselogel et al. 1991; Flagler and McKinley 1992; Shafer et al. 1993). All plant component parts examined (roots, foliage, and stems) exhibited decreases in growth. Results were variable depending on family and exposure concentrations used. These results are summarized in more detail in Flagler and Chappelka (1995); color prints of typical ozone-induced chlorotic mottle of needles of the southern pines are also presented.

Horton et al. (1990) found that the potting media used had a large influence on loblolly pine response to ozone. They found that root biomass was decreased more than above-ground biomass, but the response varied with the growing media used. The decrease in root biomass was greater when the seedlings were grown in a peat-vermiculite commercial growth media than in mineral soil. The decrease in above-ground biomass was not significantly affected by the growing media used.

Hogsett et al. (1985) subjected two geographical varieties of slash pine (*Pinus elliottii* Englem. vars. *densa* and *elliottii*) to continuous exposures of ozone. Although visible injury, manifested by mild forms of chlorotic mottle and spotting, was minimal, reductions in growth were observed with increasing ozone exposures. Root biomass was the most severely affected plant component by ozone; seed germination was unaffected by ozone exposures.

Spence et al. (1990) exposed loblolly pine seedlings to either charcoal-filtered air or 0.120 μl l^{-1} ozone 7 h day^{-1}, 5 days $week^{-1}$ for 12 weeks. No visible injury was observed. In order to characterize changes in carbon allocation seedlings were labeled with ^{11}C. Ozone-treated pines exhibited a 16% reduction in CO_2 assimilation, and 11% exhibited a decrease in the speed of phloem transport. A 40% reduction in the amount of photosynthate in the phloem and a 45% decrease in transport of photosynthate to the roots was observed. In addition, accumulation of photosynthate not transported to the roots was observed in the stems.

Since the early 1980s many studies have been conducted using open-top chambers to establish a cause–effect relationship between ozone exposure and individual tree growth. The majority of these studies have been conducted within the framework of large multi-institutional co-operative studies and carried out under ideal growing conditions, i.e., adequate moisture and free from competition with other plants (Flagler 1992; Kelly et al. 1993; Flagler and Chappelka 1995).

Kelly et al. (1993) summarized published results from a multi-disciplinary research program the overall goal of which was to determine the response of loblolly pine to ozone, acidic deposition and soil Mg. The studies were conducted in open-top chambers using loblolly pine seedlings grown in large pots in native soil with either adequate or deficient amounts of Mg in the soil. Plants were exposed to subambient, ambient, and twice the level of ambient ozone, and were also exposed to different rain acidities. Rain acidity and/or soil Mg status had little effect on loblolly pine growth in these studies. With respect to ozone, there were no significant differences in diameter and height growth detected among treatments after 3 years, although family responses were different (Adams et al. 1990a). Biomass did however decrease with increasing ozone levels during each of the 3 years of study. The largest decreases observed were in fine root biomass. This correlated with ^{14}C studies (Adams et al. 1990 b; Edwards et al. 1992) in which allocation of photosynthate to fine roots was reduced in the presence of elevated ozone levels.

In a long-term study with loblolly pine in North Carolina, (Shafer et al. 1987; Shafer and Heagle 1989) loblolly pine exhibited decreasing growth and biomass production as ozone concentrations increased. The magnitude of the response varied between families and reductions in growth were not correlated with foliar symptoms of ozone exposure.

Through the auspices of the Southern Commercial Forest Research Cooperative (Fox et al. 1992), a series of studies were developed to determine the effects of acidic precipitation and/or ozone on the growth and productivity of three commercially grown southern pine species: loblolly pine, slash pine, and shortleaf pine (*P. echinata*). In an effort to provide a regional perspective, five intensive field sites were located at different locations with representatives of the range of southern pines. Trees were grown in situ in native soils and exposed to varying concentrations of ozone and acidic rain for 3 years. Results from these studies are summarized in Flagler (1992) and Flagler and Chappelka (1995). In general, increasing ozone concentrations resulted in decreases in height and diameter growth for all three species investigated. The magnitude of response varied by species, genotype, levels of ozone exposures, and location. Height growth response was more variable, and less consistent decreases in growth were observed than with diameter growth. No study from this cooperative research program could find any direct evidence of deleterious effects of acidic rain on tree growth or productivity.

Premature senescence of foliage with increasing ozone concentrations was the most commonly observed response (Allen et al. 1992; Stow et al. 1992). This resulted in decreases in foliage area with increasing ozone levels in all three of the southern pine species investigated (Boutton and Flagler 1990; Allen et al. 1992; Byres et al. 1992). Although the greatest impacts of foliage loss were with the above-ambient ozone treatments, some decreases were observed at ambient levels of ozone.

Since the trees were grown directly in the ground it was difficult to conduct studies investigating below-ground responses to ozone. Root data was, however, obtained from two studies. Faulkner et al. (1991) using soil cores, found a 50% decrease in root biomass of loblolly pine after 2 years of exposure to three times the level of ambient ozone in North Carolina. No differences were observed with any other exposure level. Qiu et al. (1992) grew loblolly pine in the ground in root exclusion tubes and exposed them for 1 year to different concentrations of ozone. They found no differences in root biomass among ozone treatments, but a significant linear decrease in root surface area with increasing ozone levels in the ozone-sensitive genotype. No differences in growth were observed with an ozone-tolerant family.

Flagler et al. (1992) reported on data combined by genotype and species for growth and biomass from five open-top chamber studies. The most severe decreases observed were with foliage biomass. At ambient levels of ozone relative to subambient (charcoal-filtered) levels there were 7.5, 1.5, and 1% decreases observed in foliage and stem biomass and diameter growth, respectively.

In an attempt to gain a better understanding of the response of loblolly pine to ozone in a competitive environment, Barbo et al. (1995) exposed 1-year-old seedlings grown in native soil in open-top chambers to varying concentrations of ozone throughout the growing season. Weeds were not removed and understory vegetation was allowed to regenerate and grow naturally in competition for resources. The first year results indicate that seedlings exposed to twice the level of ambient ozone exhibited reductions in height and diameter growth; however, trees grown in the presence of subambient levels of ozone exhibited similar decreases in growth. These reductions were attributed to the severe competition for resources with understory vegetation in the charcoal-filtered (subambient) chambers, demonstrated by an increased number of understory species and in creased percent cover in these chambers. Before definitive conclusions can be made regarding this study at least 1 more year of exposure is necessary.

Investigations of ozone responses in deciduous tree species are somewhat limited; however, a body of research is beginning to emerge. In some early investigations in the Shenandoah National park, Virginia, Duchelle et al. (1982) exposed eight tree species to ambient and subambient levels of ozone in open-top chambers for 2 years. Tree species exposed included: *L. tulipifera*, *Liquidambar styraciflua*, *Fraxinus pennsylvanica* Marsh., *Robinia pseudoacacia* L., *P. pungens* (Chapm.) Vasey, *P. strobus*, *P. virgininana*, Mill. and *Tsuga canadensis* (L.) Carr. All species exhibited reduced height growth in ambient air compared with those grown in subambient levels of ozone. Reductions in growth, however, were not necessarily correlated with visible foliar symptoms of ozone injury. Ozone-induced symptoms manifested as purple stippling on the adaxial leaf surface developed only on the yellow poplar and green ash seedlings grown in the open air or nonfiltered chambers.

Neufeld et al. (1992) exposed 39 different plant and tree species native to Great Smoky Mountains National Park in open-top chambers. They reported that approximately 33% of the species fumigated were classified as sensitive to ozone on the basis of visible foliar injury. Some of these species also exhibited reductions in growth and biomass; however other species underwent these reductions in the absence of visible injury.

Samuelson (1994a) exposed black cherry and red maple (*Acer rubrum*, L.) seedlings to subambient, ambient, and twice the ambient concentrations of ozone (ozone exposure data are presented in Samuelson 1994a) in open-top chambers for one growing season. Seedlings were grown in pots in native soil and in reduced light (25% ambient) to simulate regeneration in the understory of mature trees. No reductions in height growth of red maple were observed across all three treatments; however net photosynthesis was decreased to 25% while visible foliar injury, expressed as percent leaf surface exhibiting general chlorosis, was observed at 23% for seedlings at twice the ambient exposures when compared with the subambient treatments. Black cherry subjected to twice the level of ambient ozone exhibited a reduction in height growth of 29%, a decrease in photosynthesis to 23%, and an increase in visible foliar injury with 2.5% of the leaf surface with stipple compared with the subambient exposure.

In an effort to determine if results obtained from seedlings studies can be compared with those from large trees, several studies were conducted with northern red oak (*Quercus rubra*) in Tennessee (Samuelson and Edwards 1993; Edwards et al. 1994; Samuelson 1994b). Individual, 30-year-old red oak trees were grown in large open-top chambers (Edwards et al. 1994) and exposed to subambient, ambient and twice the ambient concentrations of ozone. Northern red oak seedlings in pots, growing in native soil from the same location were exposed simultaneously to the same concentrations of ozone in smaller open-top chambers. After 2 years of exposure, the large trees exhibited decreases in leaf dry weight and decreased net photosynthesis in the chambers with twice the level of ambient ozone. Lower canopy leaves were more responsive to ozone exposure than upper canopy leaves. Leaf physiology, growth, and biomass of the seedlings were unaffected by any ozone treatment and no visible symptoms were described (Samuelson and Edwards 1993). Although direct comparisons could not be made, these results indicate that ozone may be affecting trees and seedlings differently.

In an effort to determine the reasons behind these differences in response to ozone, Samuelson (1994a) placed northern red oak seedlings in different canopy positions within the crowns of the large trees and allowed them to be exposed to the same ozone treatments for one growing season. Microclimate influenced the growth and physiology of the seedlings, but no differences in response to ozone were observed. Larger trees were more deleteriously affected by the ozone exposures than seedlings. These results indicate that other physiological and biochemical factors may be more important in determining differences in response between seedlings and mature trees than microclimate.

Since ozone is ubiquitous and tree response is altered by many factors (light, nutrition, moisture), it has proven difficult to determine whether ozone significantly affects tree growth and productivity in the field. At present, there is little research being done on ozone-induced effects on forested ecosystems. The majority of published research is correlative and based on visible symptoms of foliar ozone injury.

The effect of oxidant air pollution on the radial growth of eastern white pine differing in ozone sensitivity (visible symptoms) in the Blue Ridge Mountains of Virginia was evaluated (Benoit et al. 1983). Reduced radial growth was observed for trees in all sensitivity classes during the period of 1955–1978. Mean annual growth of the ozone-sensitive trees was significantly less than for exhibiting no visible symptoms.

Swank and Vose (1991) observed chlorotic mottling and tipburn on white pine in a 13.4 ha watershed located in the southern Appalachian Mountains after several ozone episodes (0.08–0.120 $\mu l\, l^{-1}$) during the 1994 growing season. They observed premature senescence of foliage, small increases in NO_3–N and K, and a reduced basal area increment, and related this to the unusually high ozone levels observed.

Winner et al. (1989) surveyed five native plant species in the Blue Ridge Mountains and found a relationship between elevation and injury, with more injury occurring at the higher elevations. Since these plots were not near ozone monitors, no direct relationships could be determined.

3.6 Recent Findings from Great Smoky Mountains National Park

In a survey of sensitive plants in Great Smoky Mountains National Park, Chappelka et al. (1994) observed visible symptoms of ozone injury throughout the Park on black cherry seedlings. Forty seven percent of all trees examined exhibited some amount of foliar injury (approximately 1600 seedlings sampled).

In a study paralleling the investigations of Hildebrand (1994), Hildebrand et al. (1996) and Chappelka et al. (1992, 1994) reported on a 3-year (1991–1993) study in Great Smoky Mountains National Park designed to determine the relationship between visible ozone injury and ozone exposure on mature trees. Three different tree species at three sites were colocated within 4 km and 150 m elevation of an ozone monitoring station. The species investigated, black cherry, yellow poplar and sassafras (*Sassafras albidum* L.) were all mature trees (>5 m in height). Field plots were established in 1991 at Cove Mountain (1243 m), Look Rock (824 m), and Twin Creeks (598 m). All sites were located on the northern periphery of the park (Tennessee).

In general, the incidence of ozone injury was greater and more consistent on black cherry and sassafras than on yellow poplar. With these species, ozone injury was greatest at the highest elevation (Cove Mountain) and least at the lowest elevation (Twin Creeks) in all 3 years of the study. Ozone concentrations were also the highest at Cove Mountain and the lowest at Twin Creeks. Data for yellow poplar was not as consistent which may be related to microsite requirements of the species. Overall, visible ozone injury was present on approximately 50% of all trees over the 3 years of the study.

It is important to note that in all years of observation at both the Shenandoah and Great Smoky Mountains National Parks significant amounts of ozone-induced foliar injuries were observed at ozone exposures below those designated as protective by the US NAAQS. Results from studies with southeastern tree species indicate that ozone may injure foliage, modify physiology, and affect productivity of individual trees.

3.7 General Conclusions

Tropospheric ozone is an air pollutant of continuing concern to the health and productivity of the forests of eastern United States. Most certainly, numerous factors affect the ozone exposures and the physiological and visible responses of individual trees, species, stands, and larger scale forested areas. When high relative humidity, mesophytic sites, and proven species sensitivity to ozone are combined with often repeated mesoscale transport of ozone across large forested areas in eastern United States, there is much cause for concern.

Specifically, ozone has been demonstrated to cause foliar marking on many common forest tree species; within a species there is a differential sensitivity regarding ozone effects, with some genotypes being very sensitive and others apparently tolerant. Convincing evidence of detrimental physiological changes in the presence of ambient ozone exposures has been gathered in recent years, with a few studies investigating mature canopy tree responses. Exposures may reduce net photosynthesis and alter the transport of carbon to the roots; root growth appears to be a very sensitive indicator of ozone stress.

Preliminary results indicate that these effects may be altered when competition with other plant species is present. Research with a few tree species indicates that response to ozone may vary depending upon the respective ages (seedlings vs mature canopy) of the trees being studied. More research with many other tree species is needed in this area, and aggressive research needs to be conducted in the field under realistic ozone exposures and in competitive surroundings.

Over the past decades, we have come to realize that understanding forest function, let alone valuing or protecting it, is a difficult problem indeed. For instance, were we interested only in timber production, it might be quite straightforward to determine ozone-induced effects by simply asking whether productivity has changed significantly relative to known changes in ozone concentration, over some period of time. While this would not provide a definitive answer, it might at least lead us to some assurance that the net effect of ozone has thus far been minimal. We might gain some comfort from experimentally defining the response of major forest species to current or projected concentrations of ozone. We might feel even more comfortable if we use large trees and incorporate additional stress factors in our studies.

But what about functions that depend on unknown or poorly understood relationships? How might changing nutrient deposition or climate alter the system response to ozone? An interaction of stress factors, over hundreds of years (the time scale of interest in many eastern United States forests) might lead to an unforseen, yet important change in ecosystem function. However, it is not yet possible to predict what long-term, continuous exposure to ozone, in concert with other potential stresses, might do to ecosystem structure and function.

References

Adams MB, Kelly JM, Taylor GE, Edwards NT (1990a) Growth of five families of *Pinus taeda* L. during three years of ozone exposure. New Phytol 116:689–694

Adams MB, Edwards NT, Taylor GE Jr, Skaggs BL (1990b) Whole-plant ^{14}C-photosynthate allocation in *Pinus taeda*:seasonal patterns at ambient and elevated ozone levels. Can J For Res 20:152–158

Allen HL, Stow TK, Chappelka AH, Kress LW, Teskey RO (1992) Ozone impacts on foliage dynamics of loblolly pine. In:Flagler RB (ed) The response of southern commercial forests to air pollution. TR-21. Air and Waste Manage Assoc, Pittsburgh, pp 149–162

Alscher RG, Amundson RG, Cumming JR, Fellows S, Fincher J, Rubin G, van Leuken P, Weinstein LH (1989) Seasonal changes in the pigments, carbohydrates and growth of red spruce as affected by ozone. New Phytol 113:211–223

Amundson RG, Kohut RJ, Laurence JA (1990) Mineral nutrition, carbohydrate content, and cold tolerance of foliage of potted red spruce saplings exposed to ozone and simulated acidic precipitation treatments. Water Air Soil Pollut 54:175–182

Amundson RG, Alscher RG, Fellows S, Rubin G, Fincher J, van Leuken P, Weinstein LH (1991) Seasonal changes in the pigments, carbohydrates, and growth of red spruce as affected by exposure to ozone for two growing seasons. New Phytol 118:127–137

Aneja VP, Li Z (1992) Characterization of ozone at high elevation sites in the eastern United States:trends, seasonal variations, and exposure. J Geophys Res 97 (DG) pp 9873–9898

Barbo DN, Chappelka AH, Stolte K (1995) Ozone effects on productivity and diversity of an early successional forest community. In:Edwards MB (ed) Proc 8th Biennial Southern Siliviculture Res Conf, Nov 1–3, 1994, Auburn, Alabama, USDA For Serv Gen Tech Rep SAS-1, pp 291–298

Barnard JE, Lucier AH, Johnson AH, Brooks RT., Karnosky DF, Dunn PH (1989) Changes in forest health and productivity in the United States of America. NAPAP State Sci Tech Rep, No 16. National Acidic Precipitation Assessment Program, Washington, DC

Bennett JP, Rassat P, Berrang P, Karnosky DF (1992) Relationships between leaf anatomy and ozone sensitivity of *Fraxinus pennsylvanica* Marsh and *Prunus serotina* Ehrh. J Exp Bot 32:33-41

Benoit LF, Skelly JM, Moore LD, Dochinger LS (1983) Radial growth reductions of *Pinus strobus* L. correlated with foliar ozone sensitivity as an indicator of ozone induced losses in eastern forests. Can J For Res 12:673–678

Boutton TW, Flagler RB (1990) Growth and water-use efficiency of shortleaf pine as affected by ozone and acid rain. In:Proc 83rd Annu Meet Air and Waste Manage Assoc, June 24–29, 1990. Pittsburgh, preprint No 90–187.7

Byres DP, Dean TJ, Johnson JD (1992) Long-term effects of ozone and simulated acid rain on the foliage dynamics of slash pine (*Pinus elliottii* var. *elliottii* Engelm). New Phytol 120:61–67

Chappelka AH, Chevone BI (1992) Tree responses to ozone. In:Lefohn AS(ed) Surface level ozone and their effects on vegetation. CRC Press, Boca Raton, pp 271–324

Chappelka AH, Hildebrand EH, Skelly JM, Mangis D, Renfro JR (1992) Effects of ambient ozone concentrations on mature eastern hardwood trees growing in Great Smoky Mountains National Park and Shenandoah National Park. In:Proc 85th Annu Meet, June 21–26, 1992, Kansas City, Missouri, preprint 92–150.04

Chappelka AH, Renfro JR, Somers GL (1994) Visible ozone injury on native plant species within Great Smoky Mountains National Park. In:Proc 87th Annu Meet Air and Waste Manage Assoc, June 19–24, 1994, Cincinnati, Ohio, preprint 94–TA36.04

Comrie AC (1992) A synoptic climatology of ozone concentrations in the forests of Pennsylvania. PhD Thesis The Pennsylvania State University, University Park

Cregg BM, Halpin JE, Dougherty PM, Teskey RO (1989) Comparative physiology and morphology of seedling and mature forest trees. In:Noble D, Martin JL, Jensen KF (eds) Proc of the US-USSR Symp on Air pollution effects on vegetation, 18–22 Sept 1988, Raleigh, North Carolina, pp 111–118

Davis DD, Skelly JM (1992) Foliar sensitivity of eight eastern hardwood species to ozone Water Air Soil Pollut 62:269–277

Dreisbach TA (1989) A needle blight of *Pinus strobus* in Acadia National Park:associated fungi and pathological anatomy. MS Thesis The Pennsylvania State University, University Park, 35 pp

Duchelle SF, Skelly JM, Chevone BI (1982) Oxidant effects on forest tree seedlings in the Appalachian Mountains. Water Air Soil Pollut 12:363–373

Eckert RT, Kohut RJ, Laurence JA, King PS (1994) An approach to detecting ozone injury in remote areas using controlled exposure studies and indicator species. In:Proc 2nd Int Conf on Science and the Manage of Protected Areas. May 16–20, 1994. Halifax NS Science and Manage of Protected Areas Assoc, Wolfville, Nova Scotia, 349–356

Edwards GS, Friend AL, O'Neill EG, Tomlinson PT (1992) Seasonal patterns of biomass accumulation and carbon allocation in *Pinus taeda* seedlings exposed to ozone, acidic precipitation, and reduced soil Mg. Can J For Res 22:640–646

Edwards GS, Wullschleger SD, Kelly JM (1994) Growth and physiology of northern red oak:preliminary comparisons of mature tree and seedling responses to ozone. Environ Pollut 83:215–221

Enders G, Teichmann U, Kramm G (1989) Profiles of ozone and surface layer parameters over a mature spruce stand. In:Georgii HW(ed) Mechanisms and effects of pollutant transfer into forests. Kluwer, Dordrecht, pp 21–35

Faulkner P, Schoeneberger MM, Kress LW (1991) Below ground changes in loblolly pine as indicators of ozone stress. In:Coleman SS, Neary DG (eds) Proc 6th Biennial So Silvic Res Conf, USDA-For Serv Gen Tech Rep, SE-70, pp 332–341

Fincher J, Cummings JR, Alscher RG, Rubin G, Weinstein LH (1989) Long-term ozone exposure affects winter hardiness of red spruce (*Picea rubens* Sarg.) seedlings. New Phytol 113:85–96

Flagler RB (ed) (1992) The response of southern commercial forests to air pollution. Air and Waste Manage Assoc TR-21 Pittsburgh

Flagler RB, Chappelka AH (1995) Growth response of southern pines to acidic precipitation and ozone. In:Fox S, Mickler RA (eds) impact of air pollutants on southern pine forests. Ecological Studies, vol 118. Springer, Berlin Heidelberg New York, pp 388–424

Flagler RB, McKinley CR (1992) Response of two loblolly pine ecotypes to ozone. Bull Ecol Soc Am 73:175

Flagler RB, Spruill SE, Chappelka AH, Dean TJ, Kress LW, Reardon JC (1992) Growth of three southern pine species as affected by acid rain and ozone: a combined analysis. In: Flagler RB (ed) The response of southern commercial forests to air pollution, TR-21. Air and Waste Manage Assoc, Pittsburgh, pp 207–224

Fox S, Mickler RA (eds) (1995) Impact of air pollutants on southern pine forests. Ecological Studies vol 118. Springer, Berlin Heidelberg New York, 118 p

Fox S, Joyner KC, Bartuska AM (1992) Overview of the Southern Commercial Forest Research cooperative. In: Flagler RB (ed) the response of southern commercial forests to air pollution, TR-21. Air and Waste Manage Assoc, Pittsburgh, pp 1–15

Frederickson TS, Joyce BJ, Skelly JM, Steiner KC, Kolb TE, Kouterick KB, Savage JE, Snyder KR (1995) Physiology, morphology, and ozone uptake of black cherry seedlings, saplings, and canopy trees. Environ Pollut 89:273–283

Fredericksen TS, Kolb TE, Skelly JM, Steiner KC, Joyce BJ, Savage JE (1996a) Light environment alters ozone uptake per net photosynthetic rate in black cherry trees. Tree Physiol 16:485–490

Frederickson TS, Skelly JM, Steiner KC, Kolb TE, Kouterick KB (1996b) Size-mediated foliar response to ozone in black cherry trees. Environ Pollut 91:53–63

Fuentes JD, Gillespie TJ, den Hartog G, Neumann HH (1992) Ozone deposition onto a deciduous forest during dry and wet conditions. Agric For Meteorol 62:1–18

Galloway JM, Skelly JM (1978) An air pollution episode in Virginia. VA Climate Advisory 3:4–6

Gilliam FS (1989) Atmospheric deposition and its potential significance in southern Appalachian hardwood forest ecosystems. Proc U K Conf on Appalachia

Gilliam FS, Sigmon JT, Reiter MA, Krovetz DO (1989) Spatial variations in ozone concentrations in the Virginia Blue Ridge Mountains: implications for forest exposures. Can J For Res 19:422-426

Hayes EM, Skelly JM (1977) Transport of ozone from the northeastern United States into Virginia and its effects on eastern white pine. Plant Dis Rep 61:778–782

Hepting GH (1971) Diseases of forest and shade trees of the United States. USDA-For Serv Handbook No 386

Hildebrand ES (1994) Incidence of ozone-induced injury on sensitive hardwood species in Shenandoah National Park, VA. MS Thesis The Pennsylvania State University, University Park, 198 pp

Hildebrand ES, Skelly JM, Fredericksen TS (1996) An analysis of 3-year (1991–1993) trend plot data on the foliar response of ozone-sensitive hardwood tree species in the Shenandoah National Park, Virginia. Can J For Res 26:658–669

Hogsett WE, Plocher M, Wildman V, Tingey DT, Bennett JP (1985) Growth response of two varieties of slash pine seedlings to chronic ozone exposures. Can J Bot 63:2369–2376

Horton SJ, Reinert RA, Heck WW (1990) Effects of ozone on three open pollinated families of *Pinus taeda* L. grown in two substrates. Environ Pollut 65:279–292

Joseph DB, Flores MI (1993) Statistical summary of ozone measurements in the National Park System 1981–1991. Quick look annual summary statistics reports, vol 1. US Dept Int Tech Rep NPS/NRAQD/NRTR-93/110. Denver, 301 pp

Karnosky DF (1991) Ecological genetics and changes in atmospheric chemistry:the application of knowledge. In:Taylor GE Jr, Pitelka LF, Clegg MT (eds) Ecological genetics and air pollution. Springer, Berlin Heidelberg New York

Keller T (1977) The effect of long during, low SO_2 concentrations upon photosynthesis of conifers. Proc IV Int Clean Air Congr Tokyo, pp 81–83

Kelly JM, Taylor GE Jr, Edwards NT, Adams MB, Edwards GS, Friend AL (1993) Growth, physiology and nutrition of loblolly pine seedlings stressed by ozone and acid precipitation:a summary of the Ropis-South project. Water Air Soil Pollut 69:363–391

Kohut RJ, Laurence JA, Amundson RG, Raba RM, Melkonian JJ (1990) Effects of ozone and acidic precipitation on the growth and photosynthesis of red spruce after two years of exposure. Water Air Soil Pollut 51:277–286

Kohut RJ, Eckert RT, Laurence JA, King PS (1992) Sensitivity to ozone of vegetation of Acadia National Park. In: The effects of air pollution on terrestrial and aquatic ecosystems in New England and New York. US Forest Service, Durham, pp 31–33

Kramer PJ (1983) Water Relations of Plants. Academic Press, Orlando

Kramer PJ, Kozlowski TT (1979) Physiology of woody plants. Academic Press, New York

Kress LW, Skelly JM (1982) Response of several eastern forest tree species to chronic doses of ozone and nitrogen dioxide. Plant Dis 66:1149–1152

Kress LW, Skelly JM, Hinkelmann KH (1982) Relative sensitivity of 18 full-sib families of *Pinus taeda* to O_3. Can J For Res 12:203–209

Laurence JA, Kohut RJ, Amundson RG (1989) Response of red spruce seedlings exposed to ozone and simulated acidic precipitation in the field. Arch Environ Contam Toxicol 18:285–290

Laurence JA, Kohut RJ, Amundson RG, Weinstein DA, Fahey TJ (1993) An evaluation of the effects of ozone and acidic precipitation on the nutrition, growth, and physiology of red spruce and sugar maple. Final Report of Project RP2799-1 to the Electric Power Research Institute, Palo Alto, 104 pp

Laurence JA, Amundson RG, Temple PJ, Pell EJ, Friend AL (1994) Allocation of carbon in plants under stress:an analysis of the ROPIS experiments. J Environ Qual 23:412–417

Lee WS, Chevone BI, Seiler JR (1990) Growth response and drought susceptibility of red spruce seedlings exposed to simulated acidic rain and ozone. For Sci 36:265–275

Lefohn AS (1992) The characterization of ambient ozone exposures. In: Lefohn AS (ed) Surface level ozone exposures and their effects on vegetation. CRC Press, Boca Raton, pp 31–92

Lefohn AS, Pinkerton JE (1988) High resolution characterization of ozone data for sites located in forested areas of the United States. J Air Pollut Control Assoc 38:1504–1511

Linzon SN (1967a) Histological studies of semimature-tissue needle blight of eastern white pine. Can J Bot 45:133–143

Linzon SN (1967b) Ozone damage and semimature-tissue needle blight of eastern white pine. Can J Bot 45:2047–2061

Manion PD, Lachance D (1992) Forest decline concepts:an overview. In:Manion PD, Lachance D (eds) Forest Decline Concepts. American Phytopathological Society, St Paul, pp 181–190

McLaughlin SB (1985) Effects of air pollution on forests:a critical review. J Air Pollut Control Assoc 35:512–534

McLaughlin SB, Adams MB, Edwards NT, Hanson PJ, Layton PA, O'Neil EG, Roy WK (1988) Comparative sensitivity, mechanisms, and whole plant physiological implications of responses of loblolly pine genotypes to ozone and acid deposition. ORNL/TM-10777, Environ Sci Div Publ No 3105. Oak Ridge National Laboratory, Oak Ridge

Merrill W (1994) Pathological anatomy of eastern white pine needles injured by exposure to known dosages of ozone and to ambient conditions in Acadia National Park. Final Report National Park Service, Mid-Atlantic Region, University Park, Pennsylvania

Miller PR, Taylor OC, Wilhour RG (1982) Oxidant air pollution effects on a western coniferous forest ecosystem. EPA Rep No, EPA-600/D-82-276. Corvallis Environ Res Lab, Corvallis, Oregon

Millers I, Shriner DS, Rizzo D (1989) History of hardwood decline in the eastern United States. USDA-For Serv Gen Tech Rep NE-126, Broomall, Pennsylvania

National Research Council (1991) Rethinking the ozone problem in urban and regional air pollution. National Academy Press, Washington, DC

Neufeld HS, Renfro JR, Hacker WD, Silsbee D (1992) Ozone in Great Smoky Mountains National Park:dynamics and effects on plants. In:Berglund R (ed) Tropospheric ozone and the environment II. Air and Waste Manage Assoc, Pittsburgh, pp 594–617

Patton RL, Jensen KF, Schier GA (1991) Responses of red spruce seedlings to ozone and acid deposition. Can J For Res 21:1354–1359

Pell EJ, Temple PJ, Friend AL, Mooney HA, Winner WE (1994) Compensation as a plant response to ozone and associated stresses:an analysis of ROPIS experiments. J Environ Qual 23:429–436

Pinkerton JE, Lefohn AS (1987) The characterization of ozone data for sites located in forested areas of the eastern United States. J Air Pollut Control Assoc 37:1005–1010

Poulida O, Dickerson RR, Doddridge BG, Holland JZ, Wardell RG, Watkins JG (1991) Trace gas concentrations and meteorology in rural Virginia 1. Ozone and carbon monoxide. J Geophys Res 96 (D-12):22461–22475

Pye JM (1988) Impact of ozone on the growth and yield of trees: a review. J Environ Qual 17:347–360

Qiu Z, Chappelka AH, Somers GL, Lockaby BG, Meldahl RS (1992) Effects of ozone and simulated acidic precipitation on above- and below-ground growth of loblolly pine *Pinus taeda* L. Can J For Res 22:582–587

Rebbeck J, Jensen KF, Greenwood MS (1992) Ozone effects on the growth of grafted mature and juvenile red spruce. Can J For Res 22:756–760

Reich PB (1987) Quantifying plant response to ozone:a unifying theory. Tree Physiol 3:63–91

Reinert RA, Shafer SR, Eason G, Horton SJ, Schoeneberger MM, Wells C (1988) Responses of loblolly pine half-sib families to ozone. In:Proc 81st Annu Meet Air Pollut Control Assoc, Dallas, preprint 88–125.5

Rennenberg H, Polle A (1989) Effects of photo-oxidants on plants. In:Georgii HW (ed) Mechanisms and effects of pollutant transfer into forests. Kluwer, Dordrecht, pp 251–258

Samuelson LJ (1994a) Ozone exposure-responses of black cherry and red maple seedlings. Environ Exp Bot 34:355–362

Samuelson LJ (1994b) The role of microclimate in determining the sensitivity of *Quercus rubra* L. to ozone. New Phytol 128:235–241

Samuelson LJ, Edwards GS (1993) A comparison of sensitivity to ozone in seedlings and trees of *Quercus rubra* L. New Phytol 125:373–379

Sanchini PJ (1986) Ozone injuries to *Pinus strobus* on permanent pine plots at Acadia National Park. 1985 Survey Results, Final Report, Natl Park Service, Acadia Natl Park, Maine

Shafer SR, Heagle AS (1989) Response of loblolly pine to ozone over three growing seasons. Can J For Res 19:821–831

Shafer SR, Heagle AS, Camberato DM (1987) Effects of chronic doses of ozone on field-grown loblolly pine:seedling responses in the first year. J Air Pollut Control Assoc 37:1179–1184

Shafer SR, Reinert RA, Eason G, Spruill SE (1993) Analysis of ozone concentration-biomass response relationships among open-pollinated families of loblolly pine. Can J For Res 23:706–715

Shaver CL, Tonnessen KA, Maniero TG (1994) Clearing the air at Great Smoky Mountains National Park. Ecol Appl 4:690–701

Sheffield RM, Cost ND (1987) Behind the decline. J For 85:29–33

Simini MJ, Skelly JM, Davis DD, Savage JE, Comrie AC (1992) Sensitivity of four hardwood species to ambient ozone in northcentral Pennsylvania. Can J For Res 22:1789–1799

Skelly JM, Innes JL (1994) Waldsterben in the forests of Central Europe and eastern North America:fantasy or reality? Plant Dis 78:1021–1032

Skelly JM, Snyder KR, Merrill W (1988) The response of two "SNB" symptomatic white pine clones taken from Acadia National Park to ozone. Phytopathology 78:1511

Skelly JM, Hildebrand ES, Fredericksen TS, Chappelka AH (1996) Ambient ozone exposures in the Appalachian Mountains and their effects on hardwood tree species under natural forest conditions. In:Vostal JJ (ed) Tropospheric ozone:critical issues in the regulatory process. Proc Air and Waste Manage Assoc, Pittsburgh, pp 256–276

Spence RD, Rykiel EJ, Sharpe PJH (1990) Ozone alters carbon allocation in loblolly pine:assessment with carbon-11 labeling. Environ Pollut 64:93–106

Stoklasa J (1923) Die Beschädigung der Vegetation durch Rauchgase und Fabriksexhalationen. Urban und Schwarzenberg, Berlin

Stow TK, Allen HL, Kress LW (1992) Ozone impacts on seasonal foliage dynamics of young loblolly pine. For Sci 38:102–119

Swank WT, Vose JM (1991) Watershed-scale responses to ozone events in a *Pinus strobus* L. plantation. Water Air Soil Pollut 54:119–133

Taylor GE Jr, Norby RJ, McLaughlin SB, Johnson AH, Turner RS (1986) Carbon dioxide assimilation and growth of red spruce (*Picea rubens* Sarg.) seedlings in response to ozone, precipitation chemistry and soil type. Oecologia 70:163–171

Taylor GE Jr, Johnson DW, Anderson CP (1994) Air pollution and forest ecosystems:a regional to global perspective. Ecol Appl 4:662–689

Thornton FC, Pier PA, McDuffie C Jr (1992) Plant and environment interactions:red spruce response to ozone and cloud water after three year exposure. J Environ Qual 21:196–202

Tingey DT, Taylor GE Jr (1982) Variation in plant response to ozone:conceptual model of physiological events. In:Unsworth MH, Ormrod DP (eds) Effects of gaseous air pollutants on agriculture and horticulture. Butterworth, London, pp 113–138

Treshow ME (1984) Establishment of white pine biomonitoring plots in Acadia National Park. Final Report to US National Park Service, Air Quality Division, Lakewood

USDA Forest Service Northeastern Area (1992) Northeastern Area Forest Health Report. NA-TP-03-93, Radnor, Pennsylvania

USEPA (1986) Air quality criteria for ozone and other photochemical oxidants, vol II of V. EPA-600/8 84-020A, Research Triangle Park, North Carolina

Volin JC, Tjoelker MG, Oleksyn J, Reich PB (1993) Light environment alters response to ozone stress in seedlings of *Acer saccharum* Marsh and hybrid *Populus* L. II Diagnostic gas exchange and leaf chemistry. New Phytol 124:637–646

Winner WE (1994) Mechanistic analysis of plant responses to air pollution. Ecol Appl 4:651–661

Winner WE, Lefohn AS, Cotter IS, Greitner CS, Nellessen J, McEvoy LR Jr, Olsen RL, Atkinson CJ, Moore LD (1989) Plant responses to elevational gradients of O_3 exposures in Virginia. Proc Natl Acad Sci 86:8828–8832

Wiselogel AE, Bailey JK, Newton RJ, Fong F (1991) Growth responses of loblolly pine (*Pinus taeda* L.) seedlings to ozone fumigation. Environ Pollut 71:43–56

Wolff GT, Lioy PJ, Wight RE, Meyers RE, Cederwall RJ (1977) An investigation of long range transport of ozone across the midwestern and eastern United States. Atmos Environ 11:797–802

Wolff GT, Kelly NA, Ferman MA (1982) Source regions of summertime ozone and haze episodes in the eastern United States. Water Air Soil Pollut 18:65–81

Woodbury PB, Amundson RG, Kohut RJ, Laurence JA (1991) Stomatal conductance, whole-tree water use, and foliar carbohydrate status of red spruce saplings exposed to ozone, acid precipitation, and drought. In: Berglund R (ed) Trans 1991 Air and Waste Manage Assoc Int Symp on Tropospheric ozone and the environment. II, TR-20, Air and Waste Manage Assoc, Pittsburgh, pp 618–628

Zahner R, Saucier JR, Myers RK (1989) Tree-ring model interprets growth decline in natural stands of loblolly pine in the southeastern United States. Can J For Res 19:612–625

4 Ozone and the Forests in Austria and Switzerland

R. Matyssek, W.M. Havranek, G. Wieser and J.L. Innes

4.1 Introduction

Following reports during the early 1980s in Germany about a spreading extensive dieback of forests (e.g., Prinz et al. 1982) occurring throughout central Europe, *Waldsterben* was believed to be a danger to the forests of Austria and Switzerland (see Schwarzenbach 1991; Amt der Tiroler Landesregierung 1985–1995). As this scenario did not materialize, the concept of "*Neuartige Waldschäden*" was introduced (FBW 1989), whereby forest status was classified by crown transparency (leaf loss) and foliage yellowing. From the outset, attention focused on ozone (O_3) as the causal agent rather than on acidic deposition (Arndt et al. 1982; Prinz 1984), especially because many soils in alpine regions possess a high buffering capacity (Bucher 1982). Moreover, high impacts of SO_2 appeared to have been short-lived and were not sufficiently widespread to account for the supposed forest decline. It also appeared relevant to consider ozone as a constraint on forests in mountainous areas, since ozone increases in concentration with altitude, due to both anthropogenic activities and natural causes (Krupa and Manning 1988). In Austria, especially in Tyrol, the O_3 hypothesis was supported by the crown transparency of trees, which increased in parallel with the O_3 concentration from 1984 through 1987 (Amt der Tiroler Landesregierung 1985–1995: Fig. 4.1).

Although most studies on the impact of ozone on trees and forests referred to scenarios specific to North America (Miller 1973, 1984; Blanchard et al. 1979; Smith 1984), or to experimental regimes unrepresentative of actual site conditions in Europe (see Reich 1987), it was tempting to relate such knowledge about "photo-oxidative injury" to field observations in central Europe. However, it was always doubtful whether estimations of crown transparency and foliage yellowing (see UN/ECE 1992) were appropriate measures of forest decline (e.g., Kandler 1985). In fact, unambiguous evidence of O_3 injury did not exist at forest sites, even

Ecological Studies, Vol. 127
Sandermann et al. (eds) Forest Decline and Ozone

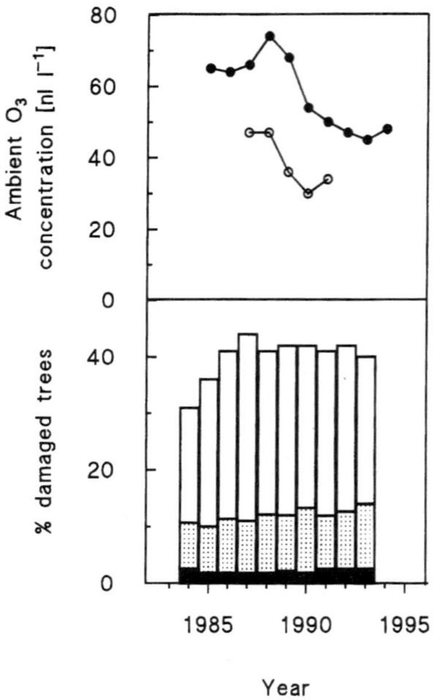

Fig. 4.1. Variation in the mean O_3 concentration of the vegetation period (May through October) 1960 m a.s.l. at the timberline near Innsbruck (closed circles) and 1000 m a.s.l. in Ziller Valley (open circles) compared with changes in the percentage of damaged trees (below) in high-altitude protection forests in Tyrol. Damage is expressed by the degree of crown transparency: *Open bars* 10–25%; *dotted bars* 26–60%; *black bars* 61–99% needle loss and dead trees

at the timberline where the O_3 concentrations are higher than at low-elevation sites (Smidt 1993).

Given such uncertainties, pressure to clarify the existence and causes of the supposed forest decline increased (Arndt et al. 1982; Bucher 1982). Investigations of the distribution of ozone in Austria and Switzerland and its dependence on altitude were initiated. In both countries, national research programs assessed the impact of air pollutants on trees and forests under field conditions, and have attempted to uncover the mechanisms of O_3 injury in forest decline. Simultaneously, the analytical approach of controlled experiments in exposure chambers was intensified.

This review reports on research from Austria and Switzerland aimed at answering the following questions: How do forest trees respond to the O_3 regimes of field conditions? Do mechanisms of O_3 injury observed under controlled experiments apply to findings from forest sites? To what extent are distribution patterns of ozone consistent with the occurrence of forest decline phenomena?

4.2 O_3 Levels and Areas at Risk in the Alps of Central Europe

Mean annual concentrations of ozone tend to increase with altitude in central Europe. In the alpine regions of Austria and Switzerland means around or above 40 nl O_3 l^{-1} were prevalent at elevations above 1500 m; at these elevations, maximum annual means could even be as high as 60 nl O_3 l^{-1} (Smidt and Gabler 1994). Annual O_3 levels around 50 nl l^{-1} coincide with the altitude of montane and subalpine forests in the Alps of central Europe (timberline at around 2000 ± 100 m a.s.l.). Mean annual O_3 levels of around 25 nl l^{-1} and less were found mainly at elevations lower than 650 m a.s.l.

Comparisons of O_3 levels between summary reports are difficult to make because of the lack of a standardized reporting format and the uncertainty about biologically meaningful exposure indices. Complications also arise with respect to the altitude-dependent influence of temperature and air pressure on the volumetric unit "$\mu g\ m^{-3}$" and its relation to temperature- and pressure-independent units such as the mole fraction "ppb" (which means nl l^{-1} = nmol mol^{-1}); the precise methods used in the conversions are rarely indicated[1].

One exposure index introduced recently is the AOT40 value (i.e., the accumulated exposure over a threshold of 40 nl O_3 l^{-1}; UN/ECE 1994). Only the part of the hourly O_3 concentration above 40 nl l^{-1} is considered relevant for plant development (e.g., biomass production). For forests, the calculation of AOT40 is made for the period from 1 April through 30 September (24 h/day). Based on findings from open-top chamber (OTC) experiments, a critical AOT40 level (CL), which may reduce the annual biomass increment of forest trees by 10%, in the range of 10 $\mu l\ l^{-1}$ h (UN/ECE 1994) has been (vaguely) proposed. In comparison, an evaluation in Austria of measurements carried out in about 45 O_3-monitoring field stations relevant to forest sites resulted in AOT40 values significantly higher than the proposed CL at nearly all stations during the period of 1993–1995 (Schneider et al. 1996; Table 4.1).

AOT40 was also calculated for forest and crop areas in Austria on a 2.75 × 2.75 km grid basis, using a time- and altitude-fitted O_3 distribution model together with the O_3 records of 110 stations (Schneider et al. 1996). This calculation suggested that the CL of 10 $\mu l\ l^{-1}$ h was exceeded in almost all forest areas. On average, Austrian forests were exposed in 1993, a rainy and cool year, to AOT40 values 2.3 times higher than the CL. The AOT40 was lower than the CL on only 1.8% of the grid area. In 1994, a year with generally high O_3 levels, CL was exceeded by a factor of four to

[1]Based on the mean annual temperature and pressure conditions of altitudinal gradients in central Europe, the conversion into "μg O_3 m^{-3}" must ensure that the multiplier of measured "nl O_3 l^{-1}" decreases from 2.05 at sea level to 1.42 at 3500 m a.s.l. (see Bucher et al. 1986).

Table 4.1. AOT40 values (ppm h = $\mu l\ l^{-1}$ h) and exceedance (%) above the "critical AOT40 level" (10 $\mu l\ l^{-1}$ h); CL of forests according to UN/ECE 1994 as derived from forest-relevant, O_3- monitoring stations in Austria. (Schneider et al. 1996)

Year	Number of stations	AOT40 (ppm h = $\mu l\ l^{-1}$ h)					
		Mean ($\mu l\ l^{-1}$h)	Exceedance (%)	Minimum ($\mu l\ l^{-1}$h)	Exceedance (%)	Maximum ($\mu l\ l^{-1}$ h)	Exceedance (%)
1993	43	28.1	181	12.9	29	49.6	396
1994	45	32.6	226	13.1	31	55.3	453
1995	48	27.5	175	0	0	49.3	393

five, with the highest AOT40 reaching 70 $\mu l\ l^{-1}$ h (in the western part of Austria). Considerably higher AOT40 values occurred in mountainous regions (due to high O_3 levels prevailing during the night) than in the lower forests of eastern Austria, where O_3 reductions at night are responsible for a decrease in AOT40 down to 50% of the high-altitude levels. However, the highest 0.5-h O_3 maxima (100–133 nl l^{-1}) occurred in eastern Austria, mainly in the plume of industrial or urban agglomerations (e.g., Linz or Vienna; Spangl 1995).

In Switzerland, the highest 0.5-h maxima have been reported for urban areas, particularly in the southern part of the country (e.g., Lugano in the canton of Ticino: 170 nl O_3 l^{-1}; BUWAL/NABEL 1995). As in Austria, CL was exceeded in the forests of the rural areas of Switzerland. However, AOT40 values seem to be influenced by local factors since no common temporal trend was found for the years 1990–1995 (Table 4.2), when several rural stations were compared in Austria (Schneider et al. 1996) and Switzerland (BUWAL/NABEL 1995). For Austrian conditions, AOT40 (y) can be approximated from half-year (April–September) mean O_3 concentrations (x) according to the equation y ($\mu l\ l^{-1}$ h) = $42.18\ x^2 - 1676\ x + 27359$ (cf. Schneider et al. 1996).

4.3 Tree Responses to O_3 Levels at Forest Sites

4.3.1 Experimental O_3 Exposure of Twigs of Mature Norway Spruce and European Larch Trees in Austrian Forests

Chamber experiments with Norway spruce and beech seedlings have indicated that ozone can induce both visible needle injury and physiological impairment (reduced gas exchange rates; Krause and Prinz 1985). The

Table 4.2. Temporal development of AOT40 (ppm h = $\mu l\ l^{-1}$ h) for the period of 1990–1995 at rural stations in Austria (After Schneider et al. 1996) and Switzerland (Buwal/Nabel 1995)

	Austria						Switzerland				
	Location[a]										
	a	b	c	d	e	f	g	h	i	j	k
Altitude (m)	305	520	726	1000	1012	1960	500	680	1030	1130	1650
Year											
1990	24	34	23	66	0	70	27	42	–	–	28
1991	22	55	27	56	30	48	23	37	–	–	21
1992	26	39	32	41	38	39	22	27	37	–	24
1993	22	31	38	40	32	36	15	25	33	–	23
1994	31	44	34	55	31	43	18	34	44	44	29
1995	25	34	31	41	34	42	–	–	–	–	–

[a] a, Dunkelsteinerwald; b, Hermannskogel; c, St. Leonhard; d, Sulzberg; e, Nebelstein; f, Innsbruck-Nordkette; g, Payerne; h, Lägeren; i, Rigi; j, Chaumont; k, Davos; – no data available.

effects were reported at lower O_3 concentrations than those persisting at the timberline sites in Austria, where clear evidence of O_3 injury had been lacking. Although controlled-environment exposure systems provide good control of O_3 regimes and environmental variables, the relevance of the results is limited (Arndt and Seufert 1990; Manning and Krupa 1992).

Plants in chambers do not experience the variety of interacting stresses that commonly occur in the field. Therefore, field exposure systems such as OTCs and branch cuvettes are used to approach near-natural conditions. OTCs are suitable for dose-response studies when dealing with large numbers of small trees. However, results only relate to young trees and may be biased by microclimatic effects (temperature, air velocity, soil moisture) as well as by problems with insect or disease control. Consequently, in Austria an exposure system consisting of branch cuvettes was developed to determine the chronic impact of ambient and above-ambient O_3 levels on forest trees under site conditions.

4.3.1.1 Study Sites, Pollution Load, and Experimental Design

The study was conducted on 33- to 65-year-old trees of Norway spruce (*Picea abies*, n=7) and European larch (*Larix decidua*, n=2) at two Tyrolean forest sites, in the Ziller Valley (1000 m a.s.l.) and on Mt. Patscherkofel near Innsbruck (1950 m a.s.l., timberline). At both sites, ozone was the dominant air pollutant, and the concentrations of SO_2, NO, and NO_2 were very low. The high-elevation site was exposed to higher annual and seasonal mean O_3 concentrations than the lower site (Fig. 4.2), while 0.5-h maxima could reach 120 nl O_3 l^{-1} at both sites. At the low-elevation site, the photochemical formation and destruction of ozone led to a typical diurnal pattern with rising concentrations during the daylight hours and a decline during the night. At the timberline, the diurnal variation was small (Fig. 4.2A). 0.5-h means of O_3 levels during the assimilation period were usually within 20-39 nl l^{-1} at the lower site and 40-59 nl l^{-1} at the timberline (Fig. 4.2B). These values are typical of O_3 conditions in alpine valleys of Austria (cf. Smidt 1993).

At both study sites, scaffolding provided access to the branches in the upper shade and sun crown. Small branches were sealed into transparent fumigation cuvettes (the sun-exposed branches being climatized; Havranek and Wieser 1990, 1994) for one or two growing seasons between 1986 and 1991. The regimes of O_3 exposure, which were determined by the diurnal and seasonal variations at the site, were charcoal-filtered air (CF), ambient air (A), or the latter enriched by 30, 60, or 90 nl $O_3 l^{-1}$. Four to eight replicates were used per treatment.

The system enabled control of the ambient climatic conditions and O_3 concentrations (tracking their diurnal fluctuations) inside the cuvettes, as well as the synchronous operation of several replicates. Comparative measurements of gas exchange in spruce did not reveal differences in the light-saturated CO_2 assimilation rate and stomatal conductance (g_{H2O})

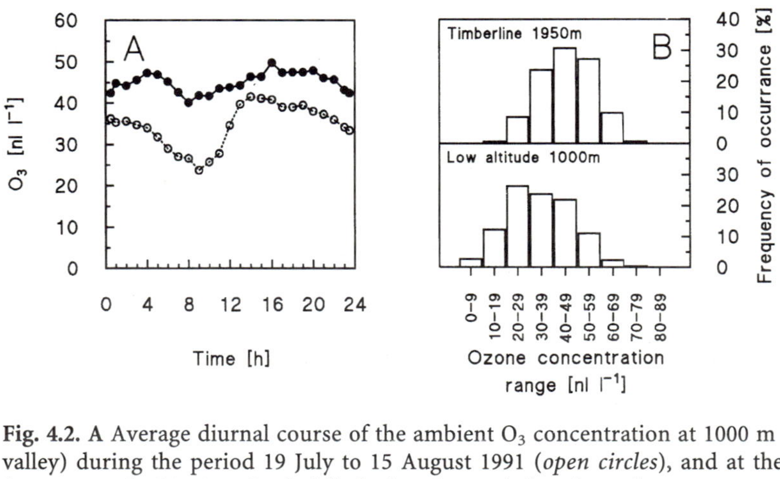

Fig. 4.2. A Average diurnal course of the ambient O_3 concentration at 1000 m a.s.l. (Ziller valley) during the period 19 July to 15 August 1991 (*open circles*), and at the timberline (1950 m a.s.l., Mt. Patscherkofel) during the period 5 through 31 August 1992 (*closed circles*). **B** Frequency distribution of 0.5-h means of O_3 concentration during the investigation periods. (After Wieser and Havranek 1994)

between twigs growing outside cuvettes and twigs exposed for 23 weeks to the ambient O_3 regime inside cuvettes (Havranek and Wieser 1994).

At the end of each fumigation period, twigs were examined for macroscopic injury and gas exchange behavior (measurements were conducted in situ in the field, but under controlled conditions of light, temperature, and humidity). Additionally, continuous measurements of the ambient O_3 concentration and the gas exchange of twigs under ambient climatic conditions were used to calculate the O_3 uptake by the needles (cf. Wieser and Havranek 1993a, 1995).

4.3.1.2 The Effects of the Various O_3 Exposure Regimes

None of the exposure regimes caused visible injury in spruce or larch needles at either study site, nor were the chlorophyll content, the dry weight of 100 needles, or the specific leaf area significantly affected relative to the corresponding control branches in filtered air. Even the highest O_3 exposure failed to cause needle loss or any injury during the subsequent winter (when cuvettes were removed from the twigs), and needle yellowing, chlorotic mottling or impaired growth of the new flush were not observed in spring (Wieser and Havranek 1994)

After fumigations throughout one assimilation period with mean O_3 concentrations ranging from near-zero to 100 nl l^{-1} (AOT40 up to 100 $\mu l\ l^{-1}h$, SUM0 up to 200 $\mu l\ l^{-1}$ h), no distinct treatment effects were detectable on the net CO_2 uptake rate and g_{H20} of current-year spruce needles (Havranek et al. 1989; Wieser et al. 1991; Wieser and Havranek 1994). Mean O_3 concentrations above 100 nl l^{-1} induced a significant decline in net photosynthesis by 20-40% in current-year needles (Fig. 4.3), whereas 1-year-old needles responded more slowly to ozone, with a decrease in photosynthesis only at AOT40 and SUM0 values above 300 and 500 ppm h, respectively. These values were generally reached during the second fumigation period. In contrast, the net photosynthesis of 2 to 4-year-old needles was increased under elevated O_3 levels (Fig. 4.3, data shown only for 2-year-old needles).

The decline in the net CO_2 uptake rate of spruce needles continuously exposed to O_3 concentrations higher than 100 nl l^{-1} was proportionally higher than the reductions in g_{H20}, revealing decreased water-use efficiency (WUE; Fig. 4.4A). These changes indicate that g_{H20} *per* se did not limit CO_2 uptake, but stomatal narrowing was probably a response to the increased CO_2 concentration in the intercellular spaces of the mesophyll tissue (c_i), the latter being caused by a decline in the carbon uptake efficiency (CE; Fig. 4.4B). Similar results were found by Reich and Amundson

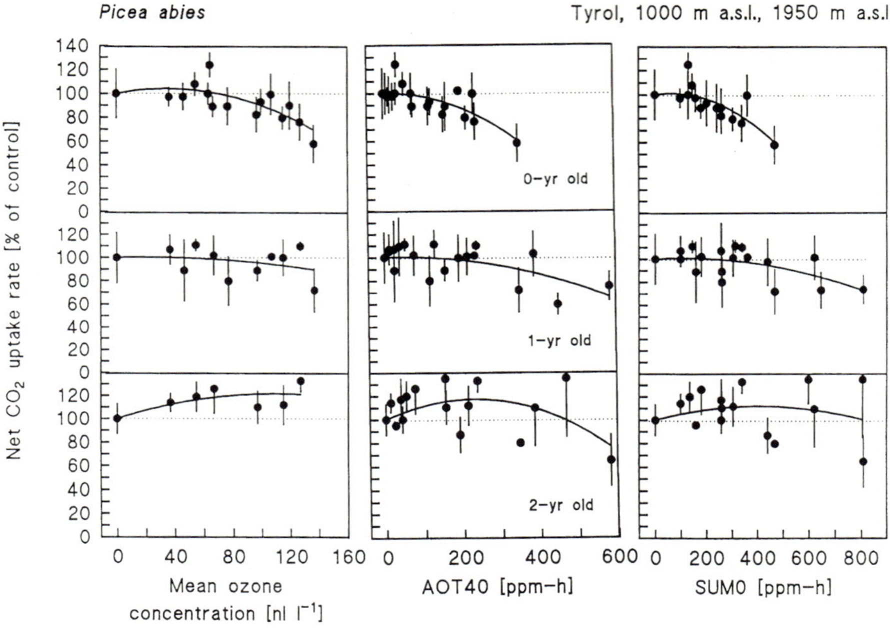

Fig. 4.3. Net CO_2 uptake rate of current. 1-and 2-year-old spruce (*Picea abies*) twigs after exposure to different O_3 treatments, related to mean O_3 concentration during one fumigation period, and AOT40 and SUM0 for one and two fumigation periods, respectively. Measurements were conducted in Tyrol in situ under controlled conditions (light saturation, 21C needle temperature; $\Delta w = 10$ Pa kPa^{-1}, 350 μmol mol^{-1} CO_2) after one or two fumigation periods, in different years and at different altitudes (1000 and 1950 m a.s.l.). Mean ambient O_3 concentrations during the fumigation periods (12–24 weeks) were 25–65 nl l^{-1}. Each data point represents the mean of 4–8 individual twigs; *bars* indicate SD (after Wieser et al. 1996)

(1987), Matyssek et al. (1991a, 1993a), Pell et al. (1992), and Wallin et al. (1992). Decreased CE may indicate inhibition of the CO_2-fixing rubisco enzyme (Farquhar and Sharkey 1982), although structural decline of the leaf mesophyll (e.g. cell collapse) can bias photosynthetic response to O_3 stress (as found in birch leaves; Matyssek et al. 1991a). However, after the second fumigation period no further decline in WUE was observed.

While Norway spruce is generally regarded to be rather O_3-tolerant, European larch has been classified as "sensitive" or at least "intermediate" relative to O_3-tolerant trees (Davies and Wilhour 1976; VDI 1989; Chappelka and Chevone 1992). However, when monitoring the gas exchange of larch twigs under conditions of diurnal and seasonal fluctuations of the microclimatic factors at the low-elevation site, it was found that ambient and twice the ambient O_3 concentrations (mean 70 nl l^{-1}, peaks >150 nl l^{-1}) did not impair photosynthesis or g_{H2O} relative to control twigs in filtered air (Havranek and Wieser 1993).

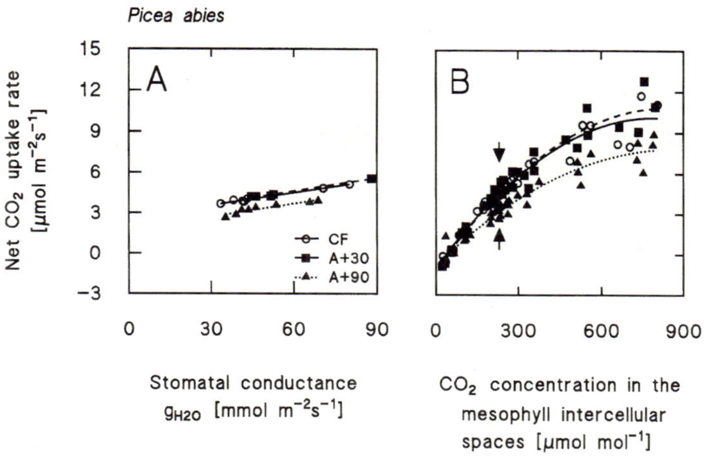

Fig. 4.4. The relationship between (**A**) the net CO_2 uptake rate and stomatal conductance (g_{H2O}) and (**B**) between the net CO_2 uptake rate and the CO_2 concentration in the intercellular spaces of the needle mesophyll ("internal CO_2 partial pressure") of current-year shoots in Norway spruce (*Picea abies*) after O_3 exposure throughout one growing season. Arrows indicate measurements under ambient CO_2 concentration. (After Wieser and Havranek 1994)

Even though ozone did not limit CO_2 uptake in spruce and larch, the responsiveness of gas exchange to fluctuations in the environmental conditions may be altered. This was examined by submitting pairs of spruce and larch twigs from exposures of ozonated or filtered air to standardized diurnal courses of increasing or decreasing irradiance, temperature, and humidity (Fig. 4.5G-H). The differently pretreated twigs were analyzed in parallel, with one gas exchange chamber each for their CO_2 uptake rate and g_{H2O} (Fig. 4.5C-F). Under the same chamber conditions, for both species there was no difference in the regulatory capacity of the stomata between control twigs and twigs that had experienced ambient or twice the ambient O_3 levels. However, regardless of the O_3 pretreatment, the gas exchange depended strongly on the microclimatic factors acting outside the gas exchange chamber (Wieser et al. 1991) and on the water status of the whole tree (Havranek and Wieser 1993).

These latter external and internal variables clearly had a stronger influence than ozone on gas exchange, as shown for larch in Fig. 4.5. After a drought period, the same pair of larch twigs had a lower net CO_2 uptake and g_{H2O} on 10 September 1991, a hot and dry day, with a high vapor pressure deficit (vpd) and a low tree water potential (Ψ), than on September 13th, when vpd was low and tree water potential was high after two preceding rainy days. In contrast, the g_{H2O} of spruce was enhanced, and the regulatory capacity of the stomata seemed to be impaired after a

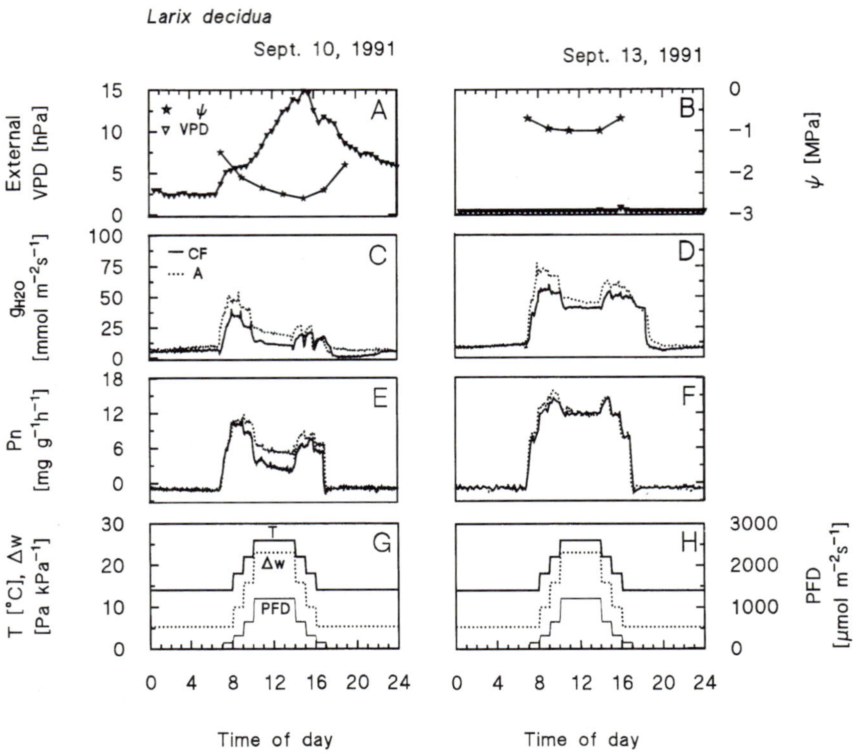

Fig. 4.5. Diurnal courses of (**A, B**) external water vapor pressure deficit (*VPD;* outside the gas exchange cuvette) and twig water potential (ψ). **C, D** Stomatal conductance (g_{H2O}); **E, F** CO_2 assimilation rate (*Pn*) of short-shoot needles of larch (*Larix decidua*) exposed to either charcoal-filtered air (*CF, solid line*) or ambient-air O_3 concentration (*A, dotted line*) at the end of a drought period 10 Sept. 1991; **A, C, E** and **G**) and after two rainy days 13 Sept. 1991; B, D, F, and H). **G, H** Experimentally standardized needle temperature (*T*), leaf/air difference in the molar fraction of water vapor (Δw), and photon flux density (*PFD*) inside the gas exchange cuvette. (After Havranek and Wieser 1994)

7-week exposure to three times the ambient O_3 concentrations (Wieser et al. 1991). Similar observations of impaired stomatal responses have been made by Keller and Häsler (1984), Maier-Maercker and Koch (1991), and Pearson and Mansfield (1993). Maier-Maercker (1989) ascribed the loss in the regulatory capacity of the stomata to an O_3–induced delignification of the walls of subsidiary and guard cells.

4.3.1.3 Stomatal Control of Ozone Uptake

The growth conditions with O_3 exposures conducted in fumigation cabinets are often nonlimiting to the plant except for the impact of the air pollutant. It is in such situations that O_3-induced plant injury has been reported (cf. Reich 1987; Pye 1988; Darrall 1989). In contrast, field sites are typified by broad natural variations in the environmental factors which often limit plant development. It may be these variations amongst others, which masked or prevented long-term effects of O_3 levels below 100 nl l^{-1} (AOT40 <150 $\mu l\ l^{-1}\ h^{-1}$, SUM0 <200 $\mu l\ l^{-1}$ h) on the leaf gas exchange of the spruce and larch trees. However, the amount of ozone taken up into the leaf, and the consequent risk that ozone poses for leaf metabolism, is determined by stomatal regulation. Thus, any factors limiting stomatal opening such as dry air or soil drought, may reduce O_3 uptake and lessen O_3 stress (Tingey and Taylor 1982; Matyssek et al. 1995b).

Variation in O_3 uptake results from changes in both the O_3 concentration of the ambient air and the leaf conductance for diffusive O_3 uptake (g_{O3}). In general, high O_3 concentrations occur during periods of high irradiance and high temperature. At both study sites, O_3 concentrations positively correlated with Δw (i.e., the difference in the mole fraction of water vapor between the leaf and air) which, like O_3 concentration, increases with irradiance and temperature (Fig. 4.6). Although more pronounced at the low-elevation site, these interrelationships seem to be of general importance (cf. Grünhage and Jäger 1994). In both spruce and larch, O_3 uptake was mainly limited by high Δw and high soil moisture deficits (Wieser and Havranek 1993a, 1995). At low Δw, however, stomata were wide open and allowed high fluxes of ozone into the needles, even at low ambient O_3 concentrations (Fig. 4.6). In spruce, stomatal narrowing caused by increasing Δw limited O_3 uptake by foliage in the shade earlier than by foliage in the sun. Larch displayed a higher maximum g_{H2O} relative to spruce, but stomatal narrowing was also more pronounced in response to increasing Δw in larch (cf. Matyssek 1985). Surprisingly, stomata were often not completely closed at night, an observation made in several other species (Benecke et al. 1981; Matyssek 1985; Zimmermann et al. 1988; Lange et al. 1989a,b; Weikert et al. 1989; Matyssek et al. 1995a). Furthermore, as shown for larch (Fig. 4.6), avoidance of O_3 uptake as the result of stomatal narrowing was more pronounced at the lower site, where water availability is generally less optimal.

Findings from chamber experiments suggest that drought protects young conifers from O_3 injury through stomatal narrowing (Freer-Smith et al. 1989; Dobson et al. 1990; Fincher and Alscher 1992; Roberts and Cannon 1992; Temple et al. 1992; Wallin and Skärby 1992; Kronfuss et al. 1996) and beech seedlings (Pearson and Mansfield 1993). Field observations in the mid-Ohio valley (USA) by Showman (1991) indicated that foliar injury of O_3-sensitive tree species (e.g., *Fraxinus americana, Lir-*

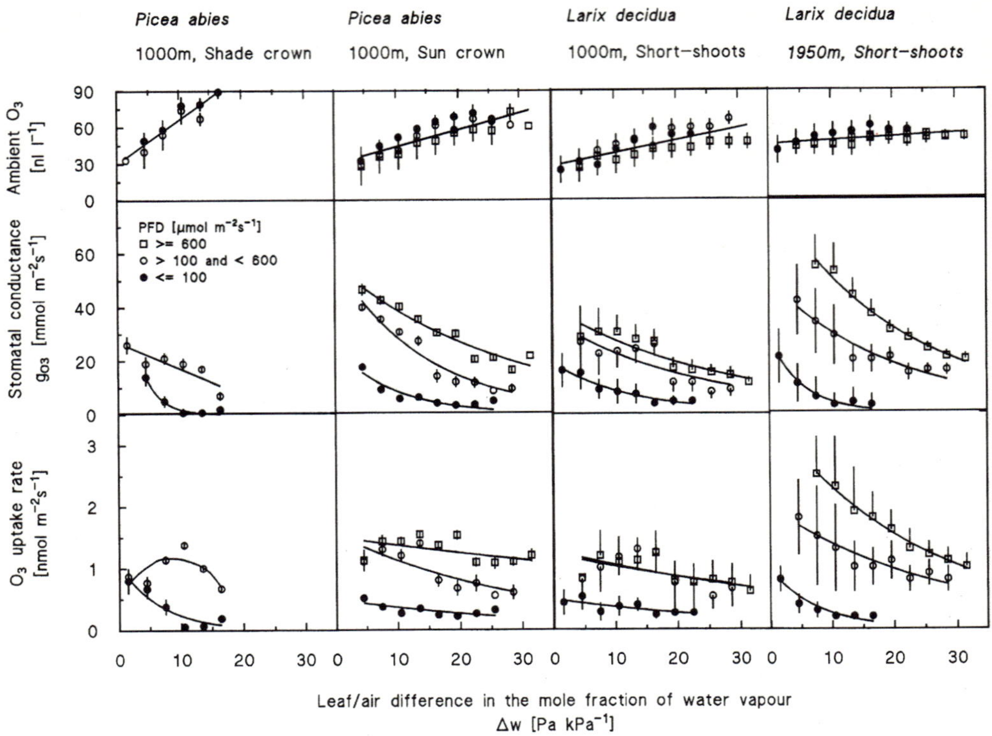

Fig. 4.6. *Top* to *bottom* Ambient O_3 concentration, stomatal conductance for O_3 (g_{O3}) and O_3 uptake in relation to the leaf/air difference in the mole fraction of water vapor (Δw) in shade and sun needles of Norway spruce (*Picea abies, left half* of figure) and in short-shoot needles of European larch (*Larix decidua*) at 1000 and 1950 m a.s.l. (*right half* of figure). Each point represents the mean of 10 to 45 0.5-h intervals ± SD. Means were selected according to the following light regimes: photon flux density (*PFD*) ≥600 (*open squares*), >100 and <600 (*open circles*) and ≤ 100 μmol m^{-2} s^{-1} (*closed circles*)

iodendron tulipifera, Prunus pennsylvanica) was more widespread in a moist year with low O_3 levels than in a dry year with elevated O_3 exposure. Paffrath and Peters (1988) and Edwards et al. (1991) also discussed the possibility of stomatal aperture controlling the occurrence of O_3 injury in the field.

Besides drought, other factors such as irradiance and nutrition affect leaf morphology and stomatal regulation and, therefore the potential for O_3 uptake (Matyssek et al. 1995b). After exposure to similar light levels, long-shoot needles of larch had a higher g_{H2O} than short-shoot needles (Wieser and Havranek 1995). The insertion of single needles on the long shoots with a lower degree of self-shading, probably favors differentiation towards sun-type foliage and, therefore, high g_{H2O} relative to short-shoot needles. Sun-exposed needles of spruce and larch are known to possess higher gas exchange rates than needles in the shade (Havranek and Wieser 1994; Matyssek 1985). Needles of fertilized trees also showed more pronounced stomatal responses to changes in Δw and irradiance than needles of nutrient-deficient trees (Wieser and Havranek 1993b). Findings about O_3 effects on trees, as part of multiple stress scenarios, are still limited and conflicting (Matyssek et al. 1995b).

4.3.1.4 Detoxification Capacity

There is evidence that not only the stomatal control of O_3 uptake but also the capability to detoxify oxygen radicals in certain exposure situations may modify the potential toxicity of the ambient O_3 concentration. For example, O_3 sensitivity of birch grown in fumigation chambers was increased at night because lower O_3 uptake (due to stomatal narrowing) led to higher reductions in biomass production as compared with the high O_3 uptake during the daylight hours (Matyssek et al. 1995a). Changes in the detoxification capacity in the field are also indicated by reported daily and annual fluctuations in the content of antioxidants in needles of spruce trees, e.g., ascorbic acid (Esterbauer et al. 1980), apoplastic ascorbate (Wieser et al. 1996) and glutathione (Grill et al. 1987; Schupp and Rennenberg 1988). Daily turnover rates in the ascorbic acid content of spruce needles are fast and hourly synthesis rates of 7% of the total content have been estimated (Esterbauer et al. 1980). Spruce seedlings also have the potential for acclimation to O_3 stress. In a chamber fumigation experiment, young spruce needles increased their antioxidant contents (ascorbate and glutathione) within a few days and were able to stop the increase of O_3 injury (i.e., the decrease in the CO_2 uptake rate) initiated by an abrupt exposure to 100 nl O_3 l^{-1} (Kronfuss 1996; Fig. 4.7).

Under field conditions, the capacity for detoxifying oxygen radicals may be greater in plants at high altitudes (Rennenberg 1988; Polle and Rennenberg 1992). High-altitude plants are generally exposed to higher oxidative stress due to low temperature, high radiation, and high O_3 concentration compared with low altitudes. The apoplastic ascorbate concentrations in spruce needles at timberline has been found to be 25% higher than in spruce needles from trees grown at 1000 m a.s.l. (Polle et al. 1995).

Differences in the levels of antioxidants, protective enzymes, and pigment contents along several altitudinal gradients in Austria were interpreted as bioindicative of oxidative stress (Bermadinger-Stabentheiner and Grill 1992; Tausz et al. 1994). However, since oxidative stress may depend on many factors varying with altitude (including nutrient defi-

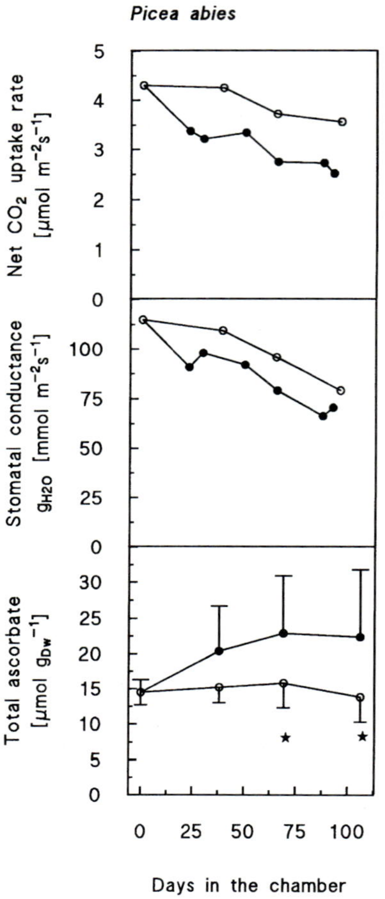

Fig. 4.7. Time course of net CO_2 uptake rate (*top*), stomatal conductance (g_{H2O}; *center*), and total needle ascorbate (*bottom*) of spruce (*Picea abies*) seedlings (n=7) exposed to charcoal-filtered air (*open circles*) and air enriched with 100 nl O_3 l^{-1} (*closed circles*) in climate chambers (* P <0.05); after Kronfuß 1996

ciency), it was difficult to trace high contents of antioxidants back to oxidative stress caused by high O_3 levels. Spruce tress from different sites actually exhibited distinctly different levels of antioxidative responses to rather similar exposures of oxidative pollutants (Tausz and Grill 1995).

Surprisingly, no O_3 effect on ascorbate or water soluble thiol contents were found in the field fumigation experiment described above (Wieser and Havranek 1994), in which twigs in the shade crown of a mature spruce tree were exposed to mean O_3 concentrations between near-zero and 130 nl l^{-1}. Chlorophyll content and the chlorophyll/carotinoid ratio were also unaffected by the O_3 treatments confirming that no chlorosis had developed. The only signs of O_3 stress to the pigments were a small decrease in α/β carotene and an increase in the xanthophyll/carotene ratio with increasing O_3 level (Bermadinger, pers. comm.).

Do single branches of a tree respond in a different way to O_3 stress than "whole plants"? The basic assumption in the use of branch cuvettes is of a certain branch autonomy, as shown for carbon uptake under different O_3 exposures (Schaap, pers. comm.). However it is unclear, whether detoxification occurs locally in the twig, or if the rest of the tree makes a contribution. Differences in O_3 uptake rate, needle morphology, light sensitivity, and – last but not least – chamber effects may also modify the capacity of protective mechanisms.

Genetic differences in O_3 sensitivity may occur within a species. In chamber experiments, young spruce (Havranek et al. 1990) and larch seedlings of low-altitude provenance (Wieser H. 1995) seemed to be more sensitive to enhanced O_3 levels than high-altitude provenances. However, further research is required to determine, whether an evolutionary adaptation to high oxidative stress, including high ambient O_3 levels, has developed at high altitudes.

4.3.2 The Swiss Research Program "Air Pollution and Forest Decline"

4.3.2.1 Effects of Air Pollutants on the Gas Exchange of Spruce

Two rural sites, Lägeren (685 m a.s.l., near Zurich) and Davos (1660 m a.s.l., eastern Switzerland), were selected in this program to examine the effects of air pollutant regimes on the gas exchange of mature spruce trees (Häsler et al. 1991). Ozone was the dominant air pollutant with monthly mean levels in summer of about 30-45 nl l^{-1} at both sites; Lägeren more often showed high O_3 episodes (381 h above 60 nl O_3 l^{-1} versus 19 in Davos; Gehrig et al. 1991).

The aim of the study was not to compare directly the gas exchange of the two individual spruce trees used in the study, but rather to analyze whether the impact of the contrasting O_3 regimes of the two sites was reflected in the gas exchange behavior of the trees. Both trees had low crown transparency. Diurnal courses of gas exchange in the sun and shade foliage were continuously monitored from 1986 through 1989. In contrast to the Austrian approach, ozone was not experimentally manipulated to reveal mechanisms of O_3 impact. Rather, the gas exchange at each site was compared between days of contrasting O_3 levels. However, even equipment with teflon tubing and nickel-coated metal surfaces for gas exchange measurement can reduce the O_3 concentration in the measuring cuvette down to 50% of the level in the ambient air (Häsler 1991b), and thus, may lower the O_3 impact during the measurement.

Given this caveat, no consistent influence of the O_3 regimes on the daily carbon gain of twigs was detectable at either site, and both high and low CO_2 uptake rates occurred under high O_3 concentrations (Fig. 4.8). In

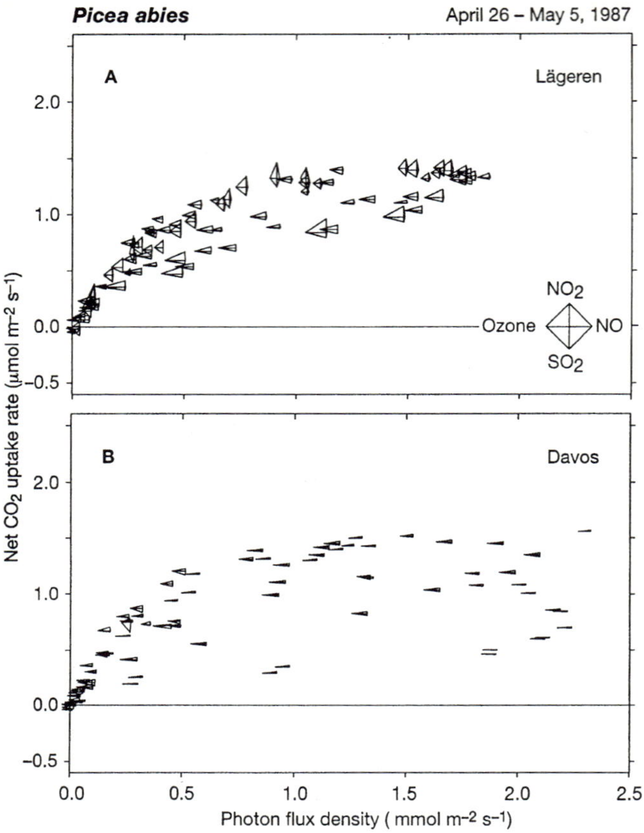

Fig. 4.8. Dependence of net CO_2 uptake rate of Norway spruce (*Picea abies*) on the photon flux density at Lägeren (**A**) and Davos (**B**). The trapezoid gives a measure of the air pollution; the scaling of the axes inside the trapezoid are 0–94 nl l^{-1} for *ozone*, 0–38 nl l^{-1} for NO_2, 0–44 nl l^{-1} for *NO*, and 0–16 nl l^{-1} for SO_2. (After Häsler et al. 1991)

addition, no distinct O_3 impact on stomatal regulation and crown transpiration was discovered (Häsler and Herzog 1991). The gas exchange appeared to be determined by the variability of the natural site factors rather than by the presence of ozone. This conclusion was supported by the use of a statistical model (Häsler and Ruckstuhl 1991) which suggests that the daily carbon gain was based solely on the natural factors irradiance, leaf and soil temperature, air humidity, and soil water tension. No further "driving variable", such as ozone, was required to explain the measured carbon uptake at either site. Consequently, it was inferred that the daily O_3 regimes at the sites were unlikely to affect the gas exchange behavior. However, O_3 effects could not be ruled out completely since a "clean-air control" was not available in this field approach.

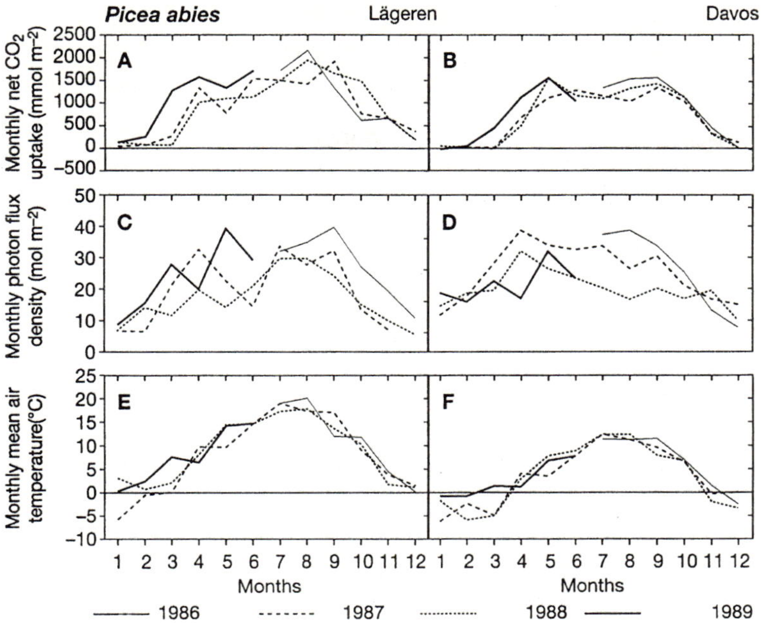

Fig. 4.9. Annual courses of monthly net CO_2 uptake (**A, B**), monthly photon flux density (**C, D**), and monthly mean air temperature (**E, F**) in Norway spruce (*Picea abies*) at Lägeren (**A, C,** and **E**) and Davos (**B, D** and **F**) throughout 1986–1989. (After Häsler 1991a)

The lack of a control made the assessment of the long-term effects of ozone difficult. Would photosynthesis decline during the course of the 3-year observation period or with increasing needle age (i.e., exposure time; Häsler 1991a)? A gradual limitation of the carbon uptake was not detected when comparing, within the 3-year period, the gas exchange of "current-year" needles formed each year in the sun crown (Fig. 4.9). The photosynthesis of shade needles (determined under the saturating irradiance of an artificial light source) up to an age of 7 years did not show any decline either, although shading may increase the susceptibility to ozone (Tjoelker et al. 1993; Matyssek et al. 1995a). Moreover, the CO_2 uptake rates of spruce at both sites were, within the range of natural variability, comparable to data obtained before this period of research (Pisek and Tranquillini 1954; Fuchs et al. 1977; Matyssek 1986).

4.3.2.2 Analysis of the Needle Wax Layer, Stem Production, and Fungal Impact

Although ozone did not measurably affect needle gas exchange, the possibility existed that the needle surfaces, which represent the ultimate tree/air interface, might show evidence of O_3 impact. The structure and chemical composition of the needle wax layer were therefore analyzed (Günthardt-Goerg 1991). Needle samples were taken from mature spruce trees in both Lägeren and Davos, including the trees used in the gas exchange study. Degradation of the wax structure (e.g., fusing and flattening of wax tubes) increased with needle age and crown transparency. This tendency was more pronounced at Davos than at Lägeren. However, crown transparency was unrelated to the amount and chemical composition of the wax and the length, dry weight, and water content of the needles. Most importantly, exposure of young cloned spruce plants to 50 or 150 nl O_3 l^{-1} during daylight hours throughout one growing season (45 h/week; the 150 nl l^{-1} O_3 exposure resulted in higher O_3 doses than at the field sites) in the exposure chambers of the Birmensdorf experimental site failed to reproduce the wax degradation found in plants of the same clone grown at the forest sites and in the mature trees.

Only the chemical differentiation of the wax was slightly delayed. Plants subjected to soil drought or SO_2 differed in their wax structure from the individuals growing at the two forest sites. The effects of all experimental treatments on the amount and composition of needle wax remained within the range of the variability found in the foliage of the mature spruce trees. Again, ozone could not be identified as the main driving factor of this variability; rather, wax degradation was determined by natural causes (e.g., genotype, exposure to climatic factors in the canopy, or altitude). This conclusion is consistent with needle wax being largely resistant to chemical impact by ozone or SO_2 (Hoigné 1988). A detailed overview of this analysis has been published recently (Günthardt-Goerg et al. 1994).

Tree-ring analysis revealed stable stem production since 1820 in spruce trees at the Davos site, whereas those at Lägeren exhibited a gradual decrease, beginning around 1960 (Joos 1991). Beech trees (*Fagus sylvatica*) at the latter site did not show this decline. Although ozone may limit the carbon allocation to stem and root (see Sect. 4.4), such conclusions cannot be drawn about the spruces at Lägeren as no experimental analysis of tree-ring formation under the given site conditions is available. Progressive canopy closure by beech had probably caused competitive stress on spruce during recent decades, and may be responsible for the growth reduction.

Fungal infections of spruce studied at many sites in Switzerland have not revealed any relation to air pollution or crown transparency (Sieber 1991). The extent of needle infection by endophytic fungi (*Lophodermium piceae* and *Thiarosporella parca*) varied strongly between sites and individual trees. Nevertheless, these fungi were excluded as the cause of needle losses involved in forest decline or of yellowish spot-like needle discolorations, elsewhere ascribed to O_3 stress (Lang and Holdenrieder 1985).

4.3.2.3 Experiments on the O_3 Sensitivity of Forest Trees

Within the Swiss research program, experiments on the O_3 sensitivity of forest trees were aimed at determining what impact O_3 exposure regimes, relevant to Swiss conditions, may have on important forest tree species (Landolt and Lüthy-Krause 1991). In a classic fumigation experiment, young individuals of *Picea abies*, *Pinus sylvestris*, *Abies alba*, and *Fagus sylvatica* were grown in charcoal-filtered air, or in 50 or 100 nl O_3 l^{-1} in the Birmensdorf exposure chambers[2] and in ambient air outside the chambers. The study looked at the similarities of O_3 responses of individuals exposed at the sites of Lägeren and Davos in OTCs to ambient or filtered air. The species comparison was extended to plants known for their O_3 sensitivity, namely *Populus x euramericana* (var. *Dorskamp*; Mooi 1981) and the herbaceous clover *Trifolium pratense* (var. *Lucrum*; EPA 1978).

At Birmensdorf, pine developed the most macroscopical leaf injury (yellowing and loss of needles) in response to ozone. Beech exposed to 100 nl O_3 l^{-1} developed brownish leaf necroses, whereas fir and spruce did not exhibit any injury during the 15-week exposure. In spruce, only extended O_3 exposure (20 weeks) caused needle yellowing. The biochemical responses of the foliage were more pronounced in pine than in spruce or beech, whereas fir did not respond. Specifically, the concentrations of malate, citrate, and pinitol (in pine; Landolt et al. 1989), and PEP carboxylase activity were increased. The levels of shikimic and china acid, inositol and malate dehydrogenase activity were lowered under O_3 stress.

It was found that the chamber environment itself (perhaps including the soil medium used) affected leaf biochemistry in ways similar to O_3 stress. Pine exposed to ambient air inside chambers differed from plants grown outside in terms of lowered inositol and elevated pinitol levels; in a similar manner plants fumigated in the chambers (with 50 or 100 nl O_3 l^{-1}) differed from their corresponding chamber control (filtered air; Fig. 4.10). Apparently, the "stress" caused by the chamber environment was not clearly "distinguished" by the biochemical response mechanisms from

[2]The OTC system required a shading roof (automatically applied at a photosynthetic photon flux density, i.e., PPFD >600 µmol m^{-2} s^{-1}) to compensate for temperature increase in full sunlight. The roof was not used at lower irradiance. Thus, plants were "moderately" shade-adapted, but much less so than inside or below forest canopies.

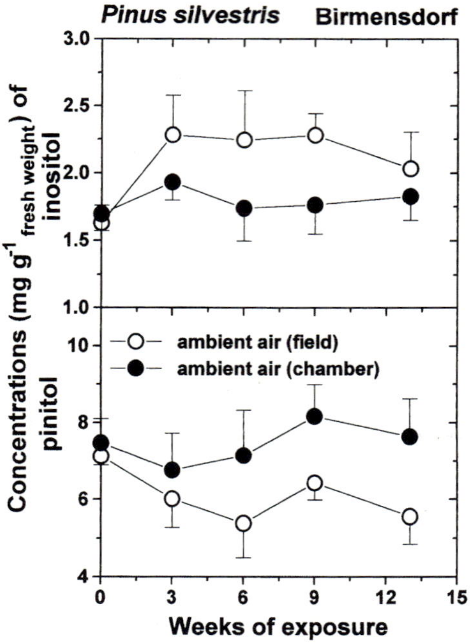

Fig. 4.10. Time course of the inositol and pinitol concentrations in the needles of *Pinus sylvestris* exposed to ambient air either inside or outside the field exposure chambers in Birmensdorf. (After Landolt and Lüthy-Krause 1991)

that caused by ozone. This means that under field conditions leaf biochemical responses may be rather unspecific for ozone and easily masked by the impact of other environmental factors.

It is therefore not surprising that at both Lägeren and Davos, the plants grown in OTCs with ambient air did not consistently differ from those exposed to filtered air in terms of biochemical and macroscopical responses. Only the O_3-sensitive clones of poplar and clover displayed leaf injury in the ambient air treatment. Whenever the mean monthly O_3 concentration exceeded 25 nl l^{-1}, leaf shedding occurred in poplar (Fig. 4.11). In clover, the proportions of necrotic leaves exponentially increased with increasing 0.5-h O_3 mean, beginning at 30 nl l^{-1}. The same poplar clone grown at a sun-exposed field site in Birmensdorf in ambient air with similar O_3 levels as the shaded forest sites did not display injury (Landolt 1992), underlining once more the effect of shading and perhaps OTC conditions on O_3 sensitivity. It was concluded that site factors strongly bias the plant's response to ozone, even in O_3-sensitive species. Overall, the findings of the Swiss research program suggested that natural site factors rather than pollutants determined tree performance. However, the experimental approaches could not rigorously clarify whether the O_3 regimes in the field adversely affect tree development in the long-term.

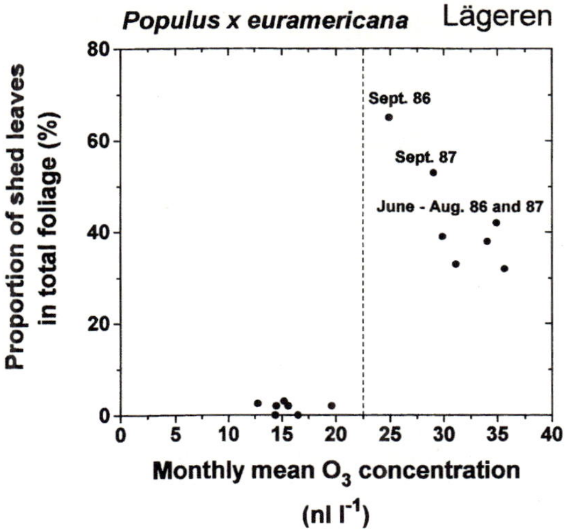

Fig. 4.11. Proportion of shed leaves in the total foliage of *Populus x euramericana* in relation to the mean monthly O_3 concentration in the open-top chambers installed at Lägeren. Pooled data from 1986 and 1987, with each *point* representing a mean of five plants. (After Landolt and Lüthy-Krause 1991)

4.3.3 Other Field Research on Trees in Switzerland

Another research program started in 1984 resulted in two "Forest Damage Reports" (*Waldschaden-Berichte*; Flückiger und Braun 1990, 1994). The emphasis was on beech (*Fagus sylvatica*), and two different approaches were employed:

1. Beech seedlings were fumigated in OTCs with either filtered or ambient air at field sites at Schönenbuch (400 m a.s.l., near Basel) and Zugerberg (1000 m a.s.l., near Zug). Again, ozone was the dominant air pollutant with pronounced daily fluctuations (peak concentrations of up to 120 nl O_3 l^{-1}) but lower seasonal mean levels at the lower site.
2. In parallel, beech trees, approximately 70 to 120 years old, were studied at about 50 forest sites (each 0.5 ha in size) in northwestern (Basel area) and eastern Switzerland (cantons of Zurich and Zug).

The beech seedlings grown in the OTCs with ambient air tended to display a reduced biomass production and root/shoot biomass ratio, and an increased water loss from detached leaves was reported. In contrast, Norway spruce seedlings exposed together with the beech did not exhibit limited biomass production in ambient air. However, beech seedlings in ambient air were prone to lethal frost injury and more intense attack by the leaf sucker *Phyllaphis fagi* relative to the control. The attractiveness of plants grown in ambient air to this insect was ascribed to enhanced concentrations of amino acids found in the phloem (Bolsinger and Flückiger 1989).

At the forest sites, the susceptibility of the mature trees to parasites correlated with an increasing N/K ratio of the foliage. Lowered levels of K, Mg, and P were attributed to the impact of acidic deposition (as concluded from laboratory experiments with seedlings), while increased leaf nitrogen was viewed in the context of increasing N deposition (>30 kg N ha^{-1} a^{-1}).

Spatial and temporal changes in crown transparency did not show an overall consistency during the 10-year observation period. However, initial assumptions that ozone determined the appearance and "status" of beech stands were eventually replaced by the realization that seasonal water availability, especially in calcareous soils, and fructification were the dominant factors.

Underlining the role of water availability, a study on beech trees in northwestern Switzerland by Saurer et al. (1995a) showed that seasonal precipitation probably determined the $\delta^{13}C$ value in the cellulose of developing tree rings. Both drought and ozone can raise $\delta^{13}C$ to less negative levels in the plant biomass (Farquhar et al. 1989; see Sect. 4.4), but it was specifically at edaphically water-limited sites that dry, warm summers caused $\delta^{13}C$ to increase. Although warm summers may be accompanied by elevated O_3 levels, the influence of dry, warm weather, and thus perhaps ozone, on $\delta^{13}C$ was unclear at edaphically moist sites (even though open stomata may allow high O_3 uptake at such sites). It appears, therefore, that water availability rather than O_3 stress determines the ecophysiology of beech in Switzerland, and that the long-term variation in annual precipitation may be reconstructed from the $\delta^{13}C$ analysis of tree rings.

4.4 Mechanisms of O_3 Impact on Trees Under Controlled Exposure Conditions

The field investigations reported above did not reveal that the prevalent O_3 regimes dominate the ecophysiology of forest trees. Tree behavior in the field is governed by a number of environmental and physiological factors which can either mask O_3 responses or impede their determination. This contributes to the different O_3 sensitivities of plants growing under field or controlled O_3 exposures, one of the shortcomings of chamber experiments. Controlled exposures yield insights into the mechanisms of O_3 effects, although they are typically restricted to young and/or small trees. Whole-plant analyses are feasible and may reveal those O_3 responses which *can* become effective and perhaps occur in the field, although the ecological relevance of the responses would need to be validated for any given tree and site. The following account is based on controlled O_3 exposures in the Birmensdorf field fumigation chambers (see Sect. 4.3.2.3), conducted with O_3–sensitive clones of birch (*Betula pendula*), poplar, and aspen (*Populus x euramericana, P. tremula*). What may be inferred about the long-term performance of forest trees.?

When ozone is taken up through the stomata, forest trees may be physiologically "challenged", although production – the more so in the absence of a "clean air" control – may not be measurably affected. Carbon cost for accumulating and maintaining a functional foliage may be enhanced by repair and detoxification processes (Reich 1983; Wolfenden and Mansfield 1991; Polle 1995) to compensate for the O_3 attack. Dark respiration rates indicated raised metabolic activity in O_3-exposed birch leaves (Matyssek et al. 1991a; Maurer and Matyssek, unpubl.), the loss of which perhaps was delayed. Leaves that developed under continued O_3 exposure resisted shedding longer than leaves formed in "clean" air and then subjected to O_3 stress later in the season (Günthardt-Goerg et al. 1993). The leaf formation rate was unchanged under O_3 stress (Mooi 1980; Matyssek et al. 1993a,b, 1995a).

Raised levels of reduced ascorbate in the cells of O_3-exposed birch leaves suggested acclimation to oxidative stress (Polle, pers. comm.). The elevated (less negative) $\delta^{13}C$ value in the biomass of O_3-exposed plants (Greitner and Winner 1988; Becker et al. 1989; Saurer et al. 1991) is also likely to be the result of a raised substrate turnover in the leaves. In birch exposed to ozone, elevated $\delta^{13}C$ of leaf and stem cellulose was related to the stimulated activity of the CO_2-binding phosphoenolpyruvate-carboxylase (PEPC; Matyssek et al. 1992; Saurer et al. 1995b) rather than to stomatal limitation of photosynthetic CO_2 uptake (Farquhar et al. 1989). This PEPC response, which favors ^{13}C incorporation relative to photosynthesis, was also found in poplar and pine exposed to ozone (Lüthy-Krause et al. 1990; Landolt et al. 1994). PEPC stimulation is seen as a part of an "anaplerotic" pathway that feeds the oxalacetate pool in C_3 plants, which have a high demand for substrate and energy (Wiskich and Dry 1985). Thus, ozone may affect the metabolism in ways which do not necessarily represent "injury" (i.e., metabolic breakdown or structural destruction) but rather reflect the capacity to cope with O_3 stress (acclimation).

The excess carbon consumed by O_3-stressed leaves to ensure their maintenance apparently cannot be allocated to the non-green organs. In young birch and poplar plants under O_3 stress, changed whole-plant carbon allocation is reflected in a reduced root/shoot ratio (R/S; Matyssek et al. 1992, 1993b; in young conifers: Schier et al. 1990; Mahoney et al. 1985); this change may also occur in mature trees growing in polluted air (McLaughlin et al. 1982). More generally, the biomass, which has accumulated over decades, may mask current and subtle changes in the carbon allocation of large forest trees. Nevertheless, it is likely that the carbon allocation of large trees responds to O_3 stress, as plants tend to maintain an equilibrium between their internal fluxes of carbon and nutrients ("set-point argument"; Mooney and Winner 1991). As the impact of ozone on leaves tends to limit the carbon flux in the plant, carbon allocation re-adjusts towards maintaining the foliage, and thus, the carbon-gaining capacity of the plant. By this, the set-point between carbon and nutrient flux is maintained. Such an acclimation to O_3 stress can limit the decline in biomass production, but lowers the assimilate supply to the non-green organs. It is not known, whether reduced, belowground carbon allocation may affect long-term root function, e.g., making the tree more susceptible to nutrient and water limitation (Pell and Dann 1991). In young birch plants experiencing low nutrient supply, O_3 stress prevented the occurrence of high R/S ratios of carbon allocation as established under clean air conditions (Maurer and Matyssek, unpubl.). In aspen, nutrient deficiency developed during the second year of continued O_3 exposure, despite a nonlimiting nutrient supply (Matyssek et al. 1993a).

Acclimation to ozone may also be reflected in changed leaf differentiation, which perhaps contributes to the delay in leaf loss. Birch leaves growing in ozonated air displayed reduced leaf area, widths, and number of cells, but increases in stomatal density, leaf mass per area, and inner leaf air space (Günthardt-Goerg et al. 1993). After growth completion, leaf photosynthesis and g_{H2O} did not differ from the control; stomatal narrowing apparently compensated for the raised stomatal density occurring under O_3 exposure conditions (Matyssek et al. 1991a). Needle differentiation of conifers resulted in reduced length when developing in ozonated air (McLaughlin et al. 1982; Schier et al. 1990).

Leaf photosynthesis declines if the intensity of O_3 stress begins to override the acclimative capacity. This was observed in the spruce trees of the Austrian experimental sites (see Sect. 4.3), although the O_3 levels required to induce the changes were higher than in ambient air. The decline pattern resembled that found in O_3-sensitive birch plants, when they could no longer maintain their leaves later in the season. The CO_2 assimilation rate decreased proportionally more than g_{H2O}, thereby lowering the water-use efficiency (Matyssek et al. 1991a).

Acclimative responses may resemble incipient O_3 injury. Altered leaf differentiation may also result from disturbed membrane formation and, thereby, enhanced cellular water loss during early stages of leaf growth under O_3 impact. Slight changes in the water potential or turgor yield threshold of cell walls may limit the extension growth of cells and tissues (Boyer 1985; Matyssek et al. 1991b). In fact, the differentiation of O_3-exposed birch leaves resembled development under "moderate" drought. In O_3-sensitive plants, such as birches and poplars, assimilate transport can be inhibited by cell collapse in the mesophyll and by phloem malfunction, which contributes to raised sugar concentrations in leaves and starch accumulation along the leaf veins (Matyssek et al. 1992; Günthardt-Goerg et al. 1993; Landolt, pers. comm.). This also limits the carbon flux to the root (cf Spence et al. 1990). Even the formation of latewood or of whole branches can be suppressed (Matyssek et al. 1992, 1993a).

Can knowledge gained from controlled experiments contribute to the assessment of risks which ozone poses for forest trees? Based on studies with sensitive plants, ozone can influence the resource acquisition and allocation, but its significance for forest trees is still uncertain. Only few attempts have been made for "scaling" branch or young-tree responses to ozone up to the level of forest trees, and no clear-cut trends have emerged; tree behaviour under multiple stress scenarios is also unclear (Matyssek et al. 1995b). Nevertheless, chronic changes in allocation, if they occur, are important as they determine, via the structure and function of root and crown, the competitiveness of the trees (Schulze et al. 1986; Matyssek and Schulze 1987; Tremmel and Bazzaz 1995). This mechanism may change the dominance of tree species in forest ecosystems in the long-term. However, this possibility is mainly of academic interest, as in most forests of central Europe species composition is determined primarily by forest practices. Relevance may apply, however, to the long-term requirements of sustaining managed mixed forests. Assertions that chronic O_3-induced shifts in carbon allocation weaken the mechanical stability of branches, stems and roots and, through this, contribute to wind and snow damage, must remain speculative for now.

4.5 Evidence of O_3 Injury on a Regional Scale?

4.5.1 Methods of Estimating Forest Condition

Throughout Europe, annual surveys of forest condition describe the appearance of tree crowns within a systematic sampling framework. The methods used in the inventories have an important bearing on the potential to identify O_3 injury. Visible symptoms of foliar injury by ozone include stippling, flecking, bleaching, and bifacial necrosis as well as chlorosis and premature senescence (Skelly et al. 1987). In all cases, there is a possibility that the symptoms are confused with those caused by other pathogenic agents, although there is a tendency for some O_3 symptoms to occur first on the upper surface of needles and leaves which can be a useful aid in diagnosis. The inventories of crown condition are based on visual assessment of crown transparency and discoloration, made from the ground using binoculars. Therefore, judgement is based on what is visible from underneath, thereby missing any axial discoloration. The latter is assessed as a single value for the whole tree, and is not recorded at proportions of less than 10% of the foliage.

4.5.2 Spatial Associations Between the Occurrence of Ozone and Crown Conditions

In Switzerland and Austria, the most important species involved in the annual inventories are *Picea abies, Abies alba, Pinus sylvestris* and *Fagus sylvatica.* With each of these, the amount of leaf or needle loss and the extent of discolored foliage is estimated. In 1993, only 7.5% of the trees in Austria were classified as "damaged" (suggesting a loss of >25% of the crown foliage; Kristöfel and Neumann 1993); in Switzerland 15.4% of trees were considered damaged. Experimental work with these species has not revealed specific symptoms of O_3 injury caused by the O_3 concentrations experienced today (see Sect. 4.3). Consequently, an alternative approach must be taken for the assessment of O_3 impact, based on the evaluation of correlations between the distribution of ozone and the distribution of forest damage.

A number of studies have attempted to relate the occurrence of crown transparency to the occurrence of raised levels of ozone. For example, Ammer et al. (1988) associated peaks in transparency of trees at Wank in the Bavarian Alps with supposed peaks in O_3 concentration. They suggested that the peaks in spruce crown transparency between 1000 and 1200 m a.s.l. and between 1400 and 1600 m a.s.l. coincided with higher O_3 concentrations associated with inversion layers, although the study was not supported by on-site measurements of O_3 concentration. However, another study (Paffrath and Peters 1988) suggested that a peak in O_3 concentration did indeed occur at such altitudes. These studies differ from other investigations of the vertical distribution of O_3 concentrations in the Alps, in which Mößmer et al. (1990) found that the vertical distributions were variable from one year to the next. In all these studies it is unclear whether weighted data were used when looking at the altitudinal distributions of ozone. Further doubts arise over the relationship between crown transparency and O_3 concentration when the result of experimental fumigations under forest conditions are examined (see Sect. 4.3).

Studies of the vertical distribution of defoliation have been carried out in Austria (Neumann 1990) and Switzerland (Innes et al. 1994). In Austria, there was no clear trend of increasing crown transparency with increasing altitude, particularly once the higher proportions of older trees at higher altitudes had been accounted for. In Switzerland, although there was some suggestion of a trend with for example *Picae abies*, the scatter was so great that a simple relationship was impossible to identify. Subsequent multivariate analyses (Webster et al. 1996) have revealed that while there is no indication of a relationship between crown transparency and altitude for *Abies alba* and *Fagus sylvatica*, such a relationship does exist for *Picea abies*, but only with those growing on skeletal soils. However, the latter relationship appears to be controlled more by the susceptibility of the soils to drought than by the distribution of ozone, providing a link to the findings described in Section 4.3.

The problems of linking crown condition to vertical O_3 distributions are illustrated by data from the *Höhenprofil Zillertal* (Puxbaum et al. 1991; Kaiser 1994). These show the altitudinal dependence of hourly O_3 concentrations at some sites. For example, at 600 m a.s.l., a marked diurnal trend is apparent. This trend steadily becomes less clear with increasing altitude until virtually no such trend is apparent at 1950 m a.s.l. This pattern is repeated elsewhere in the Alps, although the nature of changes at intermediate altitude is strongly dependent on the dynamics of the mixing layer (e.g., Paffrath et al. 1986). However, the changes in O_3 exposure patterns of the altitude gradient in the Ziller Valley were not accompanied by corresponding changes in crown transparency. There has been a suggestion that the differential exposure to ozone with altitude might be reflected in the antioxidative potential of the trees; differences in the needle contents of thiols and ascorbic acid with altitude have been documented repeatedly at similar altitudes along several altitudinal gradients in Austria (e.g., Bermadinger-Stabentheiner 1984; Grill et al. 1988). However, it is not clear whether such differences are controlled by exposure to ozone or by other environmental factors.

In other Austrian research areas (e.g., Böhmerwald/Schöneben, Northern Limestone Alps/Tyrol), forest damage and crown transparency are more pronounced than in the Central Alps (Ziller Valley) and are accompanied by needle yellowing. Since both areas are exposed to rather high O_3 concentrations, damage patterns may be significantly influenced by ozone (Grill et al. 1994). However, at Schöneben, needle yellowing is caused mainly by Mg deficiency, and disappears after fertilization (Jandl and Katzensteiner 1992). Similarly, in plots of the Northern Limestone Alps, strong nutrient deficiencies, root rot, and skeletal soils have been assumed to have a greater impact on trees than O_3 exposure. The degree of damage seemed to be based on local site conditions rather than on O_3 exposure regimes, the latter not being expected to vary dramatically between plots of the same altitude and exposition (Krehan 1991).

A regional statistical analysis of the Austrian survey data revealed that bedrock, soil type, and soil water relations, as well as altitude, age, and other stand characteristics had a significant influence on the degree of crown transparency (Neumann 1990), mirroring the Swiss survey analyses (Webster et al. 1996). In contrast, the levels of ozone, as estimated with passive samplers (Werner 1988) in more than 100 different survey plots in the northeastern part of the country (Leitner 1991), did not correspond to published maps of crown transparency (Kristöfel and Neumann 1993). Special phytopathological investigations in plots with elevated O_3 levels (all sites situated above 1000 m a.s.l.) have not revealed poor crown condition. Symptoms could only be tentatively ascribed to O_3 impact in a few cases (Krehan 1992).

A similar assessment has not been undertaken in Switzerland, and the distribution of active O_3 monitoring stations is such that interpretation of the data with respect to particular forest stands is extremely difficult. Nevertheless, it has been shown recently that protective alpine forests in Switzerland, which appeared to be heavily damaged, recovered after fertilization (Flückiger and Braun 1995).

4.5.3 Associations of Temporal Trends in O_3 Concentration and Crown Condition

Assessments of crown condition have been made over a number of years in Austria and Switzerland (Kristöffel and Neumann 1993; Innes et al. 1994). The distribution of crown transparency reveals that the majority of trees (and year-to-year changes) are at the lower ("denser") end of the transparency scale and that there has been little change in the properties of trees with higher levels of crown transparency. Mortality of trees in Switzerland has remained more or less constant between 1985 and 1994, ranging from about 0.3 to 0.5% per annum. Mortality of trees in the Austrian inventory has been lower (Stemberger 1991), with a maximum of 0.12% occurring between 1992 and 1993 (Kristöfel and Neumann 1993). The different mortality rates probably reflect differences in the sampling strategy, with only dominant and codominant trees being selected in Austria and all social classes of trees being used in Switzerland (suppressed trees probably have a greater risk of mortality).

The trend analysis through time suggests that since 1989, when the current methodology was adopted, there has been a general reduction in crown transparency in Austria (Kristöfel and Neumann 1993). In Switzerland, there has been a more complex pattern, with an apparent increase in transparency since 1985. However, changes in methodology cannot be excluded as a possible cause of this trend.

Ground-level O_3 concentrations are known to have increased over the last 100 years in Europe (e.g., Anfossi et al. 1991; Low et al. 1991), and it is likely that in the long-term, a similar trend will emerge in the Alps. However, some sites have shown no trends over the last 10 years (e.g., Bucher 1994), or even a distinct decrease in maximum 0.5-h or annual mean O_3 concentrations (Austrian side of the Northern Limestone Alps, between 1988 and 1994; Amt der Tiroler Landesregierung 1985–1995). Thus, generalization is difficult, and given the large interannual fluctuations in both crown condition and O_3 concentrations, the time period for which data are available is much too short to draw any confident conclusions about temporal associations between O_3 concentrations and crown condition.

4.5.4 Is there any Unambiguous Evidence of O_3 Injury to Trees in Switzerland and Austria?

The O_3 levels in parts of Switzerland suggest that at least the more sensitive individuals of the more sensitive species should show signs of foliar injury. This appeared to be confirmed in 1993 when symptoms typical of O_3 injury (cf. Davis and Skelly 1992, Simini et al. 1992) were found in the southern part of the country on *Prunus serotina* (reddish stippling on the upper surface of leaves; Fig. 4.12, see Appendix. These observations were followed by a survey of the symptoms in 1994, which revealed that similar symptoms were present at 14 sites (from a total of 77 where *Prunus serotina* was found) in canton Ticino.

At the same time, seedlings grown from symptomatic and asymptomatic trees were exposed in continuously stirred tank reactors to varying levels of ozone (30, 60, 90, or 120 nl l^{-1}, 7 h day^{-1}, 5 days/week, during the period 22 July – 25 August, representing an exposure of 140 h for each O_3 concentration). Symptoms developed within 4 days at higher O_3 concentrations (90/120 nl l^{-1}) but took 2 weeks at the lower concentrations (Skelly in prep.). For comparison, in Lugano, the number of hours with O_3 levels exceeding 60 nl l^{-1} was 752 (!) in 1994, and the maximum 1-h mean was 170 nl O_3 l^{-1}. The level of symptom development in the chambers was in accordance with that seen on the mother trees in Switzerland. In addition, seedlings derived from the asymptomatic trees generally showed no symptoms, except towards the end of the fumigation period in the highest O_3 treatment. These findings demonstrate the importance of genetically based differences in the O_3 sensitivity of individuals within a species. The findings also suggest that with a reasonable degree of confidence the symptoms observed on *Prunus serotina* in southern Switzerland can be attributed to O_3 injury. This species is not native to Switzerland, having been introduced from North America. Similar symptoms have also been observed in other woody species, although these observations are as yet unsupported by chamber studies.

4.6 Conclusions

The findings presented for Austria and Switzerland show that tropospheric ozone presently does not represent a dominating stress on trees and forests. The percentage of trees classified as severely damaged (with needle loss >60 %) by the national inventories has been low, and the mortality rate is low and stable. Some areas are affected more than others but no clear association with O_3 exposure has been found. Nevertheless, elevated O_3 levels cannot be excluded as a risk to forest development. Ozone can temporarily reach levels that are known to adversely affect trees at sites in North America or in controlled chamber experiments.

However, findings from chamber experiments leave us with an inherent problem, namely the difficulty of simulating the environmental conditions of forest sites, where a number of factors interact, including competition between the trees. Often, edaphic and climatic factors (e.g., nutrient deficiency and drought) have been identified in the field as the cause of discoloration or needles loss, and these may predispose trees to biotic stress (fungi, insects). The given site conditions can strongly bias O_3 responses even in sensitive plants. Therefore, altered plant behaviour cannot be readily ascribed to ozone as long as the physiological mechanisms of O_3 impact in concert with those of other site factors are not thoroughly understood. Given this restriction, the potential limitations imposed by ozone appear to be of a similar magnitude to those exerted by the other environmental factors in the Swiss and Austrian forests. It is pointless to speculate whether tree production, which is currently increasing in Europe (Innes 1993; Kuusela 1994; Neumann and Schadauer 1995; Nicolussi et al. 1995), might be even greater, if preindustrial O_3 levels existed today. The difficulties in defining critical O_3 levels for tree production in Europe are still evident (UN/ECE 1994). Until now, ozone injury has only been demonstrated in the field and experimentally verified with sensitive genotypes of *Prunus serotina*, a species introduced to Europe.

The extent to which genetic variation may be responsible for the avoidance and tolerance of O_3 stress within species and for discrepancies between findings from (often clonal) experimental plants and field plots is unknown. Ecotypes native to high altitudes may be efficiently adapted to the elevated O_3 background levels. In sensitive rather than in tolerant plants, raised O_3 levels may influence – through acclimation or incipient injury – the resource acquisition and allocation in such a way as to change the competitiveness of the plants. Some genotypes may be lost in the long-term. This deserves attention, as O_3 stress is not the only anthropogenic constraint acting on allocation. Increases in O_3 levels have been accompanied by a steady increase in atmospheric CO_2 concentration and enhanced nitrogen deposition in many parts of central Europe.

Tree behaviour is difficult to predict in situations where factors have contrasting effects on carbon allocation – as might be expected with both raised ozone and CO_2 levels. Elevated CO_2 may favor allocation of carbon to belowground organs, but it is speculative – given the uncertainties of CO_2 effects on trees – as to whether a high CO_2 supply compensates for root-limiting O_3 effects. High nitrogen deposition may direct carbon into foliage growth, perhaps supporting acclimation to ozone, but the low assimilate transport to the root relative to the aboveground carbon flux may be exacerbated by the O_3 stress. Similar "conflicts" in allocation may occur between limitation of root growth by ozone and stimulation by water or nutrient shortage or by the carbon demand of mycorrhizae. There is a lack of information, especially on forest trees, concerning the interactions between ozone and other abiotic and biotic factors (Matyssek et al. 1995b). Understanding the control of allocation is important not only in relation to competition in natural ecosystems, but also with respect to the long-term stability and management of forests.

Ozone is not causing a significant decline of Swiss and Austrian forests at the moment. However, uncertainty remains about the extent to which tree metabolism is influenced by present O_3 scenarios and, along with other types of stress, about their long-term consequences for tree and forest development. Linked to controlled chamber experiments, case studies in the field, which must include the experimental manipulation of site conditions, need to be intensified in order to demonstrate and confirm the action of ozone on the ecophysiology of forest trees. Further research is necessary to determine the risks that ozone presents to trees in a changing environment.

Acknowledgements. The critical comments by Drs. J.B. Bucher, W. Landolt, and M. Neumann on the manuscript are highly appreciated.

References

Alscher RG, Amundson RG, Cumming JR, Fellows S, Fincher J, Rubin G, van Leuken P, Weinstein LH (1989) Seasonal changes in pigments, carbohydrates and growth of red spruce as affected by ozone. New Phytol 113:211–223

Ammer U, Burgis M, Koch B, Martin K (1988) Untersuchungen über den Zusammenhang zwischen Schädigungsgrad und Meereshöhe im Rahmen des Schwerpunktprogramms zur Erforschung der Wechselwirkungen von Klima und Waldschäden. Forstwiss Centralbl (Hamb) 107:145–151

Amt der Tiroler Landesregierung (1985–1995) Untersuchungen über den Waldzustand und die Immissionsbelastung. Berichte an den Tiroler Landtag 1985–1995, Innsbruck

Anfossi D, Sandroni S, Viarengo S (1991) Tropospheric ozone in the nineteenth century: the Montcalieri series. J Geophys Res 96D:17349–17352

Arndt U, Seufert G (1990) Introduction to the Hohenheim long-term experiment. Environ Pollut 68:195–204

Arndt U, Seufert G, Nobel W (1982) Die Beteiligung von Ozon an der Komplexkrankheit der Tanne (*Abies alba* Mill.) – eine prüfenswerte Hypothese. Staub - Reinhalt Luft 42:243–247

Becker K, Saurer M, Egger A, Fuhrer J (1989) Sensitivity of white clover to ambient ozone in Switzerland. New Phytol 112:235–243

Benecke U, Schulze ED, Matyssek R, Havranek WM (1981) Environmental control of CO_2-uptake and leaf conductance in *Larix decidua* Mill. I. A comparison of contrasting natural environments. Oecologia 50:54–61

Bermadinger-Stabentheiner E (1994) Untersuchungen zum antioxidativen Schutzsystem von Fichten an den "Höhenprofilen Achenkirch". In: Herman F, Smidt S (eds) Ökosystemare Studien im Kalkalpin. Höhenprofile Achenkirch. Ergebnisse aus dem Bereich Phyllosphäre. FBVA Ber 78:73–82

Bermadinger-Stabentheiner E, Grill D (1992) Pflanzenphysiologische Untersuchungen am Höhenprofil Zillertal. Forstliche Bundesversuchsanstalt Wien. FBVA-Ber 67:87–93

Blanchard RO, Baas J, van Cotter H (1979) Oxidant damage to eastern white pine in New Hampshire. Plant Dis Rep 63:177–182

Bolsinger M, Flückiger W (1989) Ambient air pollution induced changes in amino acid pattern of phloem sap in host plants – relevance to aphid infestation. Environ Pollut 56:209–216

Boyer JS (1985) Water transport. Annu Rev Plant Physiol 36:473–516

Bucher JB (1982) Physiologische Veränderungen und ökotoxikologische Wirkmechanismen. Probleme der Differentialdiagnose. In: Waldschäden durch Immissionen? Vorträge der Informationstagung. Gottlieb Duttweiler-Institut (GDI) Rüschlikon/Zürich, pp 91–109

Bucher JB (1994) Die Ozonbelastung als Risikofaktor für den Wald. In: Innes JL, Böhm JP, Bucher JB, Dobbertin M, Jansen E, Kull P, Rigling A, Walthert L, Zimmermann S (eds) Sanasilva-Bericht 1993. Der Zustand des Schweizer Waldes. Berichte 334 der Eidgenössischen Forschungsanstalt für Wald, Schnee und Landschaft, Birmensdorf, pp 48–49

Bucher JB, Landolt W, Bleuler P (1986) Ozonmessungen auf dem Rötliboden ob Göschenen UR. Schweiz Z Forstwes 137:607–621

BUWAL/NABEL (1995) Luftbelastung 1994. Schriftenreihe Umwelt Nr 24, Bundesamt für Umwelt, Wald und Landschaft, Bern

Chappelka AH, Chevone BI (1992) Tree response to ozone. In: Lefohn AS (ed) Surface level ozone exposures and their effects on vegetation. Lewis, Chelsea, pp 271–324

Darrall NM (1989) The effect of air pollutants on physiological processes in plants. Plant Cell Environ 12:1–30

Davis DD, Skelly JM (1992) Foliar sensitivity of eight eastern hardwood tree seedlings exposed to ozone, acidic precipitation, and sulfur dioxide. J Air Waste Manage Assoc 42:309–311

Davis DD, Wilhour R (1976) Susceptibility of woody plants to sulphur dioxide and photochemical oxidants. A literature review. EPA 600/3-76-102, Environ Res Lab Corvallis

Dobson MC, Taylor G, Freer-Smith PH (1990) The control of ozone uptake by *Picea abies* (L.) Karst and *P. sitchensis* (Bong.) Carr. during drought and interacting effects on shoot water relations. New Phytol 116:465–474

Edwards PJ, Wood F, Kochenderfner JN (1991) Characterisation of ozone during consecutive drought and wet years at a rural West Virginia site. J Air Waste Manage Assoc 41:1450–1453

EPA (US Environmental Protection Agency) (1978) Diagnosing vegetation injury caused by air pollution. Applied Science Assoc, Corvallis, Oregon

Esterbauer H, Grill D, Welt R (1980) The annual rhythm of the ascorbic acid system in needles of *Picea abies*. Z Pflanzenphysiol 98:393–402

Farquhar GD, Sharkey TD (1982) Stomatal conductance and photosynthesis. Annu Rev Plant Physiol 33:317–345

Farquhar GD, Ehleringer JR, Hubick KT (1989) Carbon isotope discrimination and photosynthesis. Annu Rev Plant Physiol Plant Mol Biol 40:503–537

FBW (Forschungsbeirat Waldschäden/Luftverunreinigungen) (1989) Dritter Bericht, Kernforschungszentrum Karlsruhe, Karlsruhe, ISSN 0931-7905

Fincher J, Alscher RG (1992) The effect of long-term ozone exposure on injury of seedlings of red spruce (*Picea rubens* Sarg.) New Phytol 120:49–59

Flückiger W, Braun S (1990) Waldschadensbericht, Untersuchungen in Buchenbeobachtungsflächen 1984–1988. Institut für Angewandte Pflanzenbiologie, Schönenbuch, 57 pp

Flückiger W, Braun S (1994) Waldschaden-Bericht, Untersuchungen in Buchenbeobachtungsflächen 1984–1993. Institut für Angewandte Pflanzenbiologie, Schönenbuch, 54 pp

Flückiger W, Braun S (1995) Revitalization of an alpine protective forest by fertilization. Plant Soil 168–169:481–488

Freer-Smith PH, Dobson M, Taylor G (1989) Factors controlling the rates of O_3 uptake by spruce and beech. In: Bucher JB, Bucher-Wallin I (eds) Air pollution and forest decline. Proc 14th Int Meet for specialists in air pollution effects on forest ecosystems, IUFRO P2.05, Interlaken, 2–8 October 1988. EAFV Birmensdorf, pp 407–409

Fuchs M, Schulze E-D, Fuchs MI (1977) Spatial distribution of photosynthetic capacity and performance in a mountain spruce forest of northern Germany. II. Climatic control of carbon dioxide uptake. Oecologia 29:329–340

Gehrig R, Jeannet P, Wunderli S (1991) Höhenabhängigkeit von Schadstoffimmissionen. In: Schüpbach E (ed) Meteorologic und Luftchemie in Waldbeständen - Lufthaushalt, Luftrerschmutzung und Waldschäden in der Schweiz, Vol 4. Verlag der Fachvereine Zürich/Switzerland, pp 71–101

Greitner CS, Winner WE (1988) Increases in $\delta^{13}C$ values of radish and soybean plants caused by ozone. New Phytol 108:489–494

Grill D, Pfeifhofer H, Esterbauer H (1987) Further investigations on the thiol content of Norway spruce needles. Phyton 27:311–317

Grill D, Pfeifhofer H, Tschulik A, Hellig K, Holzer K (1988) Thiol content of spruce needles at forest limits. Oecologia 76:294–297

Grill D, Tausz M, Bermadinger-Stabentheiner E, Edl M, Gailhofer M, Halbwachs G, Havranek WM, Kromp-Kolb H, Müller M, Nemetz C, Puchinger L, Ruppert W, Scardelli U, Strohl A, Wagner K, Wieser G, Wimmer R, Zellnig G (1994) Physiological and biochemical bioindication and its application to the FIW II Project 'Schöneben'. Forstl Schriftenreihe Univ Bodenkultur Wien 7:123–145

Grünhage L, Jäger HL (1994) Influence of the atmospheric conductivity on the ozone exposure of plants under ambient conditions: considerations for establishing ozone standards to protect vegetation. Environ Pollut 85:125–129

Günthardt-Goerg MS (1991) Die Einwirkung von Luftschadstoffen und Klimafaktoren auf die Wachsschicht von Fichtennadeln. In: Stark (ed) Luftschadstoffe und Wald-Lufthaushalt, Luftverschmutzung und Waldschäden in der Schweiz, vol 5. Verlag der Fachvereine Zürich pp 107–118

Günthardt-Goerg MS, Matyssek R, Keller T (1993) Birke (*Betula pendula*): Mögliche Akklimatisierung an Ozon-Streß. Forstwiss Centralbl 112:46–49

Günthardt-Goerg MS, Keller T, Matyssek R, Scheidegger C (1994) Structural and chemical variation in the wax of Norway spruce needles, from mature trees at two forest sites and from young plants experimentally exposed to O_3 or SO_2. Eur J Plant Pathol 24:92–111

Häsler R (1991a) Vergleich der Gaswechselmessungen der drei Jahre (Juli 1986 - Juni 1989). In: Stark (ed) Luftschadstoffe und Wald-Lufthaushalt, Luftverschmutzung und Waldschäden in der Schweiz, vol 5. Verlag der Fachvereine Zürich, pp 177–184

Häsler R (1991b) Einfluß der Gaswechselmeßkammer auf Luftschadstoffe am Beispiel von Ozon. In: Stark M (ed) Luftschadstoffe und Wald-Lufthaushalt, Luftverschmutzung und Waldschäden in der Schweiz, vol 5. Verlag der Fachvereine Zürich, pp 261–265

Häsler R, Herzog K (1991) Wasserhaushalt der Fichte. In: Stark (ed) Luftschadstoffe und Wald-Lufthaushalt, Luftverschmutzung und Waldschäden in der Schweiz, vol 5. Verlag der Fachvereine Zürich, pp 97–103

Häsler R Ruckstuhl A (1991) Modellierung der Photosynthese der Fichte unter dem Einfluß natürlicher Parameter. In: Stark (ed) Luftschadstoffe und Wald-Lufthaushalt, Luftverschmutzung und Waldschäden in der Schweiz, vol 5. Verlag der Fachvereine Zürich, pp 169–175

Häsler R, Savi C, Herzog K (1991) Photosynthese und stomatäre Leitfähigkeit der Fichte unter dem Einfluß von Witterung und Luftschadstoffen. In: Stark (ed) Luftschadstoffe und Wald-Lufthaushalt, Luftverschmutzung und Waldschäden in der Schweiz, vol 5. Verlag der Fachvereine Zürich, pp 143–168

Havranek WM, Wieser G (1990) Research design to measure ozone uptake and its effects on gas-exchange of spruce in the field. In: Payer HD, Pfirrmann T, Mathy P (eds) Environmental research with plants in closed chambers. Air pollution research report 26. Commission of the European Communities, Brussels, pp 148–152

Havranek WM, Wieser G (1993) Zur Ozontoleranz der Lärche (*Larix decidua* Mill.). Forstwiss Centralbl 112:56–64

Havranek WM, Wieser G (1994) Design and testing of twig chambers for ozone fumigation and gas exchange measurements in mature trees. Proc R Soc Edinb Sect B 102:541–546

Havranek WM, Wieser G, Bodner M (1989) Ozone fumigation of Norway spruce at timberline. Ann Sci For 46 Suppl: 581s–585s

Havranek WM, Pfeifhofer H, Grill D (1990) Pigmentgehalte und Gaswechsel von Tief- und Hochlagenfichten nach chronischer Ozonbelastung. Forstwiss Centralbl 109:200–209

Hoigné J (1988) Bildung und Chemische Bedeutung von Photooxidantien in der wässerigen Phase. In: Schulte-Hostede S (ed) Proc Int Symp Garmisch-Partenkirchen 1988. GSF-Ber München 17:166–175

Innes JL (1993) Air pollution and forests - an overview. In Schlaepfer R (ed) Long-term implications of climate change and air pollution on forest ecosystems. Progress report of the IUFRO task force 'Forest, Climate Change and Air Pollution'. IUFRO World Series, vol 4, Vienna, pp 77–100

Innes JL, Böhm JP, Bucher JB, Dobbertin M, Jansen E, Kull P, Rigling A, Walthert L, Zimmermann S (1994) SanasilvaBericht 1993. Der Zustand des Schweizer Waldes. Berichte 334 der Eidgenössischen Forschungsanstalt für Wald. Schnee und Landschaft, Birmensdorf

Jandl R, Katzensteiner K (1992) Düngungsversuche mit Magnesitdüngern zu Fichte im Magnesiummangelgebiet Schöneben. In: Glatzel G, Jandl R, Sieghardt M, Hager H (eds) Magnesiummangel in mitteleuropäischen Waldökosystemen, vol 5. Forstl Schriftenreihe Univ Bodenkultur Wien pp 152–170

Joos K (1991) Jahrringanalysen auf den Beobachtungsflächen Davos, Alptal und Lägeren. In: Stark (ed) Luftschadstoffe und Wald-Lufthaushalt, Luftverschmutzung und Waldschäden in der Schweiz, vol 5. Verlag der Fachvereine Zürich, pp 45–52

Kaiser A (1994) Projekt "Höhenprofil Zillertal". Analyse der vertikalen Temperatur- und Windstruktur und ihr Einfluß auf die Immissionskonzentrationen. FBVA Ber 77, Forstliche Bundesversuchsanstalt, Wien

Kandler O (1985) Immissions- versus Epidemie-Hypothesen. In: von Kortzfleisch G (ed) Waldschäden, Oldenbourg München, pp 19–59

Keller T, Häsler R (1984) The influence of a fall fumigation with ozone on the stomatal behaviour of spruce and fir. Oecologia 64:284–286

Krause GHM, Prinz B (1985) Zur Wirkung von Ozon und saurem Nebel (einzeln und in Kombination) auf phänomenologische und physiologische Parameter an Nadel- und Laubgehölzen im kombinierten Begasungsexperiment. Spez Ber der Kernforschungsanlage Jülich 369:208–221

Krause GHM, Prinz B, Jung KD (1983) Forest effects in West Germany. In: Davis DD, Miller AA, Dochinger L (eds) Air pollution and the productivity of the forest. Izak Walton League of America, Washington, DC, pp 297–332

Krause GHM, Arndt U, Brandt CJ, Bucher J, Kenk G, Matzner E (1986) Forest decline in Europe: development and possible causes. Water Air Soil Pollut 31:647–668

Krehan H (1991) Forstpathologische Sondererhebungen im Rahmen der österreichischen Waldzustandsinventur 1984-1988. Mitt Forstl Bundesversuchsanst Wien 166:133–161

Krehan H (1992) Ergebnisse forstpathologischer Untersuchungen. Forstliche Bundesversuchsanstalt, FBVA-Berichte 71:85–92

Kristöfel F, Neumann M (1993) Ergebnisse der terrestrischen Kronenerhebungen 1993. Österr Forstz 103 (12):10–12

Kristöfel F, Neumann M, Pollanschütz J (1989) Ergebnisse der Kronenzustandserhebungen 1989 - Es war schon schlechter. Österr Forstz 11/1988:51–53

Kronfuß G (1996) Einfluß von Ozon und mildem Trockenstreß auf Gaswechsel, Antioxidantien und Schutzenzyme bei Fichte (*Picea abies* L. Karst.). Dipl Thesis, Universität Innsbruck

Kronfuß G, Wieser G, Havranek WM, Polle A (1996) Ozone and drought effects on quaiacol peroxidase and lipid peroxidation. J Plant Physiol 148:203–206

Krupa SV, Manning WJ (1988) Atmospheric ozone: formation and effects on vegetation. Environ Pollut 50:101–137

Kuusela K (1994) Forest resources in Europe. Cambridge University Press, Cambridge, 168 pp

Landolt W (1992) Methoden und Probleme der experimentellen Ursachenforschung. In: Schlaepfer R (ed) Forum für Wissen 1992, Waldschadenforschung in der Schweiz:

Stand der Kenntnisse. Eidg Forschungsanstalt für Wald. Schnee u Landschaft (WSL), Birmensdorf, pp 73–79

Landolt W, Lüthy-Krause B (1991) Wirkungen umweltrelevanter Ozon-Konzentrationen auf verschiedene Pflanzen. In: Stark (ed) Luftschadstoffe und Wald - Lufthaushalt, Luftverschmutzung und Waldschäden in der Schweiz, vol 5. Verlag der Fachvereine Zürich, pp 127–134

Landolt W, Pfenninger I, Lüthy-Krause B (1989) The effect of ozone and season on the pool sizes of cyclitols in Scots pine (*Pinus sylvestris*) Trees 3:85–88

Landolt W, Günthardt-Goerg MS, Pfenninger I, Scheidegger C (1994) Ozone-induced microscopical changes and quantitative carbohydrate contents of hybrid poplar (*Populus x euramericana*). Trees 8:183–190

Lang KJ, Holdenrieder O (1985) Nekrotische Flecken an Nadeln von *Picea abies* - Ein Symptom des Fichtensterbens? Eur J For Pathol 15:52–58

Lange OL, Heber U, Schulze E-D, Ziegler H (1989a) Atmospheric pollutants and plant metabolism. In: Schulze ED, Lange OL, Oren S (eds) Forest decline and air pollution. A study of spruce (*Picea abies*) on acidic soils. Ecological studies, vol 77. Springer, Berlin Heidelberg New York, pp 238–273

Lange OL, Weikert RM, Wedler M, Gebel J, Heber U (1989b) Photosynthese und Nährstoffversorgung von Fichten aus einem Waldschadensgebiet auf basenarmem Untergrund. Allg Forstztg 3:55–64

Leitner J (1991) Immissionskundliche Untersuchungen integrierende Luftschadstoffmessungen. In: Österreichisches Waldschaden-Beobachtungssystem. Ziele, Methoden und erste Ergebnisse. FBVA Berichte 49:77–87

Low PS, Davies TD, Kelly PM, Reiter R (1991) Uncertainties in the surface ozone trend at Hohenpeißenberg. Atmos Environ 25A:511–515

Luethy-Krause B, Pfenninger I, Landolt W (1990) Effects of ozone on organic acids in needles of Norway spruce and Scots pine. Trees 4:198–204

Mahoney MJ, Chevone BI, Skelly JM, Moore LD (1985) Influence of mycorrhizae on the growth of loblolly pine *Pinus taeda* seedlings exposed to ozone and sulphur dioxide. Phytopathology 75:679–682

Maier-Maercker U (1989) Delignification of subsidiary and guard cell walls of *Picea abies* (L.) Karst. by fumigation with ozone. Trees 3:57–64

Maier-Maercker U, Koch W (1991) Experiments on the control capacity of stomata of *Picea abies* (L.) Karst. after fumigation with ozone and in environmentally damaged material. Plant Cell Environ 14:175–184

Manning WJ, Krupa SV (1992) Experimental methodology for studying the effects of ozone on crops and trees. In: Lefohn AS (ed) Surface level ozone exposures and their effects on vegetation. Lewis, Chelsea, pp 93–156

Matyssek R (1985) Der Kohlenstoff-, Wasser-, und Nährstoffhaushalt der wechselgrünen und immergrünen Koniferen Lärche, Fichte, Kiefer. PhD Thesis, University Bayreuth, 224 pp

Matyssek R (1986) Carbon, water and nitrogen relations in evergreen and deciduous conifers. Tree Physiol 2:177–187

Matyssek R, Schulze E-D (1987) Heterosis in hybrid larch (*Larix decidua x leptolepis*) II. Growth characteristics. Trees 1:225–231

Matyssek R, Günthardt-Goerg MS, Keller T, Scheidegger C (1991a) Impairment of the gas exchange and structure in birch leaves (*Betula pendula*) caused by low ozone concentrations. Trees 5:5–13

Matyssek R, Maruyama S, Boyer JS (1991b) Growth-induced water potentials may mobilize internal water for growth. Plant Cell Environ 14:917–923

Matyssek R, Günthardt-Goerg MS, Saurer M, Keller T (1992) Seasons growth, $\delta^{13}C$ in leaves and stem, and phloem structure of birch (*Betula pendula*) under low ozone concentrations. Trees 6:69–76

Matyssek R, Keller T, Koike T (1993a) Branch growth and leaf gas exchange of *Populus tremula* exposed to low ozone concentrations throughout two growing seasons. Environ Pollut 79:1–7

Matyssek R, Günthardt-Goerg MS, Landolt W, Keller T (1993b) Whole-plant growth and leaf formation in ozonated hybrid poplar (*Populus x euramericana*). Environ Pollut 81:207–212

Matyssek R, Günthardt-Goerg MS, Maurer S, Keller T (1995a) Night time exposure to ozone reduces whole-plant production in *Betula pendula*. Tree Physiol 15:159–165

Matyssek R, Reich PB, Oren R, Winner WE (1995b) Response mechanisms of conifers to air pollutants (review article). In: Smith WK, Hinckley TM (eds) Physiological ecology of coniferous forests. Physiological Ecology Series, Academic Press, New York, pp 255–308

McLaughlin SB, McConathy RK, Duvick D, Mann LK (1982) Effects of chronic air pollution stress on photosynthesis, carbon allocation and growth of white pine trees. For Sci 28:60–70

Miller PR (1973) Oxidant-induced community change in a mixed conifer forest. In: Naegele JA (ed) Air pollution damage to vegetation. Advances in chemistry series No 122. American Chemical Society, Washington, DC, pp 101–117

Miller PR (1984) Ozone effects in the San Bernardino National Forest. In: Davis DD, Miller AA, Dochinges L (eds) Symp on air pollution and the productivity of the forest. Izaak Walton League of America, Arlington, pp 161–197

Mooi J (1980) Influence of ozone on growth of two poplar cultivars. Plant Dis 64:772–773

Mooi J (1981) Influence of ozone and sulphur dioxide on defoliation and growth of poplars. Mitt Forstl Bundes-Versuchsanst Wien 137:47–51

Mooney HA, Winner WE (1991) Partitioning response of plants to stress. In: Mooney HA, Winner WE, Pell EJ (eds) Response of plants to multiple stresses. Academic Press, San Diego, pp 129–141

Mößmer R, Tränkner H, Maier G, Burgis M, Grad M, Marin K (1990) Schwerpunktprogramm Erforschung der Wechselwirkungen von Klima- und Waldschäden am Schwerpunktstandort Kalkalpen. In: Waldschadenserhebung 1989 – Luftbildauswertung – terrestrische Schadansprache. Bayerische Forstliche Versuchs- und Forschungsanstalt, München

Neumann M (1990) Zu Fragen des Einflusses von Standorts- und Bestandesfaktoren auf den Kronenzustand. In: Neumann M, Schadauer K (eds) Waldzustandsinventur. Methodische Überlegungen und Detailauswertungen. FBVA Berichte 45:53–83

Neumann M, Schadauer K (1995) Die Entwicklung des Zuwachses in Österreich an Hand von Bohrkernanalysen. Allg Forst Jagdz 166:230–235

Nicolussi K, Bortenschlager S, Körner C (1995) Increase in tree-ring width in subalpine *Pinus cembra* from the central Alps that may be CO_2-related. Trees 9:181–189

Paffrath D, Peters W (1988) Betrachtung der Ozon-Vertikalverteilung im Zusammenhang mit den neuartigen Waldschäden. Forstwiss Centralbl 107:151–159

Paffrath D, Peters W, Stangl H (1986) Airborne measurements of ozone and other pollutants north and south of the Alps. Proc 7th Clean Air Congr, Sydney, vol 3, pp 194–203

Pearson M, Mansfield TA (1993) Interacting effects of ozone and water stress on the stomatal resistance of beech (*Fagus sylvatica* L.) New Phytol 123:351–358

Pell EJ, Dann MS (1991) Multiple stress-induced foliar senescence and implications for whole-plant longevity. In: Mooney HA, Winner WE, Pell EJ (eds) Response of plants to multiple stresses. Academic Press, San Diego, pp 189–204

Pell EJ, Eckhardt N, Enyedi AJ (1992) Timing of ozone stress and resulting status of ribulose bisphosphate carboxylase/oxigenase and associated net photosynthesis. New Phytol 120:397–405

Pisek A, Tranquillini W (1954) Assimilation und Kohlenstoffhaushalt in der Krone von Fichten- (*Picea excelsa* Link.) und Rotbuchenbäumen (*Fagus silvatica* L.). Flora 141:237–270

Polle A (1995) Protection from oxidative stress in trees as affected by elevated CO_2 and environmental stress. In: Mooney HA, Koch GW (eds) Terrestrial ecosystem response to elevated CO_2. Physiological Ecology Series, Academic Press, New York, pp 299–315

Polle A, Rennenberg H (1992) Field studies on Norway spruce trees at high altitudes: II. Defense systems against oxidative stress in needles. New Phytol 121:635–642

Polle A, Wieser G, Havranek WM (1995) Quantification of ozone influx and apoplastic ascorbate content in needles of Norway spruce trees (*Picea abies* L., Karst.) at high altitude. Plant Cell Environ 18:681–688

Prinz B (1984) Woran sterben unsere Wälder? Umschau 18:544–549

Prinz B, Krause GHM, Stratmann H (1982) Waldschäden in der Bundesrepublik Deutschland. LIS-Bericht 28, Essen, 154 pp

Puxbaum H, Gabler K, Smidt S, Glattes F (1991) A one-year record of ozone profiles in an alpine valley (Zillertal, Tyrol, Austria, 600-2000 m a.s.l.). Atmos Environ 25A:1759–1765

Pye JM (1988) Impact of ozone on the growth and yield of trees: a review. J Environ Qual 17:347–360

Reich PB (1983) Effects of low concentrations of O_3 on net photosynthesis, dark respiration, and chlorophyll contents in aging hybrid poplar leaves. Plant Physiol 73:291–296

Reich PB (1987) Quantifying plant response to ozone: a unifying theory. Tree Physiol 3:63–91

Reich PB, Amundson RG (1987) Ambient levels of ozone reduce net photosynthesis in tree and crop species. Science 230:566–570

Rennenberg H (1988) Wirkung von Photooxidantien auf Pflanzen. In: Schulte-Hostede S, Kirchner M, Reuter M (eds) Int Symp Verteilung und Wirkung von Photooxidantien im Alpenraum. GSF Bericht 17/88, pp 360–370

Roberts BR, Cannon WN Jr (1992) Growth and water relationship of red spruce seedlings exposed to atmospheric deposition and drought. Can J For Res 22:193–197

Saurer M, Fuhrer J, Siegenthaler U (1991) Influence of ozone on the stable carbon isotope composition, $\delta^{13}C$ of leaves and grain of spring wheat (*Triticum aestivum* L.). Plant Physiol 97:313–316

Saurer M, Siegenthaler U, Schweingruber F (1995a) The climate–carbon isotope relationship in tree rings and the significance of site conditions. Tellus 47B:320–330

Saurer M, Maurer S, Matyssek R, Landolt W, Günthardt-Goerg MS, Siegenthaler U (1995b) The influence of ozone and nutrition on $\delta^{13}C$ in *Betula pendula*. Oecologia 103:397–406

Schier GA, McQuattie CJ, Jensen KF (1990) Effects of ozone and aluminium on pitch pine (*Pinus rigida*) seedlings growth and nutrient relations. Can J For Res 20:1714–1719

Schneider J, Loibl W, Spangl W (1996) Kumulative Ozonbelastung der Vegetation in Österreich. Berechnung und Darstellung nach dem Konzept der kritischen Belastungsgrenzen ('Critical Levels'). Reports UBA-96-127, Bundesministerium für Umwelt, Wien, 74 pp

Schulze E-D Küppers M, Matyssek R (1986) The notes of carbon balance and branching pattern in the growth of woody species. In: Girnish TJ (ed) On the economy of plant form and function. Cambridge University Press, Cambridge, pp 585–602

Schupp R, Rennenberg H (1988) Diurnal changes in the glutathione content of spruce needles (*Picea abies* L.). Plant Sci 57:113–117

Schwarzenbach FH (1991) Vorgeschichte und Planung des Nationalen Forschungsprogrammes NFP14+ 'Waldschäden und Luftverschmutzung in der Schweiz'. In: Stark M (ed) Luftschadstoffe und Wald. Verlag der Fachvereine Zürich, pp 1–11

Showman RE (1991) A comparison of ozone injury to vegetation during moist and drought years. J Air Waste Manage Assoc 41:63–64

Sieber TN (1991) Bedeutung von Pilzen in epigäischen Pflanzenteilen für die Fichtenschäden. In: Stark (ed) Luftschadstoffe und Wald-Lufthaushalt, Luftverschmutzung und Waldschäden in der Schweiz, vol 5. Verlag der Fachvereine Zürich, pp 119–126

Simini M, Skelly JM, Davis DD, Savage JE, Comrie AC (1992) Sensitivity of four hardwood species to ambient ozone in northcentral Pennsylvania. Can J For Res 22:1789–1799

Skelly JM, Davis DD, Merill W, Cameron EA, Brown HD, Drummond DB, Dochinger LS (1987) Diagnosing injury to eastern forest trees. US Dep Agric, For Serv, For Pest Manage and Pennsylvania State University

Smidt S (1993) Die Ozonsituation in alpinen Tälern Österreichs. Centralbl Gesamte Forstwes 110:205–220

Smidt S, Gabler K (1994) Development of SO_2, No_x, and ozone annual mean values in Austria. Centralbl gesamte Forstwes 111:183–176

Smith WH (1984) Auswirkungen von regionalen Luftschadstoffen auf die Wälder in den USA. Forstwiss Centralbl (Hamb) 103:48–61

Spangl W (1995) Ozone in Austria, Overview. In: Exceedances of critical loads and levels – background papers. Int worksh spatial and temporal assessment of air pollutant impact on ecosystems, Nov 22-24, 1995, Vienna

Spence DR, Rykiel EJ, Sharpe PJH Jr (1990) Ozone alters carbon allocation in loblolly pine assessment with carbon-11 labelling. Environ Pollut 64:93–106

Stemberger A (1991) Über Ausmaß und Verteilung der Mortalität: Gegenüberstellung von Ergebnissen der Waldzustandsinventur mit früheren Untersuchungen. In: Zusammenfassende Darstellung der Waldzustandsinventur. Mitt Forstl Bundesversuchsanst Wien 166:163–173

Tausz M, Grill D (1995) Physiological reactions of spruce trees to environmental stresses – field study results from various locations in Austria. Proc Int Workshop: Exceedances of critical loads and levels, spatial and temporal interpretation for elements in landscapes sensitive to atmospheric pollutants (Background papers) Nov 22–24, 1995, Vienna

Tausz M, Müller M, Bermadinger-Stabentheiner E, Grill D (1994) Stress-physiological investigations and chromosomal analysis on Norway spruce (*Picea abies* L., Karst.) – a field study. Phyton 34:291–308

Temple PJ, Riechers GH, Miller PR, Lennox RW (1992) Growth response of ponderosa pine to long-term exposure to ozone, wet and dry acidic deposition, and drought. Can J For Res 23:59–66

Tingey DT, Taylor GE Jr (1982) Variation in plant response to ozone: a conceptual model of physiological events. In: Unsworth HM, Ormond DD (eds) Effects of gaseous air pollution in agriculture and horticulture. Butterworth, London, pp 113–138

Tjoelker MG, Volin JC, Oleksyn J, Reich PB (1993) Light environment alters response to ozone stress in seedlings of *Acer saccharum* Marsh. and hybrid *Populus* L. I. In situ net CO_2 exchange and growth. New Phytol 124:627–636

Tremmel DC, Bazzaz FA (1995) Plant architecture and allocation in different neighborhoods – implications for competitive success. Ecology 76:262–271

UN/ECE-CEC (1992) Forest condition in Europe. Convention on long-range transboundary air pollution. International cooperative programme on assessment and monitoring of air pollution effects on forests. CEC-UN/ECE Brussels, 117 pp

UN/ECE (1994) Critical level for ozone. In: Fuhrer J, Achermann B (eds) UNECE Workshop Report, Bern 1993 Schriftenreihe FAC 16, 328 pp

VDI (1989) Maximale Immissions-Konzentrationen für Ozon. VDI Handbuch Reinhaltung der Luft. Beuth, Berlin, Blatt 6

Wallin G, Skärby L (1992) The influence of ozone on the stomatal and non-stomatal limitation of photosynthesis in Norway spruce (*Picea abies* (l.) Karst.) exposed to soil moisture deficits. Trees 6:128–136

Wallin G, Ottosson S, Sellden G (1992) Long-term exposure of Norway spruce, *Picea abies* (L.) Karst, to ozone in open top chambers. IV. Effects on the stomatal and non-stomatal limitation of photosynthesis and on carboxylation efficiency. New Phytol 121:395–401

Webster R, Rigling A, Walthert L (1996) An analysis of tree crown conditions in relation to enviornment in Switzerland. Forestry (in press)

Weikert RM, Wedler M, Lippert M, Schramel P, Lange OL (1989) Photosynthetic performance, chloroplast pigments, and mineral content of various needle age classes of spruce (*Picea abies*) with and without the new flush: an experimental approach for analysing forest decline phenomena. Trees 3:161–172

Werner H (1988) Erfahrungen mit passiven Ozonintegratoren im Freiland- und im Kammertest. Symp Verteilung und Wirkung von Photooxidantien im Alpenraum April 11–15, 1988 in Garmisch Partenkirchen. GSF Bericht 17/88, pp 346–359

Wieser G, Havranek WM (1993a) Ozone uptake in the sun and shade crown of spruce: quantifying the physiological effects of ozone exposure. Trees 7:227–232

Wieser G, Havranek WM (1993b) Einfluß der Nährstoffversorgung auf den Gaswechsel der Fichte. Centralbl Gesamte Forstwes 110:135–149

Wieser G, Havranek WM (1994) Exposure of mature Norway spruce to ozone in twig-chambers: effects on gas exchange. Proc R Soc Edinb Sect B 102:119–125

Wieser G, Havranek WM (1995) Environmental control of ozone uptake in *Larix decidua* Mill.: a comparison between different altitudes. Tree Physiol 15:128–136

Wieser G, Havranek WM, Polle A (1996) Effects of ambient and above-ambient ozone concentrations on mature conifers in the field. In: Kärenlampi L, Skärby L (eds) Critical levels for ozone in Europe: Testing and finalizing the concepts. UN-ECE Workshop Report, Kuopio University, Finland, pp 183–190

Wieser G, Weih M, Havranek WM (1991) Ozone fumigation in the sun crown of Norway spruce. In: Reuter M, Kirchner M, Kirchinger E, Reiter H, Rösel K, Pfeifer U (eds) Waldschadensforschung im östlichen Mittereuropa und in Bayern. Proc Expertentagung Schloß Neuburg/Inn bei Passau Nov 13-15 1990. GSF-Bericht 24/91, pp 567–573

Wieser H (1995) Auswirkungen chronischer Ozonbelastung auf die Gaswechselphysiologie junger Hoch- und Tieflagenherkünfte der Lärche (*Larix decidua* L.) in Klimakammern. Dipl Thesis, Universität Innsbruck

Wiskich JT, Dry IB (1985) The tricarboxylic acid cycle in plant mitochondria: its operation and regulation. In: Douce R Day DA (eds) Higher plant cell respiration. Encyclopaedia of plant physiology, new series, vol 18. Springer Berlin Heidelberg New York, pp 281–313

Wolfenden J, Mansfield TA (1991) Physiological disturbances in plants caused by air pollutants. Proc R Soc Edinb 97B:117–138

Zimmermann R, Oren R, Schulze ED, Werk KS (1988) Performance of *Picea abies* (L.) Karst. at different stages of decline II Photosynthesis and leaf conductance. Oecologia 76:518–523

5 Role of Ozone in Forest Decline on Wank Mountain (Alps)

H. RENNENBERG, A. POLLE and M. REUTHER

5.1 Introduction

From crown-thinning of spruce it has been estimated that one-third of the mountain forests in the Bavarian Alps are severely damaged (Kennel 1990). Photo-oxidants transported above inversion layers have been thought to be responsible for the damage observed in these forests (Seiler 1988; Fabian 1990). As O_3 is a key product of photo-oxidative chain reactions in the troposphere and tropospheric O_3 concentrations change with increasing altitude, O_3 has been implicated as the central cause of this forest decline (Seiler 1988). Climatic stress, nutrient deficiencies, forest management and genetic properties of the trees have also been considered as predisposing or triggering factors (Rennenberg and Reuther 1991). The general objective of the studies performed in the Bavarian Alps at the Wank mountain research site has been to test this hypothesis using an interdisciplinary approach. To meet this objective, scientists working in forestry, meteorology, air chemistry, tree physiology, soil sciences, ecology and taxonomy participated in these studies which lasted for almost a decade. The individual objectives of the research performed were (1) to characterize the particular stand, especially the atmospheric and pedospheric environment, and to assess the vitality and the nutritional status of the trees and (2) to identify and characterize those physiological parameters that exhibit changes with altitude similar to those associated with O_3 (Rennenberg and Reuther 1991). In this chapter, the results obtained in the studies at the Wank mountain research site are summarized. In addition, the question is addressed as to whether altitude-dependent changes of the parameters studied provide evidence for stress to the trees, can be connected with the damage observed, and are the result of O_3 exposure.

Ecological Studies, Vol. 127
Sandermann et al. (eds) Forest Decline and Ozone

5.2 Site Characteristics

5.2.1 The Bavarian Alps

The Bavarian Alps represent the northern rim of the Alps and stretch over more than 200 km in an east-west direction along the border to Austria (Fig. 5.1). The Bavarian Alps mainly consist of steep calcareous rock formations, skirted with a small zone of soft rock and marl in the north (Bayer. StMELF 1982). Due to its high elevation, this alpine region is characterized by relatively low temperatures (annual mean of +7 to −5°C) and high rainfall and snowfall (mean annual precipitation of 1200 to 2500 mm; Mößmer 1986; Bayer. OFoD München 1991).

Based on distinct elevation zones, the following classification of natural forest vegetation has been developed for the Bavarian Alps: at 500 to 1000 m a.s.l., submountanous and mountanous mixed alpine forests [Norway spruce (*Picea abies*), silver fir (*Abies alba*), common beech (*Fagus sylvatica*)]; at 1100 to 1400 m a.s.l., high mountanous spruce/fir forests; at 1400 m a.s.l. to the timber line, subalpine spruce forests and dwarf pine

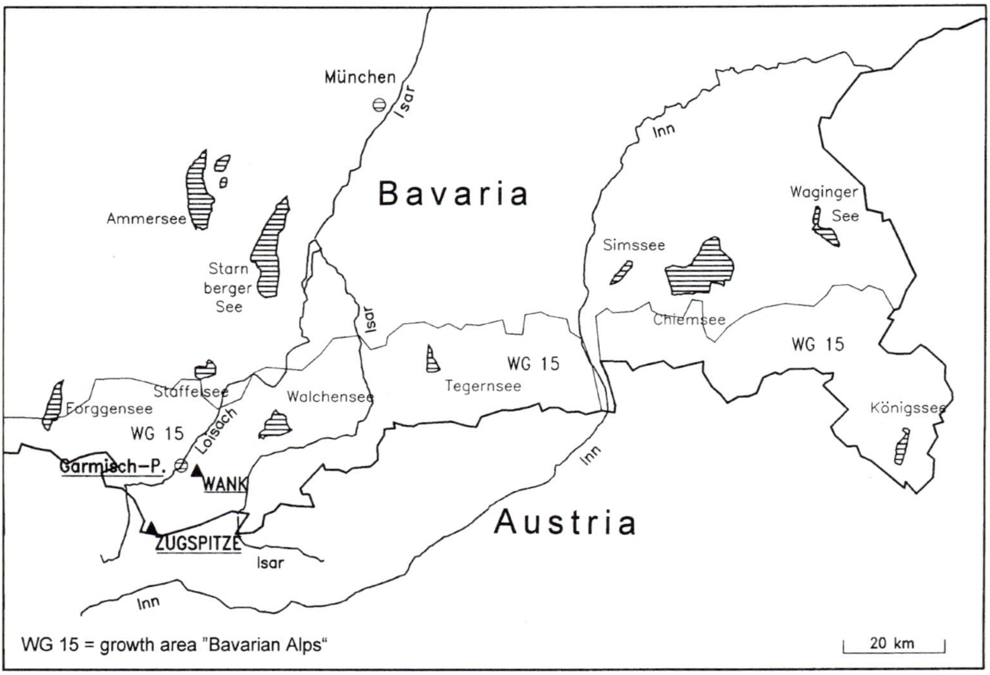

Fig. 5.1. Location of the Bavarian Alps and the region of Garmisch-Partenkirchen in southern Bavaria, Germany

(*Pinus mugo* L.). Recently planted forests in the Bavarian Alps differ considerably from natural forest vegetation. Anthropogenic activities have led to a massive decline of beech and fir, so that now spruce is the dominating tree species covering 66% of the Bavarian alpine forests (Bayer. StMELF 1982).

Despite intensive tourism, for a long time the Alps have been considered as a clean air region since they are far away from heavily populated industrial centers. This view has been confirmed for the classical gaseous air pollutants, SO_2 and NO_X, which show rather low atmospheric gas mixing ratios in the Bavarian Alps compared to other regions (PBWU 1987; Weißgerber 1987; Rennenberg 1989). In contrast, relatively high tropospheric O_3 levels are observed in the Bavarian Alps. The O_3 concentrations have increased over the last decades and also have increased with altitude at different measuring sites in German, Austrian and Swiss alpine regions (Gabler et al. 1990; Smidt et al. 1990; Seiler 1991; Sladkovic et al. 1994a,b).

5.2.2 The Wank Mountain Research Site

The Wank mountain is located in the Loisach valley close to the village of Garmisch-Partenkirchen (Fig. 5.2, see Appendix). It is part of the Bavarian Alps in the south of Germany and rises from about 700 m a.s.l. in the Loisach valley up to 1780 m a.s.l. at the summit (N 47° 30′ 36″, E 11° 8′ 39″). The geological formation of the region is dominated by calcareous rocks and thus is typical for the Bavarian Alps (Bayer. OFoD München 1991; Liu and Payer 1992). Sun exposure, steepness of the slopes, precipitation and microrelief structure determine the thickness of the soil layer which ranges from shallow rendzina soil types on sun-exposed, steep slopes to deep and fertile soils on shallow, shady slopes. Also the climate of the region with high precipitation and low temperatures is typical for the Bavarian Alps. The average precipitation per year ranges from 1300 mm at valley level to 1800 mm at the summit. Typically, maximum precipitation takes place in summer, and minimum precipitation, in winter (Polle et al. 1992). The mean annual air temperature ranges from 7 °C at valley level to 3 °C at the mountain summit. Western winds prevail, but on 30 to 40 days per year, a dry (20–40% humidity) warm wind from south/southwest (foehn conditions) is experienced. Furthermore, a daily uphill/downhill circulating wind system is superimposed on this regular wind regime. Inversion layers have been reported on about 30 days per year, most frequently in autumn and winter (Reiter and Kanter 1972).

The original natural forest vegetation at the Wank mountain research site was a mixed spruce, fir and beech forest, supplemented by Scots pine on sun-exposed, dry and shallow sites. In contrast, the present forest vegetation is dominated by spruce (about 75%) and pine (15%; Ammer et al. 1988). This difference between the natural and present forest vegetation is the consequence of intensive anthropogenic use. After the first Roman settlements in the Loisach valley, forest cropping and forest litter utilization as well as grazing by horses, sheep, goats and cattle took place. As a consequence, shortage of timber and forest devastation had occured by the end of the 15th century. When the forests were taken over in 1803 by the Bavarian state, forest utilization became regulated by law, but mismanagement, adverse climatic events such as storms, the continuation of forest use as pasture, browsing by game, game keeping and feeding led to the overmature, spruce-dominated forests presently found. Recent programmes for redevelopment and protection of the forests, as well as progressive game management, are expected to improve the forests on Wank mountain in the future.

The research site on Wank mountain used for the present study is located on the south-west slope in the direction of Garmisch-Partenkirchen. At the beginning of the research project in 1986, a metereological station was set up at middle elevation (1175 m a.s.l.) to continuously record climatic and air pollution data. This station supplemented existing metereological stations in the valley (735 m a.s.l.) and at the Wank mountain summit (1782 m a.s.l.). Since 1987, measurements of O_3, SO_2, NO, NO_y (= sum of gaseous nitrogen oxides including NO, NO_2, N_2O_5, PAN, HNO_2, and HNO_3 as major species), H_2O_2, several hydrocarbons including PAN, and various meteoreological parameters (air temperature, humidity, wind, precipitation, rain water composition, etc.) have been undertaken. The results of these measurements were used to estimate transport of air pollutants to, and production of air pollutants at the site (Seiler 1988).

For tree analysis at the Wank mountain research site, seven sampling plots of about 0.1 to 0.2 ha were chosen at different altitudes along a transect ranging from 780 m a.s.l. up to 1720 m a.s.l. with a difference in elevation of ca. 150 m between each plot (Fig. 5.3). All plots were stocked with 70 to 140 year-old spruce trees (Table 5.1). Because of heavy storms in February 1990, one plot at middle altitude had to be replaced by a younger spruce stand (ca. 40 year-old trees). In addition to this vertical transect, two sampling plots were chosen along the south-west slope of Wank mountain in 1991 at 1150 m a.s.l. in order to set up a horizontal transect with different expositions (south-west and north-west). Some characteristics of the sampling plots are shown in Table 5.1. Needles from the sun-exposed crown were sampled from up to 60 spruce trees per site

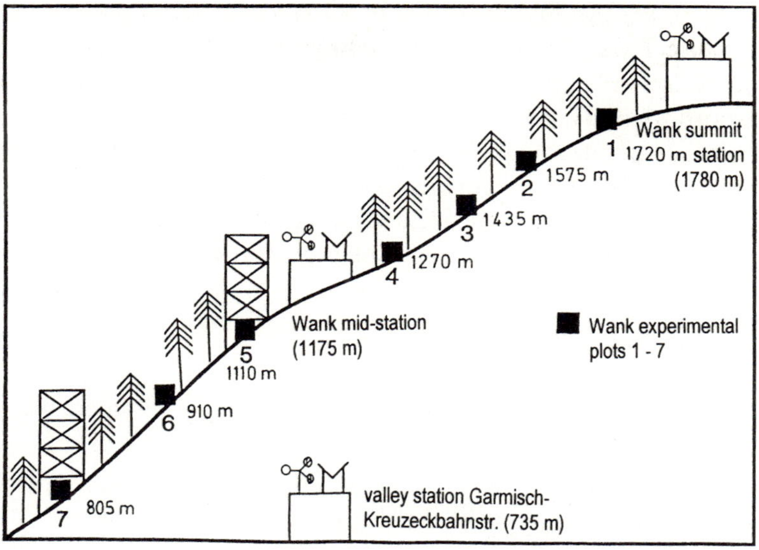

Fig. 5.3. Vertical profile of the sampling plots 1 to 7 and the meteorological stations at Wank mountain. (Liu and Payer 1992)

Table 5.1. Characteristics of the sampling plots 1 to 7 of the vertical profile at Wank mountain. (Liu et al. 1994; Mößnang and von Schönborn 1992)

Sampling site	1	2	3	4	5	6	7
Elevation m a.s.l.	1720	1575	1435	1270	1110	910	805
Exposition	S	S-SW	SW	SW	SW-W	SW-W	SW-W
Slope steepness (%)	20–30	20–30	10–30	30	20–25	25	5–10
Soil type	TfR	RTf	TfR	RTf	RTf	RTf	Tf
Long-term annual mean Temperature (°C)	+3 (Summit)						+8 (Valley)
Annual precipitation (mm/year)	1,800 (Summit)						1,300 (Valley)
Growing stock	74% spruce, 15% pine, 8% fir, 3% broad-leaf trees						
Mean age (year)	140	130	130	90	70/40	130	140
Mean tree height (m)	15	18	22	19	19	19	27
Breast height diam. (cm)	28	34	42	34	29	31	42

TF Terra fusca; RTF Rendzina–Terra fusca, TfR Terra fusca–Rendzina over several years three to five times a year. Each of the seven plots along the vertical transect was equipped with metereological measuring units for air temperature, soil temperature, and air humidity. These weather stations were also used to set up integrating monitoring systems for O_3. At two of the plots, one at valley level (780 m a.s.l.) and one at middle elevation (1110 m a.s.l.), a measuring platform was installed to allow experiments on the upper crowns of the surrounding spruce trees. These plots were used for more intensive experimental studies.

5.2.3 Tree Vitality at the Research Site

The vitality of the spruce trees at Wank mountain was estimated from coloured infrared aerial photographs of the entire research site and from terrestrial surveys on the sampling plots along the vertical profile (Ammer et al. 1988; Mößmer et al. 1991, 1993). The evaluation was based on the methods used for regular national surveys, i.e. the determination of needle loss and needle yellowing compared to representative healthy reference trees. From these estimates, the Wank mountain research site was determined to be severely damaged. As shown in Table 5.2, the percentage of needle loss ranges from 15 to 60% with a mean of ca. 30% (Sinner and Schuster 1993). Moreover, there was a strong tendency of increasing needle loss with increasing elevation (Mößmer et al. 1993) and needle loss appeared to increase from the 80th to the 90th estimate (Table 5.2). However, strong hurricanes destroyed a large number of single trees in February 1990 and sampling plots 3 and 5 were totally destroyed by storms in 1991. These events may have influenced the needle and twig mass of the remaining trees and, therefore, may be partially responsible for the high degree of needle loss found in the 90th estimate. However, considerable chlorosis of needles was only observed in 1992 (Mößmer et al. 1993).

Table 5.2. Results of the terrestrial surveys on crown classification from 1987 to 1992 at the seven sampling plots at the Wank vertical profile in percent needle loss. (Mößnang and von Schönborn 1992; Sinner and Schuster 1993)

Year	1	2	3	4	5	6	7	Mean
1987	25.5	31.0	22.9	24.5	15.6	21.3	22.0	23.5
1988	25.0	33.5	19.2	23.0	13.8	25.0	27.5	24.2
1989	25.5	32.5	20.3	24.8	15.9	22.6	25.0	24.1
1990	33.7	50.0	35.0	37.4	22.0	33.0	27.9	33.1
1991	38.1	51.9	41.7	32.4	24.2	25.0	24.4	32.1
1992	51.1	60.0	ND	45.3	ND	36.5	38.8	46.3

ND, No data available

The results from the surveys in 1986 and 1987 indicated an enhanced needle loss in two zones, at 800 to 1000 m a.s.l. and at 1300 to 1500 m a.s.l., respectively (Ammer et al. 1988; Mößmer et al. 1991). It was assumed that these zones of enhanced damage corresponded to the zones of the inversion layers in winter and summer that are connected with increased exposure to photo-oxidants. Thus, the surveys in 1986 and 1987 supported the hypothesis that air pollution by photo-oxidants could be a major stress factor for the forests in this alpine region. However, evaluations of subsequent and previous surveys with a larger data base did not confirm the existence of two distinct zones of damage. Furthermore, a common pattern of altitude-dependent differences in tree vitality was not found when compared with other inventories on forest decline in alpine regions (Kirchner et al. 1992). While the surveys from 1986 to 1989 showed an influence of exposition on tree vitality, this was no longer found in the subsequent evaluations (Mößmer et al. 1993).

In summary, the Wank mountain research site is characterized by a high degree of needle loss, but clear-cut differences are only observed between trees growing at low and high elevation. Enhanced damage to the trees at high elevation is partially due to the more unfavourable climatic conditions. As a consequence, dieback of the upper crown, crown deformations, and bark and stem injury was observed on individual trees growing at high elevation. Signs of biotic injury by insects were generally not observed in surveys performed in 1989 and 1992 (Köhler and Lang 1993, unpubl.). However, the beginning of a bark beetle (*Ips typographus* L.) calamity in 1992 might have been a consequence of the heavy loss of trees by the storm events in 1990 and 1991.

5.3 Atmospheric Exposure

Basic meteorological parameters and pollution climate were continuously recorded at three meteorological stations at valley level (735 m a.s.l.), middle elevation (1175 m a.s.l.), and the Wank mountain summit (1782 m a.s.l.). Atmospheric gas mixing ratios of SO_2 and nitrogen oxides were relatively low as expected for a region far away from highly populated industrial centres. For nitrogen oxides, annual mean values of 0.4 nl l^{-1} (NO) and 3 nl l^{-1} (NO_y) were determined. Maximum NO and NO_y concentrations (50 nl l^{-1} and 20 nl l^{-1}, respectively) observed in winter could be attributed to local sources in the valley. Annual mean values for SO_2 amounted to 0.4–1 nl l^{-1} with a decreasing trend from 1987 to 1993 at middle elevation. Peak concentrations of SO_2 (50 nl l^{-1}) were observed exclusively during episodes of advective transport of air masses from SO_2-polluted areas (Junkermann and Langer 1992; Junkermann et al. 1992; Sladkovic et al. 1994a). From these data it can be concluded that atmospheric SO_2 and NO_x gas mixing ratios do not reach values which are injurious to spruce trees of Wank mountain forests (Slovik et al. 1992).

In contrast to the low atmospheric gas mixing ratios of SO_2 and NO_x, atmospheric O_3 concentrations were relatively high at the Wank mountain research site. From 1978 to the mid 1980s, atmospheric O_3 concentrations appeared to increase, but this trend did not continue (Sladkovic et al. 1994a,b). Annual mean values of atmospheric O_3 at the middle elevation amounted to 30–50 nl l^{-1} and distinct annual fluctuations were found. Monthly maximum half-hour values of up to 100 nl l^{-1} were measured in summer (May to August); minimum values were found in winter (Junkermann and Langer 1992; Sladkovic et al. 1994a). Maximum monthly mean O_3 values ranged from 45 to 55 nl l^{-1}. Atmospheric O_3 concentrations showed a pronounced diurnal or several-day-variation, with low gas mixing ratios at night, when O_3 was consumed by its reaction with NO_x or deposited on plants and other surfaces (Junkermann and Langer 1992). In Fig. 5.4, daily and monthly mean values of atmospheric O_3 gas mixing ratios are summarized for 1990. A comparison of O_3 concentrations at valley level, middle altitude and Wank mountain summit showed increasing O_3 levels with increasing elevation (Fig. 5.5), whereas diurnal variations decreased from valley level to the mountain summit (Fig. 5.6).

So-called O_3 bulges, i.e. high concentrations at distinct elevations, reported from vertical measurements performed at other sites in the Alps (Paffrath and Peters 1988; Smidt et al. 1990), were not found at the Wank research site. This result was surprising since it was initially thought that high levels of O_3 and photo-oxidants at distinct altitudes may be a major factor involved in injury to the trees at the Wank site (PBWU 1987; Seiler 1988; Rennenberg 1989). A careful analysis of temperature profiles, however, indicated that inversion episodes are scarce during the day and can be rapidly removed by uphill–downhill wind circulations. However, the level of photo-oxidants at the Wank research site was highly influenced by advective long-distance transport of polluted air from the prealpine region (Junkermann and Langer 1992; Sladkovic et al. 1994a,b).

Given the needle loss of spruce trees and the increase in atmospheric O_3 concentrations increasing with increasing elevation, the question as to whether the needle loss observed can be attributed to the atmospheric O_3 concentrations has to be addressed. The O_3 levels determined at the Wank research site (Junkermann et al. 1992; Sladkovic et al. 1994a,b) and also at several Austrian research sites in the Alps (Smidt et al. 1991; Schneider and

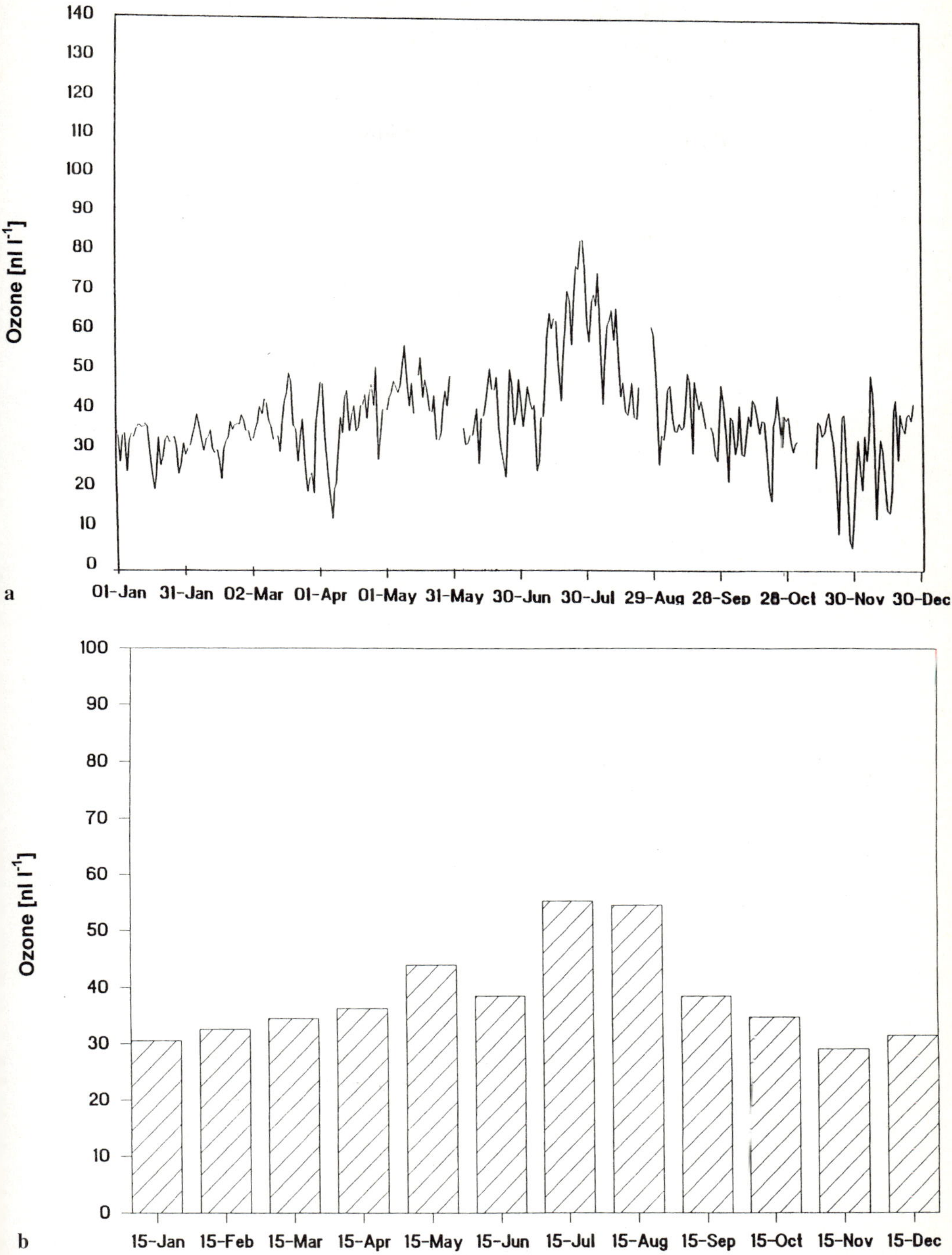

Fig. 5.4. Daily (**a**) and monthly (**b**) mean values of atmospheric O_3 gas mixing ratios at the middle elevation (1125 m a.s.l.). (Junkermann and Langer 1992)

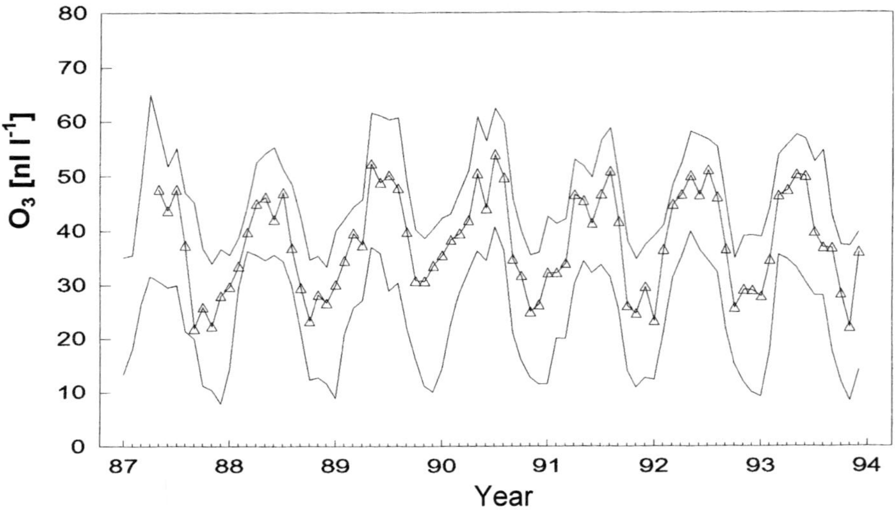

Fig. 5.5. Comparison of monthly mean values of O_3 between 1987 and 1994 at valley level (735 m a.s.l., *lower line*), middel altitude (1175 m a.s.l., *lines and symbols*) and Wank mountain summit (1782 m a.s.l., *upper line*). (Sladkovic et al. 1994a)

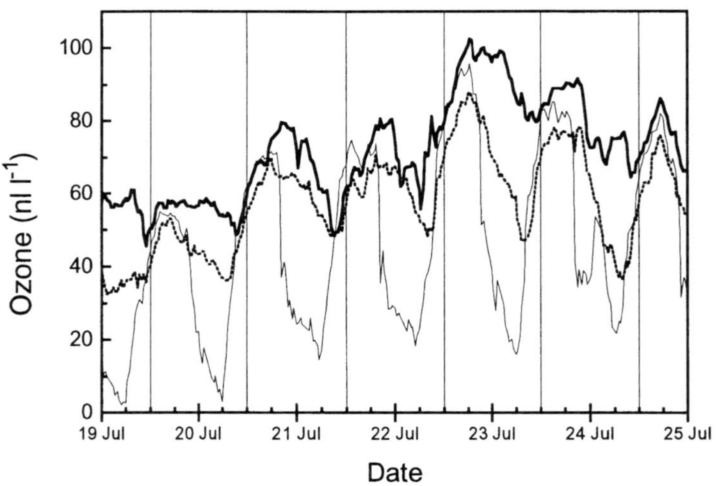

Fig. 5.6. Diurnal fluctuations of O_3 during 6 days in July 1990 at meteorological stations located at valley level in Garmisch-Partenkirchen (735 m a.s.l., *thin line*), middle elevation of Wank mountain (1175 a.s.l., *broken line*) and the summit of the Wank mountain (1782 m a.s.l., *thick line*). (Courtesy of W. Junkermann, Fraunhofer Institut, Garmisch-Partenkirchen)

Loibl 1996) exceed the existing AOT-40 threshold value of 10 $\mu l\ l^{-1}$ h (UN-ECE 1994).

5.4 Soil Characteristics and Tree Nutrition

In addition to the atmospheric environment, the physical and chemical properties of the soil along the slope of Wank mountain were considered as causal or predisposing factors for altitude-dependent symptoms of injury observed on the above-ground parts of spruce (Liu et al. 1991). In order to obtain information on the interaction of soil properties with the below-ground biomass of spruce and with tree nutrition in general, fine root biomass, distribution, and dynamics as well as elemental contents of both roots and needles (Mößnang and von Schönborn 1992; Mößnang 1992; Sandhage-Hofmann and Zech 1993; Zech and Sandhage-Hofmann 1993) were analyzed.

Between 1988 and 1992, the chemical and physical properties of the soil were studied in seven plots at different altitudes of Wank mountain (Liu et al. 1991; Liu and Payer 1992). These studies showed that differences in the geological formation from which the soil was developed and in soil morphology, as well as human activities have caused a high degree of heterogeneity of the soil along the mountain slope. The dominating soil type is a Rendzina–Terra fusca with mould as the main humus type. At least at the low elevation site, tree litter utilization has resulted in a loss of organic material from the soil of ca. 100 tonnes ha^{-1}. Moreover, despite relatively high precipitation (Table 5.1), drought stress cannot be excluded as a factor which contributes to injury of the trees at Wank mountain, since the shallow soils have a relatively high clay content (Liu and Payer 1992).

Compared to other forest sites (e.g. Kreutzer and Pröbstle 1991), total soil content of N, Mg, Ca, Mn and Fe were high on all plots, whereas total P and K content were relatively low. Optimum levels of plant-available K, Mg and Ca were found, while amounts of plant-available P in particular, but also N and Mn, were insufficient independent of total soil contents (Liu and Payer 1992). Element analysis of fine roots and needles of spruce at both the high and low elevation site, confirmed the assumptions drawn from the results of the soil analysis. With 0.3–0.7 mg P g^{-1} dry matter and 5.3–11.5 mg N g^{-1} dry matter in fine roots, and 0.8–1.0 mg P g^{-1} dry matter and 11–12 mg N g^{-1} dry matter in the needles, the content of these elements were relatively low. Slightly better N supplies were found for the needles and fine roots of trees at low elevations as compared to those higher up (Mößnang and von Schönborn 1992; Zech and Sandhage-Hofmann 1993). Apparently, N-minerlization and -release from fine root turnover is lower in the soil at the high elevation site than lower down. Part of the difference in N supply between trees growing at the two sites may also be the consequence of enhanced N deposition at the low elevation site close to Garmisch-Partenkirchen. However, the annual wet deposition of ca. 19 kg N ha^{-1} measured at Wank mountain is low when compared to other forests (Liu et al. 1994).

At both the low and high elevation sites, mycorrhizal and non-mycorrhizal fine roots of spruce were mainly located in the humus layer and the upper 10 cm of the mineral soil. With increasing soil depth, the relative portion of dead fine roots also increased. The sites differed considerably in total fine root content in the soil with values of 4550 kg ha^{-1} at the higher site, compared to 6340 kg ha^{-1} at the lower one. From 1990 to 1993, the fine root content of the soil decreased by 2570 kg ha^{-1} at the higher site and 1820 kg ha^{-1} at the lower one. The carbohydrate content of the fine roots was significantly higher in the roots of trees growing in the valley (Zech and Sandhage-Hofmann 1993). Therefore, it may be concluded that a reduced pigment content of the needles and enhanced needle loss resulted in a reduced root biomass as well as a reduced supply of assimilates to the remaining roots. Alternatively, drought and/or other stress factors may have reduced root growth and, as a consequence, may have caused enhanced needle loss and reduced pigment content of the remaining needles.

From these findings, it appears doubtful that the small differences in P and N supply along the altitudinal gradient significantly contribute to the differences in the degree of needle loss observed between trees growing at high and low elevations. However, P and N deficiencies can cause premature needle shedding (Linder 1989). Thus, it was concluded that these deficiencies contributed to the generally high degree of damage to the trees observed at Wank mountain (Polle et al. 1992). This high degree of damage cannot be attributed to acidification, since the proton burden from wet deposition is rather low (0.4 kg ha^{-1} $year^{-1}$) and the calcareous soils with soil water pH values of about 8 do not provide a considerable acidification risk (Liu et al. 1994).

5.5 Physiological Observations

5.5.1 Ozone

Typical signs of O_3 injury, such as necrotic spots, yellow bands or severe bleaching, were not found in needles of spruce trees grown along an altitudinal gradient from 805 to 1720 m a.s.l. at Wank mountain. In general, spruce needles from higher altitudes contained less chlorophyll than spruce needles from lower down (Lütz 1992; Polle et al. 1992). Since the photosynthetic pigments showed normal seasonal fluctuations at the different altitudes (Lütz 1992; Polle et al. 1992), the reduction in chlorophyll content can be interpreted as an acclimation of the needles to high light intensities and low temperatures rather than an indication of O_3-induced damage. This suggestion is supported by the observation that needles from higher altitudes contained enhanced carotenoid to chlorophyll ratios (Polle et al. 1992). Apparently, the photo-protection of the residual pigments in needles from high altitude was thereby improved (Polle et al. 1992).

When mature needles from spruce trees grown at Wank mountain were challenged in short-term experiments with extreme concentrations of O_3, reductions in photosynthesis only occurred after exposure to more than 1000 nl l^{-1} O_3 for several hours (Knoppik et al. 1992). Apparently, spruce trees at Wank mountain are well equipped to withstand significant O_3 concentrations during short-term peak episodes of this pollutant. In studies addressing chronic effects of O_3, Wieser and Havranek (1996) found reductions in photosynthesis in current year needles above SUMO 200 $\mu l\ l^{-1}$ h and in previous-year needles above SUMO 400 $\mu l\ l^{-1}$ h indicating that field-grown trees are more resistant to O_3 than young trees exposed in open top chambers (Skärby et al. 1995). In order to trace chronic O_3 effects in needles at the Wank mountain, a range of biochemical parameters of trees grown at low altitudes were compared with those of trees grown higher up. Plant hormones such as gibberellic acid (Hock 1990) and abscisic acid (Ziegler et al. 1992) did not show altitude-dependent changes. Fungal pathogens were apparently not involved in the observed needle loss at Wank mountain (Hippeli et al. 1991). Schröder et al. (1993) excluded chlorinated hydrocarbons as injurious agents at this site, but they observed higher glutathione-S-transferase activities in needles of damaged trees than in needles of healthy trees. From this finding the induction of a defence reaction in diseased trees may be concluded. However, a causal connection between the response of glutathione-S-transferase and O_3 was not established (Schröder et al. 1993). In controlled O_3 exposure experiments, Sandermann et al. (1990) identified secondary metabolities that are induced specifically by both O_3 and pathogens. Several studies have suggested that O_3 can elicit similar responses to those elicited by pathogens (Kangasjärvi et al. 1994), but these compounds were not studied at the Wank mountain (Heller et al. 1994).

Ozone is a strong oxidant and will give rise to a range of reactive oxygen species in aqueous phases (Hoigne and Bader 1975; Grimes et al. 1983). Consequently, antioxidative systems are considered especially important for protection from O_3 injury (Tanaka 1994). Superoxide dismutases, which scavenge O_2^-, are thought to play a key role in the detoxification of oxy-products of O_3 (Bennett et al. 1984). However, at Wank mountain significant altitude-dependent differences in superoxide dismutase activities were not observed in summer or fall but in winter when the needles were not protected by a snow cover (Polle and Rennenberg 1992; Kröniger 1993; cf. Fig. 5.7 A). In controlled fumigation experiments including those simulating "Wank" climate, an induction of superoxide dismutase activity in needles in response to O_3 has not been found (Hausladen et al. 1990; Polle and Rennenberg 1991; Wingsle et al. 1992; Nast et al. 1993, Polle et al. 1993, Benes et al. 1995; Schwanz et al. 1996; Polle and Baumbusch, unpubl.). These observations suggest that either the superoxide dismutase activities present in spruce needles are already high enough to counterbalance oxidative stress as imposed by O_3 or that these enzymes are not directly involved in O_3 detoxification. The latter assumption is corroborated by several studies with transgenic plants overexpressing superoxide dismutase. When these plants were exposed to O_3, in most cases an improved protection from O_3 was not observed, whereas the tolerance for other stress factors such as herbicides, chilling, etc was enhanced (for reviews see Foyer et al. 1994; Rennenberg and Polle 1994).

Along the altitudinal gardient at Wank mountain, activities of apoplastic peroxidases and concentrations of several soluble phenolic components in total extracts from spruce needles decreased significantly with increasing elevation (Polle et al. 1991; Ziegler et al. 1992). In clonal spruce trees exposed for 2 years at low, middle and high altitudes, the lowest apoplastic peroxidase activities, however, were found in needles from trees exposed at the middle altitude (Polle and Junkermann 1994a). In needles from mature spruce trees, total foliar peroxidase activity declined slightly with increasing elevation and also showed a significant drop in activity at the middle elevation (Fig. 5.7B). Levels of glutathione increased with increasing altitude but often displayed lowest amounts in needles from trees grown at the middle altitude (Polle and Rennenberg 1992, Fig. 1D). Lütz (1992) also reported a reduction in the violaxanthin to antheraxanthin ratio in needles from spruce trees grown at the middle elevation which implies a reduction in the protection from photoinhibition at this site. At middle elevation similar biochemical irregularities as found in needles from mature spruce were also observed in needles from clonal spruce planted or exposed in pots at this site (Polle and Junkermann 1994b; Polle and Rennenberg 1994). The latter results suggest that the effects on biochemical parameters observed at the middle altitude were independent of genetic or soil-borne factors and, thus, likely to be caused by air-borne pollutants. The O_3 doses increased with increasing altitude e.g. from 129 $\mu l\ l^{-1}$ h at 735 m a.s.l. to 237 $\mu l\ l^{-1}$ h at 1780 m a.s.l.

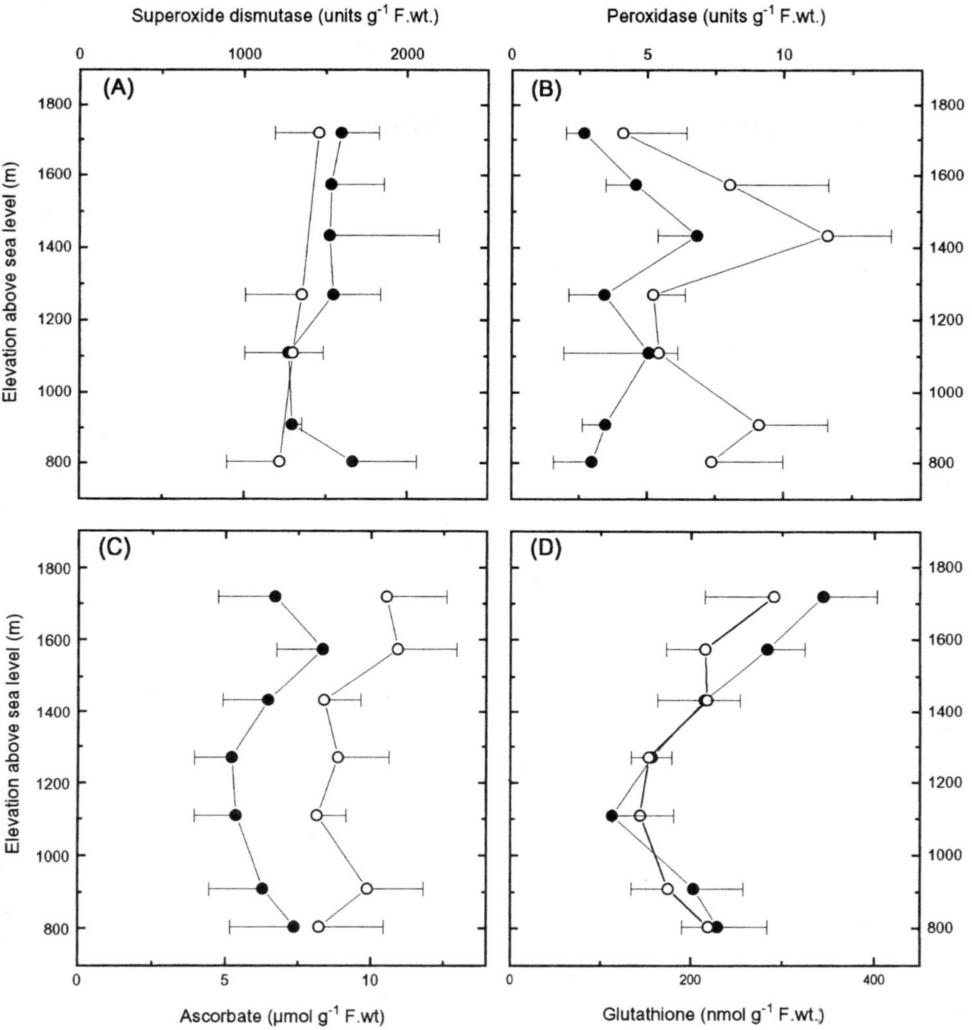

Fig. 5.7. Altitude-dependent fluctuations of antioxidative enzymes (**A** superoxide dismutase; **B** guaiacol peroxidase) and antioxidants (**C** ascorbate; **D** glutathione) in current-year (•–•) and previous-year needles (o–o) observed at Wank mountain in September 1988. F.wt. Fresh weight. Data points represent means (SD, *n*=6). The measurements were conducted as described by Polle and Rennenberg (1992)

(SUMO from April to October in 1995). Therefore, it is unlikely that O_3 alone was a causal factor for the irregularities found at middle elevation.

Foliar ascorbate content increased with increasing altitude (Polle and Rennenberg 1992). However, the extent of this increase was fairly small in some months, e.g. in September (Fig. 5.7C). Furthermore, ascorbate showed pronounced seasonal fluctuations which corresponded roughly to seasonal fluctuations of mean O_3 concentrations at the high and low elevations (Fig. 5.8). The altitude-dependent increase in ascorbate was higher in summer (May, June) than in the other months investigated (Fig. 5.7C). It has been known for a long time that high tissue concentrations of

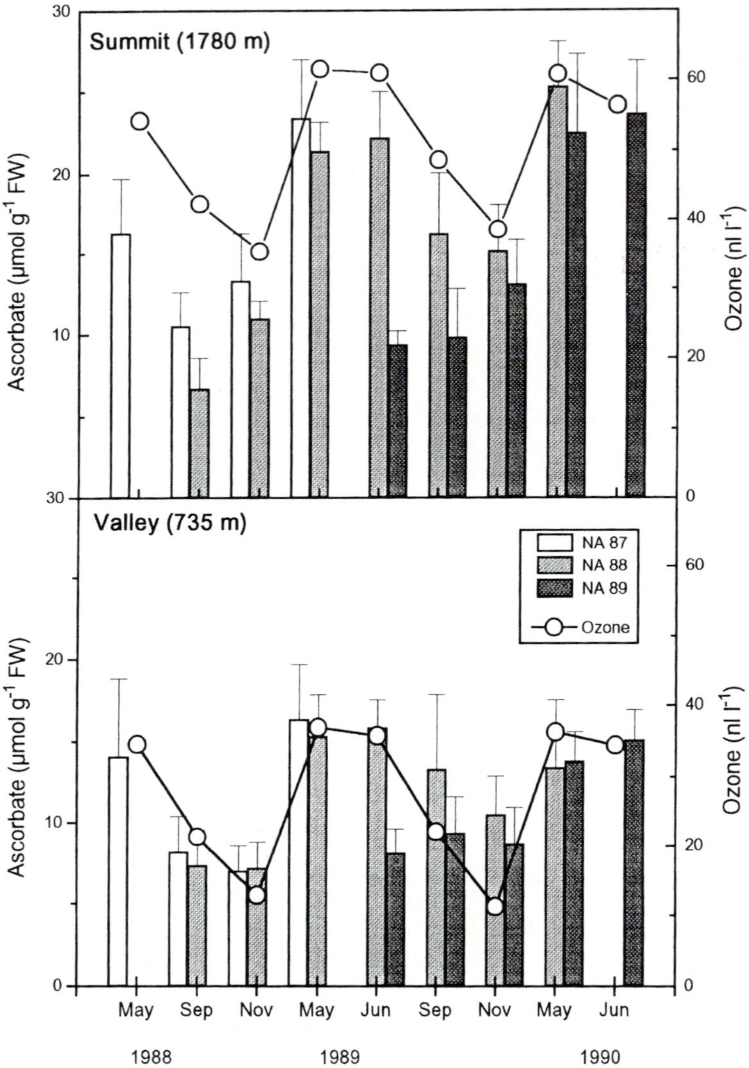

Fig. 5.8. Seasonal fluctuations of the total foliar ascorbate content (*bars*) in three consecutive years in spruce trees grown close to the summit of Wank mountain 1720 m and in spruce trees grown close to valley level at 805 m in comparison with mean O_3 concentrations at these sites (circles). *NA* Needle age class; *FW* Fresh weight; *bars* indicate means (SD, n=6). Data were taken from Polle et al. (1992) and Polle and Rennenberg (1992)

ascorbate confer enhanced O_3 tolerance (Menser 1964; Urbach et al. 1989). Since O_3 enters plants mainly via open stomata and will diffuse through the aqueous matrix of the cell wall, the primary reaction sites of O_3 are expected to be in the apoplastic space (Polle and Rennenberg 1993). Spruce needles contain considerable concentrations of apoplastic ascorbate (Castillo et al. 1987; Polle et al. 1990). Since ascorbate reacts rapidly with O_3, the apoplastic ascorbate pool might provide an important protection against the destruction of membranes and intracellular components by this strong oxidant (Chameides 1989). At Wank mountain, the concentration of apoplastic ascorbate increased with increasing altitude from about 1 mM at valley level (735 m a.s.l.) to 4 mM at high altitude (1720 m a.s.l.; Polle et al. 1995).

The crucial question that is currently being investigated is whether this pool is readily accessible and regenerated at a sufficient rate for an efficient removal of O_3. Exposure of spinach leaves to O_3 (300 nl l^{-1}) resulted in a depletion of apoplastic ascorbate within 6 h and it seemed that 90% of the O_3 taken up by the leaves had not reacted with ascorbate (Luwe et al. 1993). These results suggest that apoplastic ascorbate does not contribute much to the protection from O_3. Unlike spinach, however, exposure of spruce needles to elevated concentrations of O_3 did not cause a depletion of the apoplastic ascorbate pool but induced an increase in apoplastic ascorbate (Polle et al. 1995). This suggests that apoplastic ascorbate in spruce needles is rapidly regenerated and might constitute an important defence mechanism. Alternatively, it may be concluded that O_3 fluxes into needles do not reach the bulk of apoplastic ascorbate. Both possibilities cannot be rejected at present.

It is conceivable that inside the substomatal cavity O_3 reacts already within or in close proximity to the stomata because structural properties provide a diffusion barrier for a more even distribution of O_3 in the apoplastic space or because of the reaction with volatile hydrocarbons. This would explain why negative effects of O_3 on stomata have frequently been reported. Maier-Maercker and Koch (1991) showed that O_3 injures stomatal regulation, possibly by delignification of the subsidiary cells (Maier-Maercker 1989). Such needles generally show higher stomatal conductances and fail to close their stomata fully (Maier-Maercker and Koch 1991). Initially, such O_3-predisposed needles would photosynthesize more than healthy needles, albeit with decreased water-use efficiency. Consequently, in periods of drought stress, these needles would suffer more from drought-induced injury than needles whose stomatal regulation is functioning well (Maier-Maercker and Koch 1991). At Wank mountain a comparison of healthy and damaged trees, which suffered from 30% needle loss, revealed a significant loss of stomatal control and reduced turgor values in needles of the damaged trees (Maier-Maercker and Koch 1995). Since Maier-Maercker and Koch (1995) studied only the behaviour of the trees grown at the lower altitude, it remains unclear to what extent drought stress and O_3 interact to produce the altitude-dependent symptoms of injury observed at Wank mountain. It would be difficult to address this complex interaction in the field because periods of drought stress generally go along with episodes of elevated O_3 concentrations.

Unfortunately, measurements of diurnal fluctuations of stomatal conductances of spruce needles at different altitudes at Wank mountain are lacking. Such measurements would have allowed us to detect irregularities of stomatal regulation and to calculate effective O_3 doses, i.e. the amount of O_3 which has actually been taken up by spruce needles. So far, the discussion has been based on average O_3 concentrations measured in ambient air at different altitudes at Wank mountain. However, current concepts of O_3 effects on plants imply that the reponse to O_3 and the degree of damage is determined by the rate and the amount of O_3 taken into the tissue rather than the average external pollutant concentrations (Reich 1987; Runeckles 1992). Since cuticles provide an almost impermeable barrier to O_3 (Kerstiens and Lendzian 1989), the flux of O_3 into a leaf depends on the degree of stomatal aperture.

A rough assessment of potential O_3 fluxes and the resulting effective O_3 doses in needles from Wank mountain can be obtained from measured fluctuations of O_3 in ambient air, if we assume that the stomatal conductance of spruce needles for water vapour shows the same diurnal fluctuations as typically observed at other sites on sunny summer days (cf. Steinbrecher 1989; Lange et al. 1989, open circles in Fig. 5.9A,B). The flux of O_3 into needles at different altitudes was estimated for two O_3 scenarios observed in July 1990: (1) a day where the ambient O_3 concentration was close to average O_3 concentrations found in the vegetation period at the three different measuring sites and (B) a day where the O_3 concentrations reached peak values of about 100 nl l^{-1} (Fig. 5.9A,B). Since measurements of stomatal conductance for water vapour were not available at different elevations at the Wank mountain, we assumed an identical stomatal behaviour for needles at valley, middle and high altitudes.

Measured maximum stomatal conductances of Norway spruce needles at high altitude in Austria showed a natural variation between between 30 and 60 mmol H_2O m^{-2} s^{-1} (Havranek et al. 1989), which is close to the figures used for the simulation in Fig. 5.9. The O_3 flux (F) was calculated according to Fick's First Law: $F = g\ (c_a - c_i)$, where g indicates the stomatal conductance for O_3 (= 0.612 $g(H_2O)$; see Runeckles 1992) and c_a and c_i are the concentrations of O_3 in air and inside the needles, re-

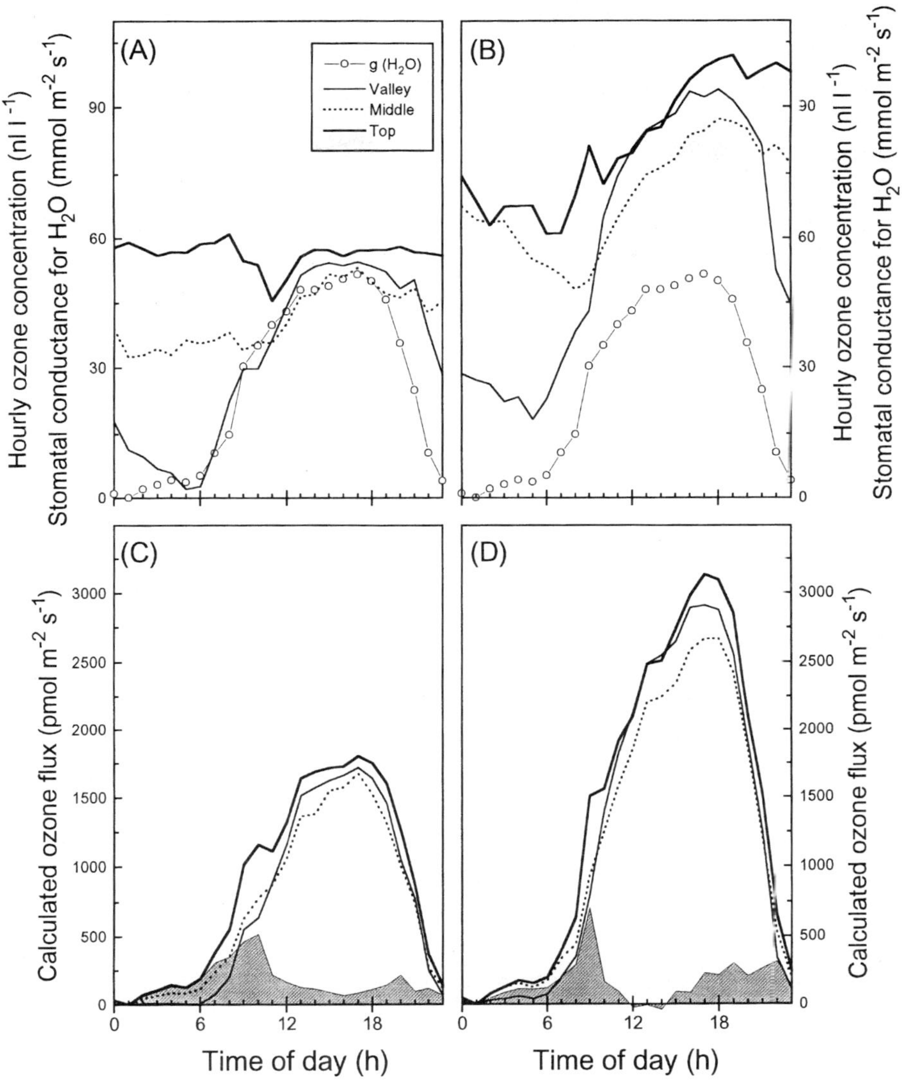

Fig. 5.9. Diurnal fluctuations of O_3 on 17 July 1990 (**A**) and 23 July 1990 (**B**) determined at meteorological stations located at valley level in Garmisch-Partenkirchen (735 m a.s.l., *thin line),* middle elevation of Wank mountain (1175 m a.s.l., *broken line*) and the summit of Wank mountain (1,782 m a.s.l., *thick line*). (Data courtesy of W. Junkermann, Fraunhofer Institut, Garmisch-Partenkirchen). Typical diurnal fluctuations of the stomatal conductance for water vapour (mmol H_2O $m^{-2}s^{-1}$) observed on sunny days in summer (o–o, after Lange et al. 1989; Steinbrecher 1989) were used to calculate the fluxes of O_3 into the foliage for a normal day (**C**) and for a day with a peak episode of O_3 (**D**). Details regarding the calculation are given in the text. The *grey underlay* shows the difference in the effective ozone doses between high and low elevations

spectively. The O_3 concentration inside the needles is assumed to be close to zero (Meyers and Baldocchi 1988; Laisk et al. 1989; Runeckles 1992; Wieser and Havranek 1993). Furthermore it was assumed that the transport of ozone to the plant surface was not hindered by gross aerodynamic and boundary layer resistances. If these assumptions are correct, the O_3 fluxes into the needles at high, middle and low altitude will not differ strongly as indicated by the three similar curves in Fig. 5.9C,D. This result may be surprising at first but is a consequence of the strong diurnal fluctuation of ozone at valley level and low fluctuations at high altitude. However, at night time the uptake of ozone is greatly reduced by low stomatal conductance.

The area under the curves in Fig. 5.9C,D indicates the effective O_3 doses for the different altitudes. The difference in the effective O_3 dose between high and low elevations obtained by this calculation was about 20% for a normal day (Fig. 5.9C, grey underlay) and 10% for a day with a peak episode of O_3 (Fig. 5.9D, grey underlay). Apparently, the effective O_3 doses to which trees were exposed at high altitude were not much higher than those experienced by trees at valley level. In line with these estimations, Knoppik et al. (1992) reported that spruce trees grown at low and middle elevations reveal no significant differences in photosynthesis, but they investigated only young needles and older needle classes suffering from loss. Simulation of the time-dependent opening of the stomata up to a maximum, which sets an upper limit to O_3 influx into the needles, reveals that significant differences in O_3 uptake occur only at night time or early in the morning, especially if the real stomatal conductances were higher during these periods than those used in the model (Fig. 5.9C,D). The simulation (Fig. 5.9) suggests that the difference in the effective O_3 doses experienced by needles of trees from different altitudes is potentially small. This might explain why typical altitude-dependent effects on biochemical parameters unequivocally attributable to O_3 were not observed. It must be noted, however, that the stress imposed by O_3 was relatively high at all sites investigated which correlated with a high degree of needles loss on all the experimental plots investigated.

5.5.2 Other Photo-oxidants

Since convincing evidence is lacking that O_3 was a primary causal agent in the site-specific forest decline phenomena at Wank mountain, the effects of other photo-oxidants have also been considered. Fels and Junkermann (1993) reported the occurrence of H_2O_2 and organic peroxides such as methylhydroperoxide (MHP) and hydroxymethyl hydroperoxide (HMHP) at Wank mountain. In summer, the gas-phase concentrations of H_2O_2 were elevated at the higher elevation compared to lower down, whereas the concentration of HMHP was enhanced at the middle compared to the higher altitude (Fels and Junkermann 1993). Polle and Junkermann (1994a) showed that spruce needles have a high intrinsic capacity for the detoxification of H_2O_2. Therefore, it is unlikely that the ambient atmospheric H_2O_2 concentrations observed at Wank mountain contribute to the damage of spruce observed. In contrast, HMHP is not detoxified but inhibits apoplastic peroxidase activities at low concentrations (Polle and Junkermann 1994b). The maximum gas-phase concentration of HMHP at the Wank mountain was 0.8 nl l^{-1} (Junkermann, pers. comm.). If the uptake of organic peroxides into spruce needles follows the same pathway as O_3, an accumulation of toxic concentrations of HMHP in the apoplastic space cannot be excluded because of its high solubility in water. In fact, significant reductions in apoplastic peroxidase activities in both mature field-grown spruce trees and in clonal spruce trees in pots were observed at the middle elevation of Wank mountain, where HMHP concentrations were greater than higher up (Polle and Junkermann 1994a; Polle et al. 1991). Atmospheric hydroperoxides are the result of gas-phase reactions of O_3 with volatile organic compounds of anthropogenic or biogenic origin (Hewitt and Kok 1991). Thus, volatile biogenic compounds such as monoterpenes or isoprene, emitted at considerable rates by spruce trees (Steinbrecher et al. 1992), may act as amplifiers in O_3-mediated injury by the production of highly toxic organic hydroperoxides (Hewitt et al. 1990).

5.6 Conclusions

In the course of the ecosystem studies at the Wank research site a massive decline of spruce was observed due to storm events. This decline mainly affected the middle altitudes of ca. 1,000 m a.s.l., apparently because of the unfavourable orographic structure and high windspeeds at this elevation. It may however be argued that spruce trees at this elevation level has already a reduced vitality and, therefore, may be more susceptible to destruction by storm events. Conventional inventories that showed enhanced injury to the trees at middle elevation only in the early surveys at Wank mountain may not be suitable because needle loss may not be an appropriate symptom of damage and only surviving trees were considered in the surveys. Therefore, the question has to be addressed as to whether the investigations on atmospheric chemistry and/or the physiological studies performed at the research site provide evidence that O_3 or other air pollutants may have contributed to a reduced vitality of trees at middle altitude.

The atmospheric concentrations of SO_2 and NO_x were generally low at the Wank mountain research site (Junkermann et al. 1992; Sladkovic et al. 1994a,b). The hypothesis that inversion layers with high atmospheric mixing ratios of photo-oxidants occur more frequently at middle altitude than at other elevation levels was not confirmed. Enhanced O_3 levels at middle elevation were not experienced. However, higher concentrations of organic peroxides that are produced in gas-phase reactions from O_3 and volatile organic compounds were found at the middle altitude than higher up (Fels and Junkermann 1993). These organic peroxides may be responsible for the biochemical irregularities in the antioxidative system of spruce needles found in the trees at the middle elevation. As these irregularities were observed with mature trees of the forest stand and also with clonal spruce trees in pots exposed at the middle elevation, genetic and soil-borne factors can be excluded (Polle and Rennenberg 1994). Thus, organic peroxides rather than O_3 or nutrient limitations may be responsible for the site-specific reduction in tree vitality at middle elevation.

Soil and needle analyses showed that the N and P supply of the spruce trees was limited and may be responsible for the generally high degree of injury at Wank mountain (Polle et al. 1992; Zech and Sandhage-Hofmann 1993). However, nutritional limitations are unlikely to account for the differences in damage between trees at valley level and those at the mountain summit. Is it, however, possible that these differences are due to the increasing levels of O_3? It is generally accepted that O_3 is primarily taken up via the open stomata and that the O_3 molecule itself will not diffuse into the interior of the cell because of its high reactivity. However, the stomatal conductance of spruce needles is much lower than that of leaves of deciduous trees or herbaceous plants. Therefore, one might expect, in line with previous observations (Reich 1987), that spruce can tolerate much higher O_3 concentrations than many other species, simply because of a lower rate of uptake. Experimental approaches indicate that O_3 itself can be detoxified by ascorbate that is present in spruce needles at high concentrations. At the Wank site, mature needles contained higher concentrations of ascorbate at higher altitudes, especially in the apoplastic space than needles from lower elevations (Polle et al. 1995). Short-term *in situ* fumigation with high O_3 concentrations did not affect photosynthesis, but significantly increased apoplastic ascorbate levels. While investigations with spinach showed a complete depletion of the apoplastic ascorbate pool by O_3 fumigation, this was not observed with spruce (Luwe et al.

1993; Polle et al. 1995). Apparently, the rate of regeneration of apoplastic ascorbate in spruce needles was high enough to maintain a sufficient detoxification potential for O_3. However, the diffusion of O_3 into the stomatal cavity has to be considered at the microscopic scale. As the free diffusion path of O_3 is not very long, O_3 may react in appreciable amounts in close proximity to the stomata. This would explain the observation that O_3 effects the regulation of the stomata and, as a consequence, can make the trees more susceptible to drought stress (Maier-Maercker and Koch 1995). Thus, despite an enhanced detoxification capacity for O_3 at high altitude, the possibility cannot be excluded that high O_3 levels at high altitude enhance the susceptibility of spruce to other stresses like drought and, hence, cause an enhanced degree of injury compared to trees growing at lower elevations. On the other hand, calculations of the effective O_3 dose indicate small differences between low and high elevations. Further experiments are required to solve this contradiction between ozone dose and degree of tree damage and to assess the importance of hydroperoxide formation.

Acknowledgements. Generous financial support of this study by Bayerisches Staatsministerium für Landesentwicklung und Umweltfragen, München and Bayerisches Staatsministerium für Ernährung, Landwirtschaft und Forsten, München, is gratefully acknowledged. The authors thank the forestry commission in Garmisch-Partenkirchen, especially Forstdirektor Dr. Gleißner, for the continuous support during this study.

References

Ammer U, Burgis M, Koch B, Martin K (1988) Untersuchungen über den Zusammenhang zwischen Schädigungsgrad und Meereshöhe im Rahmen des Schwerpunktprogramms zur Erforschung der Wechselwirkung von Klima und Waldschaden. Forstwiss Centralbl 107:145–151

Bayer OFoD (1991) Langfristige Forstbetriebsplanung (Forsteinrichtung) für das Forstamt Garmisch-Partenkirchen. München

Bayer StMELF (1982) Grundsätze für die Waldbehandlung im bayerischen Hochgebirge. München

Benes SE, Murphy TM, Anderson PD, Houpis JL (1995) Relationship of antioxidant enzymes to ozone tolerance in branches of mature ponderosa pine (*Pinus ponderosa*) trees exposed to long-term low concentration ozone fumigation and acid precipation. Physiol Plant 94:124–134

Bennett J, Lee E, Heggestad H (1984) Biochemical aspects of plant tolerance to ozone and oxyradicals: superoxide dismutase. In: Koziol M, Whatley F (eds) Gaseous air pollutants and plant metabolism. Butterworths, New York, pp 413–423

Castillo FJ, Miller PR, Greppin H (1987) 'Waldstreben': extracellular biochemical markers of photochemical oxidant air pollution damage to Norway spruce. Experientia 43:111–115

Chameides W (1989) The chemistry of ozone deposition to plant leaves: role of ascorbic acid. Environ Sci Technol 23:595–600

Fabian P (1990) Luftschadstoff-Belastung in den nördlichen Kalkalpen. In: Kommission für Ökologie der Bayerischen Akademie der Wissenschaften (Hrsg) Zustand und Gefährdung des Bergwaldes. Parey, Hamburg, pp 80–86

Fels M, Junkermann W (1993) The occurrence of organic peroxides in air at a mountain site. Geophys Res Lett 21:341–344

Foyer C, Descourvieres P, Kunert KJ (1994) Protection against oxygen radicals: an important defence mechanism studied in transgenic plants. Plant Cell Environ 17:507–523

Gabler K, Smidt S, Puxbaum H (1990) Ozone profiles in alpine valleys. Poster presented at Euroanalysis VII 1990, Vienna

Grimes H, Perkin K, Boss W (1983) Ozone degrades into hydroxyradicals under physiological conditions. Plant Physiol 72:1016–1020

Hausladen A, Madamachani N, Fellows S, Alscher R, Amundson R (1990) Seasonal changes in antioxidants in red spruce as affected by ozone. New Phytol 115:447–456

Havranek WM, Wieser G, Bodner M (1989) Ozone fumigation of Norway spruce at timberline. Ann Sci For 46 (Suppl):581–585

Heller W, Rosemann D, Sandermann H (1994) Untersuchung sekundärer Stoffwechselleistung von Fichte und Kiefer als möglicher Indikator für Schadeffekte im Höhenprofil des Wank. Abschlußbericht (PBWU, Hrsg). München. 109 pp

Hewitt CN, Kok GL (1991) Formation and occurrence of organic hydroperoxides in the troposphere: laboratory and field observations. J Atmos Chem 12:181–194

Hewitt CN, Kok G, Fall R (1990) Hydroperoxides in plants exposed to ozone mediate air pollution damage to alkene emitters. Nature 344:56–58

Hippeli S, Osswald W, Elstner EF (1991) Veränderung der Konzentration präformierter Abwehrtoxine gegenüber mikrobiellen Schaderregern in Koniferennadeln. Abschlußbericht (PBWU, Hrsg). München, 59pp

Hock B (1990) Einfluß von Luftschandstoffen und weiteren Streßfaktoren auf Phytohormonkonzentrationen in Bäumen. Abschlußbericht (PBWU, Hrsg). München, 95pp

Hoigne J, Bader H (1975) Ozonation of water: role of hydroxyl radicals as oxidixing intermediates. Science 190:782–784

Junkermann W, Langer E (1992) Betrieb einer Basisstation für die Waldschadensforschung am Forschungsschwerpunkt Kalkalpen. Abschlußbericht (PBWU, Hrsg). München, 37pp

Junkermann W, Langer E, Zardini D, Tranos S (1992) Untersuchungen luftgetragener Spurensubstanzen und photochemischer Sekundärprodukte am Schwerpunktstandort Kalkalpen. Abschlußbericht (PBWU, Hrsg). München, 95pp

Kangasjärvi J, Talvinen J, Utriainen M, Karjalainen R (1994) Plant defense systems induced by ozone: a review. Plant Cell Environ 17:783–794

Kennel E (1990) Ubersicht über neuartige Schäden im Bergwald und zeitliche Entwicklung. In: Kommission für Okologie der Bayerischen Akademie der Wissenschaften (Hrsg) Zustand und Gefährdung des Bergwaldes. Parey, Hamburg, pp 66–79

Kerstiens G, Lendzian K (1989) Interactions between ozone and plant cuticles. I. Ozone deposition and permeability. New Phytol 112:13–19

Kirchner M, Mößmer R, Reuther M (1992) Vertikale Verteilung der Waldschäden in den Alpen. Allg Forstztg 24:1283–1287

Knoppik D, Dunkl M, Knapp W, Müller T, Schinzel M (1992) Gaswechselmessungen als Methode zur Früherkennung von Waldschäden im Bergwald (Garmisch-Partenkirchen). Abschlußbericht (PBWU, Hrsg). München, 87 pp

Kreutzer K, Pröbstle P (1991) Einfluß von saurer Beregnung und Kalkung auf Ernährungszustand und Streufall von Altfichten. Forstwiss Forsch 39:35–40

Kröniger W (1993) Biochemische, physiologische und ökophysiologische Untersuchungen der Superoxiddismutasen in Fichten (*Picea abies,* L., Karst.). Thesis, TU München

Laisk A, Kull O, Moldau H (1989) Ozone concentrations in leaf intercellular air spaces is close to zero. Plant Physiol 90:1163–1167

Lange OL, Heber U, Schulze E-D, Ziegler H (1989) Atmospheric pollutants and plant metabolism. In: Schulze E-D, Lange O, Oren R (eds) Forest decline and air pollution. Ecological Studies, vol 77. Springer, Berlin Heidelberg New York, pp 238–273

Linder S (1989) Nutritional control of forest yield. Marcus Wallenberg Foundation Symp Proc 6: 62–87

Liu JC, Payer HD (1992) Untersuchungen zur Wasser- und Stoffdynamik der Böden am Wank bei Garmisch-Partenkirchen. Abschlußbericht (PBWU, Hrsg), München, 189 pp

Liu JC, Pfirrmann T, Payer HD, Runkel KH, Schramel P (1991) Wasser- und Stoffdynamik der Böden am Wank und ihr möglicher Einfluß auf die Fichten. In: 2. Statusseminar der PBWU zum Forschungsschwerpunkt "Waldschäden", GSF Bericht 26/91, Neuherberg, pp 237–251

Liu JC, Keller T, Runkel KH, Payer HD (1994) Bodenkundliche Untersuchungen zu Ursachen des Nadelverlusts der Fichten (*Picea abies* (L.) Karst.) auf Kalkstandorten der Alpen. Forstwiss Centralbl 113:86–100

Lütz C (1992) Einfluß von Höhenlage, Witterung und Jahreszeit auf das Pigmentmuster von Nadeln ausgewählter Fichten am Wank. Abschlußbericht (PBWU, Hrsg). München, 24 pp

Luwe M, Takahama U, Heber U (1993) Role of ascorbate in detoxifying ozone in the apoplast of spinach (*Spinacia oleracea* L.) leaves. Plant Physiol 101:969–976

Maier-Maercker U (1989) Delignification of subsidiary and guard cell walls of *Picea abies* (L.) Karst. by fumigation with ozone. Trees 3:57–64

Maier-Maercker U, Koch W (1991) Experiments on the control capacity of stomata of *Picea abies* (L.) Karst. after fumigation with ozone and in environmentally damaged material. Plant Cell Environ 14:175–184

Maier-Maercker U, Koch W (1995) Poor stomatal control of water balance and the abscission of green needles from a declining stand or spruce trees [*Picea abies* (L.) Karst.] from the northern Alps. Trees 10:63–73

Menser A (1964) Response of plants to air pollutants: III. A relation between ascorbic acid levels and ozone susceptibility of light pre-conditioned tobacco plants. Plant Physiol 39:564–567

Meyers TP, Baldocchi DD (1988) A comparison of models for driving dry deposition fluxes of O_3 and SO_2 to a forest canopy. Tellus 40B:270–284

Mößmer R (1986) Verteilung der neuartigen Waldschäden an der Fichte nach Bestandes- und Standortsmerkmalen in den Bayerischen Alpen. Schriftenreihe der Forstwissenschaftlichen Fakultät der Universität München und der Bayer. Forstlichen Versuchs- und Forschungsanstalt 73/1986

Mößmer R, Tränkner H, Schuster M (1991) Waldschadenserhebung am Schwerpunktstandort Wank in den Jahren 1987 bis 1990. In: 1. Statusseminar der PBWU zum Forschungsschwerpunkt "Waldschäden". GSF-Bericht 26/91, Neuherberg, pp 65–76

Mößmer R, Büchler-Kleemeier B, Bäumler A, Greune A, Martin K (1993) Höhenabhängigkeit des Kronenzustands der Fichten am Wank untersucht an Hand von

Farbinfrarot-Luftbildern 1986 und 1989. Abschlußbericht (PBWU, Hrsg). München, 36 pp

Mößnang M, von Schönborn A (1992) Elementgehalte von Fichten entlang eines Höhenprofils im bayerischen Alpenraum. Abschlußbericht (PBWU, Hrsg). München, 96 pp

Nast W, Mortenson L, Fischer K, Fitting I (1993) Effects of air pollutants on the growth and antioxidative system of Norway spruce exposed in open top chambers. Environ Pollut 80:85–90

Paffrath D, Peters W (1988) Betrachtung der Ozonvertikalverteilung im Zusammenhang mit den neuartigen Waldschäden. Forstwiss Centralbl 107:152–159

PBWU (1987) Die Rolle von Photooxidantien und anderer Spurensubstanzen in der Atmosphäre in Zusammenhang mit den neuartigen Waldschaden. GSF-Bericht 26/87, Neuherberg

Polle A, Junkermann W (1994a) Does atmospheric hydrogen peroxide contribute to damage to forest trees? Environ Sci Technol 28:812–815

Polle A, Junkermann W (1994b) Inhibition of apoplastic and symplastic peroxidase activity from Norway spruce by the photo-oxidant hydroxymethyl hydroperoxide. Plant Physiol 104:617–621

Polle A, Rennenberg H (1991) Superoxide dismutase activity in needles of Norway spruce (*Picea abies* L. Karst.) and in Scots pine (*Pinus sylvestris*, L.) under field and chamber conditions: lack of ozone effects. New Phytol 117:335–343

Polle A, Rennenberg H (1992) Field studies on Norway spruce trees at high altitudes. II. Defence systems against oxidative stress in needles. New Phytol 121:635–642

Polle A, Rennenberg H (1993) The significance of antioxidants in plant adaptation to environmental stress. In: Mansfield T, Fowden L, Stoddard F (eds) Plant adaptation to environmental stress. James and James, London, pp 263–273

Polle A, Rennenberg H (1994) Entgiftung von Peroxiden in Fichtennadeln (*Picea abies* L.) am Schwerpunktstandort "Kalkalpen (Wank)" IFU-Schriftenreihe Bd 31. Dr Maraun Wissenschafts-Verlag, Frankfurt, pp 41–63

Polle A, Chakrabarti K, Schürmann W, Rennenberg H (1990) Composition and properties of hydrogen peroxide decomposing systems in extracellular and total extracts from needles of Norway spruce (*Picea abies* L, Karst.). Plant Physiol 94:312–319

Polle A, Chakrabarti K, Rennenberg H (1991) Entgiftung von Peroxiden in Fichtennadeln (*Picea abies* L.) am Schwerpunktstandort Kalkalpen (Wank). In: 2. Statusseminar der PBWU zum Forschungsschwerpunkt "Waldschäden", GSF-Bericht 26/91, Neuherberg, pp 151–161

Polle A, Mössnang M, von Schönborn A, Sladkovic R, Rennenberg H (1992) Field studies on Norway spruce trees at high altitudes. I. Mineral, pigment and soluble protein contents of needles as affected by climate and pollution. New Phytol 121:89–99

Polle A, Pfirrmann T, Chakrabarti S, Rennenberg H (1993) The effects of enhanced ozone and enhanced carbon dioxide concentrations on biomass, pigments and antioxidative enzymes in spruce needles (*Picea abies* L.). Plant Cell Environ 16:311–316

Polle A, Wieser G, Havranek M (1995) Quantification of ozone uptake and apoplastic ascorbate in spruce needles (*Picea abies* L. Karst.) at high altitude. Plant Cell Environ 18:681–688

Reich PB (1987) Quantifying plant response to ozone: a unifying theory. Tree Physiol 3:63–91

Reiter R, Kanter HJ (1972) Das Klima am Wank. Physikalisch-Bioklimatische Forschungsstelle Garmisch-Partenkirchen. Wissenschaftliche Mitteilung Nr 7, 1972

Rennenberg H (1989) Konzept der Forschungsarbeiten am Schwerpunktstandort Kalkalpen. In: 1. Statusseminar der PBWU zum Forschungsschwerpunkt "Waldschäden". GSF-Bericht 6/89, Neuherberg, pp 351–356

Rennenberg H, Polle A (1994) Protection from oxidative stress in transgenic plants. Biochem Soc Trans 22:936–940

Rennenberg H, Reuther M (1991) Eintrag, Umwandlung und Wirkung von Luftschadstoffen in Bergwaldökosystemen der Alpenregion. Forschungskonzept des Schwerpunktstandorts Kalkalpen (Wank/Kramer bei Garmisch-Partenkirchen) der PBWU. In: 2. Statusseminar der PBWU zum Forschungsschwerpunkt "Waldschäden". GSF-Bericht 26/91, Neuherberg, pp 15–40

Runeckles V (1992) Uptake of ozone by vegetation. In: Lefohn A (ed) Surface-level ozone exposures and their effects on vegetation. Lewis, Boca Raton, pp 157–188

Sandermann H, Langebartels C, Heller W (1990) Ozonstreß bei Pflanzen. Frühe und Memory-Effekte von Ozon bei Nadelbäumen. Z Umweltchem Ökotox 2: 14–15

Sandhage-Hofmann A, Zech W (1993) Dynamik und Elementgehalte von Fichtenfeinwurzeln in Kalkgesteinsböden am Wank (Bayerische Kalkalpen). Z Pflanzenernaehr Bodenkd 156: 181–190

Schneider J, Loibl W (1996) Kumulative Ozonbelastung der Vegetation in Österreich. Umweltbundesamt Wien, UBA Bericht 96–127, 72 pp

Schröder P, Berkau C, Nathaus FJ, Weiß A, Wolf A (1993) Untersuchungen zur Aufnahme anthropogen emittierter Chlorkohlenwasserstoffe und deren Entgiftung durch Fichten vom Schwerpunktstandort Kalkalpen. Abschlußbericht (PBWU Hrsg), München, 68 pp

Schwanz P, Häberle KH, Polle A (1996) Interactive effects of elevated CO_2, ozone and drought stress on the activities of antioxidative enzymes in needles of Norway spruce trees (*Picea abies* L. Karsten) grown with luxurious N-supply. J Plant Physiol 148:351–355

Seiler W (1988) Die Waldschäden aus der Sicht der Luftchemie. In: Verteilung und Wirkung von Photooxidantien im Alpenraum, GSF-Bericht 17/88, pp 48–75

Seiler W (1991) Zeitlicher Trend des troposphärischen Ozons im Bereich der Bayerischen Kalkalpen. In: 2. Statusseminar der PBWU zum Forschungsschwerpunkt "Waldschäden". GSF-Bericht 26/91, Neuherberg, pp 41–63

Sinner HU, Schuster M (1993) Waldzustandserhebung 1992 auf den Probeflächen des Höhenprofils am Wank-terrestrische Schadansprache. Abschlußbericht (PBWU Hrsg), München

Skärby L, Wallin G, Sellden G, Karlsson PE, Ottosson S, Sutinen S, Grennfelt P (1995) Tropospheric ozone – a stress factor for Norway spruce in Sweden. Ecol Bull 44:133–146

Sladkovic R, Scheel HE, Junkermann W (1994a) Betrieb einer Basisstation für die Waldschadensforschung am Forschungsschwerpunkt Kalkalpen. Abschlußbericht (PBWU Hrsg), München, 65 pp

Sladkovic R, Scheel HE, Seiler W (1994b) Ozone climatology at the mountain sites Wank and Zugspitze. In: Borell P (ed) Proc of EUROTRAC Symp 1994. SPB Acad Publ, The Hague

Slovik S, Heber U, Kaiser W, Kindermann G, Körner C (1992) Quantifizierung der physiologischen Kausalkette von SO_2-Immissionsschäden: II Ableitung von SO_2 Immissionsgrenzwerten für chronische Schäden an der Fichte. Allg Forstztg 17:913–920

Smidt S, Gabler K, Puxbaum H (1990) Die zeitliche und vertikale Zunahme der Ozonkonzentrationen. Österr Forstzg 7/1990. Forstliche Bundesversuchsanstalt, Informationsdienst 247. Folge, Vienna

Smidt S, Gabler K, Halbwachs E (1991) Beurteilung österreichischer Ozonmeßdaten in Hinblick auf wirkungsbezogene Grenzwerte. Staub-Reinhaltung der Luft 51: 43–49

Steinbrecher R (1989) Gehalt und Emission von Monterpenen in oberirdischen Organen von *Picea abies* (L.) Karst. Thesis, Technische Universität, München

Steinbrecher R, Karbach I, Mits S, Ziegler H (1992) Isoprene emission from Norway spruce (*Picea abies* L.) and behaviour in the atmosphere of a major spruce stand. In: Borrell PM, Borrell P, Cvitas T, Seiler W (eds) Proc of EUROTRAC Symp 1992. SPB Acad Publ, The Hague, pp 286–288

Tanaka K (1994) Tolerance to herbicides and air pollutants. In: Foyer C, Mullineaux P (eds) Causes of photooxidative stress and amelioration of defense systems in plants. CRC Press, Boca Raton, pp 365–378

UN-ECE Workshop Report (1994) Critical levels for ozone. In: Fuhrer J, Achermann B (eds) Schriftenreihe der FAC, Nr 16 Liebefeld-Bern, 328 pp

Urbach W, Schmidt W, Kolbowski J, Rühmele S, Reisberg E, Steigner W, Schreiber U (1989) Wirkung von Umweltschadstoffen auf Photosynthese und Zellmembranen von Pflanzen. In: 1. Statusseminar der PBWU zum Forschungsschwerpunkt "Waldschäden". GSF Bericht 6/89, Neuherberg, pp 195–206

von Schönborn A, Mößnang M (1991) Elementgehalte von Fichten entlang eines Höhenprofils im bayerischen Alpenraum. In: 2. Statusseminar der PBWU zum Forschungsschwerpunkt "Waldschäden", GSF-Bericht 26/91, Neuherberg, pp 225–235

Weißgerber K (1987) Zur Schadstoffbelastung im bayerisch österreichischen Nordalpenraum. Mitt Geogr Ges München 72:55–69

Wieser G, Havranek W (1993) Ozone uptake in the sun and shade crown of spruce: quantifying the physiological effects of ozone exposure. Trees 7:227–232

Wieser G, Havranek W (1996) Evaluation of ozone impact on mature spruce and larch in the field. J Plant Physiol 148:189–194

Wingsle G, Mattson A, Ekblad A, Hällgren JE, Selstam E (1992) Activities of glutathione reductase and superoxide dismutase in relation to changes of lipids and pigments due to ozone in seedlings of *Pinus sylvestris* (L.). Plant Sci 82:167–178

Zech W, Sandhage-Hofmann A (1993) Wachstum und Nährstoffversorgung von Feinwurzeln unterschiedlich geschädigter Fichten auf Böden aus Kalkgestein (Wank-Massiv). Abschlußbericht (PBWU, Hrsg), München, 242 pp

Ziegler H, Kraus M, Deigele C (1992) Waldschäden: Nachweis von Phytohormonen und sekundären Pflanzenstoffen in gesunden und geschädigten Nadeln vom natürlichen Standort (Wank), bei definierter Schadstoffbelastung und definierten Infektionskrankeiten. Abschlußbericht (PBWU, Hrsg), München, 168 pp

6 Ozone Responses of Trees: Results from Controlled Chamber Exposures at the GSF Phytotron

C. Langebartels, D. Ernst, W. Heller, C. Lütz, H.-D. Payer and H. Sandermann Jr.

6.1 Introduction

Tree damage in the field is judged primarily by visual inspection (Forschungsbeirat Waldschäden 1989; Chappelka and Chevone 1992) using needle and leaf discoloration, premature abscission and crown transparency as the principle and rather nonspecific symptoms. A diagnostic field guide is available for distinguishing "novel" decline symptoms from known mineral deficiencies and frost or insect damage (Hartmann et al. 1988).

Norway spruce (*Picea abies* [L.] Karst.) is the most important conifer in Germany covering about 40% of the forested area (Forschungsbeirat Waldschäden 1989). Premature fall of older needles is the predominant symptom of decline while yellowing of older needles is a minor but characteristic feature of damage on certain field sites. This symptom of montane yellowing is restricted to the upper side of the needles which are directly exposed to sunlight (Hartmann et al. 1988) and has been thoroughly studied in terms of Mg^{2+} deficiency and nutritional imbalance (Schulze et al. 1989). In Scots pine (*Pinus sylvestris* L.), the second most important conifer species (20% of the forested area), chlorosis and premature needle loss of 2- and 3-year-old needles occur. Both European beech (*Fagus sylvatica* L., 20%) and English oak (*Quercus robur* L., 5%), two major European hardwood species, show yellowing and necrotic lesions, leaf loss, and alterations in crown structure and have a high incidence of biotic diseases.

In most species, visible damage involves a loss of chloroplast pigments and, as a result, reduced photosynthesis. Thus, any causal factor for the observed changes ultimately affects the chloroplast although this organelle is particularly well protected against oxidative stress (Elstner and Osswald 1994; Foyer et al. 1994). The mechanisms which are involved in pigment degradation are similar in various stress situations and in their light requirement (Elstner et al. 1985; Elstner and Osswald 1994). Mineral

Ecological Studies, Vol. 127
Sandermann et al. (eds) Forest Decline and Ozone

deficiencies, air pollutants, soil acidification and other climatic and biotic stresses may all produce these unspecific symptoms (Schulze et al. 1989; Matyssek et al. 1995). The exact mechanism of premature needle abscission, which is the main damage symptom of Norway spruce, is still unknown. In comparison to spruce, only few studies exist on needle and leaf symptom development in Scots pine and European beech.

Over the past 10 years, studies in the closed environment chamber system at the GSF-Forschungszentrum, Neuherberg (Payer et al. 1990), have concentrated on O_3 and acid mist effects on trees and crops. Research in this phytotron was undertaken by a large working team (Münchner Arbeitsgemeinschaft Luftschadstoffe, MAGL) which included scientists from both universities in Munich and from the universities of Bayreuth, Darmstadt, Dresden (Tharandt), Freiburg, Karlsruhe, and Würzburg as well as the GSF itself. In particular, studies have been aimed at determining biomarkers for O_3 and understanding the mechanisms of injury, tolerance and acclimation.

The O_3 concentrations and doses used in the chamber studies ranged from subambient to acutely phytotoxic. Total O_3 doses (SUMO) were between 2 and 10 $\mu l\ l^{-1} h$ (exposure for days to weeks) and between 5 and 380 $\mu l\ l^{-1}$ h (season-long exposure). The latter values are well within the range of typical cumulative doses on European and American field sites (150 to 350 $\mu l\ l^{-1}$ h per growing season; Lefohn 1992). AOT40 values, i.e., the accumulated hourly exposure over a threshold of 40 nl l^{-1} (UN-ECE 1994), were between 5 and 80 $\mu l\ l^{-1}$ h for experiments with deciduous trees and between 10 and 220 $\mu l\ l^{-1}$ h with conifers. Episodes with such high O_3 loads are rare in Europe (Grennfelt and Beck 1994). The exposure profiles represented 1.5 times to three times the level of ambient O_3.

At low O_3 levels, trees may act as a sink for O_3 due to antioxidative processes. Chronic O_3 stress, however, may cause reductions in photosynthesis and plant growth without visible, external symptoms (Guderian 1985; Reich 1987). Prolonged O_3 stress can also lead to impaired photosynthesis and carbon allocation, as well as to predisposition to insect or microbial attack (Heath 1988; Sandermann et al. 1989a, 1994; Chappelka and Chevone 1992). Typically, sub-acute O_3 exposure of conifers causes visible symptoms on older needles whereas high (acute) levels preferentially lead to visible symptoms in current-year needles (Hartmann et al. 1988; Miller 1993).

Biochemical and physiological responses to individual stress parameters are difficult or even impossible to detect in adult trees in the field. Reasons for this are genotypic variations and complex interactions between various stress factors occurring at the same time. A wide array of early biochemical responses to O_3 has, however, been detected in controlled chamber studies using the GSF phytotron (Fig. 6.1; see Appendix). It was recently established that these O_3 responses in trees and crops are surprisingly similar to those known for plant-pathogen interactions (Sandermann 1996).

6.2 Chamber Design and Performance

Indoor systems used to study the phytotoxic potential of air pollutants range from small cuvettes up to complex exposure chambers which allow realistic environmental simulations of both diurnal and seasonal changes during exposures to various pollutants (Payer et al. 1990, 1994; Thiel et al. 1996; see also Chap. 9, this Vol., Musselman and Hale). Alternatively, studies under semi-natural conditions are performed in open top chambers (OTCs) which are well suited to determine yield and biomass changes (Guderian 1985; Manning and Krupa 1992) but have limited or no control over light, humidity and temperature.

The GSF closed chamber system consists of a set of four walk-in environmental chambers, each of which can be subdivided into four subchambers and root compartments. Fig. 6.2 shows a schematic view of one of these chambers with the plant cabinet (9 m^2) at the bottom and the lamp cabinet on top. The chambers were constructed in 1985 from noncorrosive materials, e.g. stainless steel, glass and perspex. In 1992, a newly developed light system was installed which covers the visible and UV-B range of the solar spectrum with high precision (Seckmeyer and Payer 1993; Thiel et al. 1996). Glass and water filters absorb excess ultraviolet-C and infrared (IR) irradiances (Fig. 6.2; Doehring et al. 1996). Parameters such as light, air and soil temperature, air humidity, soil moisture, wind and dosing of mists or rain, irrigation and exposure to air pollutants are controlled dynamically and individually.

As shown at the top of Fig. 6.2, four types of lamps [metal halide (HQI), quartz-halogen (HAL) and blue fluorescent (TLD); 238 lamps in total], mounted at the top of the lamp cabinet, are used for the simulation of the visible and far-red components of the sunlight spectrum (Seckmeyer and Payer 1990; Thiel et al. 1996). In addition, 248 UV-B fluorescent lamps are mounted on the four walls of the lamp cabinet. The lamps at the ceiling are separated from the remaining system by a 2-cm layer of deionized water contained in a borosilicate glass basin. This constantly flowing water represents the water vapour of the atmosphere and absorbs most of the IR radiation. The lamp and plant cabinets are separated by two layers of borosilicate glass that absorb UV-C radiation and reproduce the UV

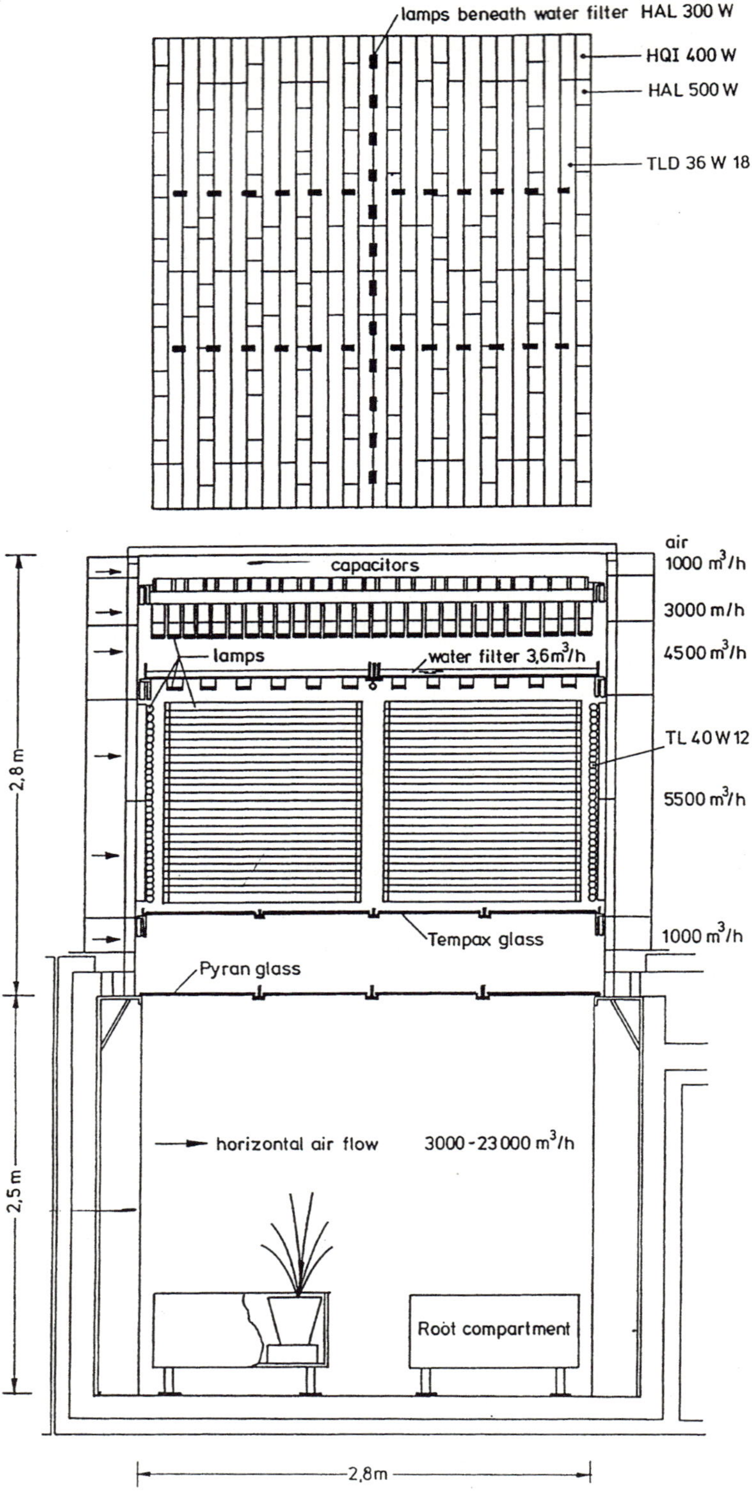
lamps beneath water filter HAL 300 W
HQI 400 W
HAL 500 W
TLD 36 W 18
air
1000 m³/h
capacitors
3000 m/h
lamps
water filter 3,6 m³/h
4500 m³/h
TL 40 W 12
5500 m³/h
Tempax glass
1000 m³/h
Pyran glass
2,8 m
horizontal air flow
3000 - 23000 m³/h
2,5 m
Root compartment
2,8 m

absorption of the atmospheric O_3 column (Payer et al. 1994). Different "cut-off" wavelengths in the UV region can be obtained by the use of various glass materials (Doehring et al. 1996). Visible and UV-B radiation are continuously recorded by pyranometers, photocells and Robertson-Berger meters which are regularly compared to spectral measurements by a double monochromator spectroradiometer (Bentham 300, UK; Seckmeyer and Payer 1990). The spectral characterization of this lighting system shows that it adequately simulates the solar spectrum (Thiel et al. 1996). UV-C radiation is below the detection limit of 10^{-6} W m^{-2} of the Bentham spectroradiometer while UV-B radiation (280–320 nm) is varied in the range of 0.5 to 6 W m^{-2} (Payer et al. 1994; Thiel et al. 1996). Total irradiance and photosynthetically active radiation ($PAR_{400-700\ nm}$) at plant canopy level can reach up to 800 W m^{-2} or 1500 μmol m^{-2} s^{-1}, respectively.

Clean air enriched with O_3 or other trace gases was used in all experiments. To this end, ambient air is first enriched with O_3 to 0.3 $\mu l\ l^{-1}$ in order to oxidize NO to NO_2. The latter compound as well as dust, SO_2 and O_3 are then absorbed by a series of filters consisting of activated carbon, permanganate-coated aluminium oxide (Purafil) and particle filters. This system is more than 99% effective for SO_2 and O_3 and removes more than 90% of NO_x (Roberts et al. 1988). Ozone is generated by silent electrical discharge in pure oxygen (Payer et al. 1986). The concentrated pollutant gases are then injected by computer- or timer-controlled mass-flow meters through Teflon lines into the incoming stream of filtered air. Air from the exposure chambers is sampled with Teflon lines at regular intervals and analyzed for O_3, SO_2, NH_3, NO, NO_2, hydrocarbons and CO_2 as described by Payer et al. (1986, 1994). The efficiency of this dynamic gas exposure system is demonstrated in Fig. 6.3 in the context of a dose-response study with European beech (Lippert et al. 1996a). In this experiment, diurnal patterns (2-h means) of O_3 levels and climate parameters from a station at Schönenbuch, Switzerland, were simulated for a total growing season. O_3 levels at Schönenbuch (ambient) were followed with high accuracy, and, in addition, increased (to 1.5 times or twice the ambient level) or decreased (to 0.15 times the ambient level) concentrations were applied.

◀

Fig. 6.2. Schematic view of one of the closed chambers of the GSF phytotron. *Top* Combination of lamp types at the top of the lamp cabinet and below the water filter. *Bottom* View of the total chamber with lamps at the top, water filter, UV-B lamps at the walls, glass filters and plant cabinet (from the *top*). Ventilation rates are indicated for each level. (Payer et al. 1994)

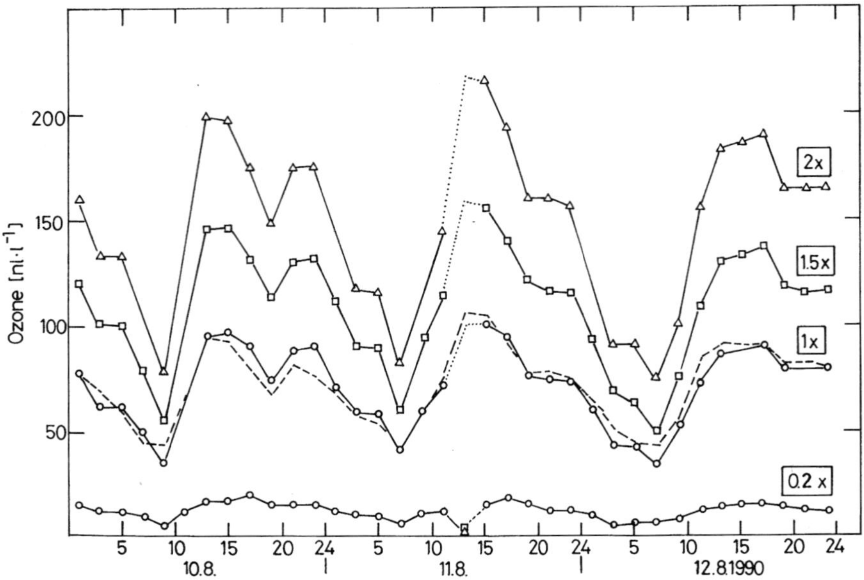

Fig. 6.3. Simulation of ozone values measured in August 1990 in Schönenbuch near Basel in the closed chamber system. 2-h means at the field station (*dotted line*) were simulated 12 days later in the closed chamber (ambient level). Additionally, ozone levels were decreased (*0.2x*) or increased (*1.5x, 2x*) in parallel treatments. Ozone exposure was interrupted on 11 August for 3 h to perform photosynthetic measurements

The GSF chambers provide a wide range of climatic conditions (Payer et al. 1994). Temperatures between −20 and +40 °C can be applied. This is especially valuable for multi-season exposure of perennial plants. For example, winter and summer maxima of -14 and +26 °C, respectively, were required for a 14-month exposure experiment with Norway spruce (Blank and Lütz 1990). The simulated temperatures usually meet the set values within a range of ± 1.5 K and parallel chambers are operated at practically identical temperatures (± 1 K).

During 10 years of operation, the GSF phytotron has been used in more than 60 individual studies which were performed in collaboration with various research groups in Germany and in other European countries. About 60% of these studies were devoted to O_3. Besides dose-effect experiments with O_3 alone, the following interactions have been studied: O_3/acidic mist, O_3/CO_2, O_3/drought, O_3/mineral deficiencies and O_3/fungal infection. More than 50% of the experiments have been performed with trees. Both short-term experiments of a few days and long-term exposures of up to two growing seasons were undertaken as described in the following sections.

6.3 Selected Exposure Experiments

6.3.1 Interactions Between Ozone and Acidic Mist

Needle loss and yellowing of Norway spruce in the Bavarian forest was thought to be due to mineral deficiencies caused by cuticular leaching through the action of O_3 and acidic mist (Bosch et al. 1983). This scenario was tested in a long-term experiment that lasted over two growing seasons involving five spruce clones that were 3 years old at the start of the experiment. Ozone (50 nl l^{-1}, with peaks of 65–180 nl l^{-1}) and acid mists (pH 3.0) were compared with control trees grown in the presence of 25 nl l^{-1} O_3 and a mist of pH 5.6. Two types of soil were used which were of acidic or calcareous nature. The results have been described in detail elsewhere (Blank and Lütz 1990) and are briefly summarized below.

Biometric effects were determined mainly by clone and soil but no clear O_3 effect was observed. Leaching during acidic mist episodes removed only small amounts of cations from the trees and was not considered to cause a significant disturbance of the nutrient balance of young spruce trees. As in a preceding long-term study (Rehfuess and Bosch 1986), the hypothesis that O_3 and acidic mist could cause damage through cuticular leaching could not be confirmed.

The winter period in this experiment also allowed the testing of the frost hardiness of the needles but the O_3 episodes neither changed the timing of development of frost hardiness nor the resistance during the dehardening phase (Senser 1990). Photosynthetic parameters, however, were influenced by the soil, provenance and needle age (Führer et al. 1990). Gas exchange measurements showed a significant depression of CO_2 fixation in 1-year-old needles of one clone after 12 months of O_3 exposure. This effect appeared to be triggered at the time of the new flush. Soil, clone and needle age all affected the gas exchange parameters. In the O_3-sensitive clone, levels of photosynthetic pigments were reduced by nearly 50%. A second clone also showed reduced pigment levels, but without a clear depression of gas exchange. Violaxanthin content increased due to O_3 exposure while lutein and β-carotene levels underwent only minor changes (Lütz et al. 1990).

Analysis of the epicuticular wax components revealed that only nonacosanol (C_{29}-ol) increased as a consequence of O_3 treatment (Lütz et al. 1990; see also Chap. 7, this Vol., Wellburn et al.). However, this change did not affect the ultrastructure of the wax layer. Ozone-induced starch accumulation in the older needles of two clones was found in both mesophyll and vascular strand cells, while granal stacking was clearly reduced in all clones in both needle age groups in O_3-exposed trees. In pulse-labeling experiments, radioactivity was incorporated preferentially into two soluble proteins of 16 and 32 kDa after O_3 exposure indicating an enhanced synthesis of defined proteins (Schmitt and Sandermann 1990). A similar result was obtained in a subsequent experiment on the D-1 thylakoid protein (Lütz et al. 1992).

It was concluded that O_3/acid mist interactions are not major factors which could explain the decline of Norway spruce on acidic field sites at higher altitudes of the Bavarian forest and similar forest areas (Blank and Lütz 1990).

6.3.2 Interactions Between Ozone and Carbon Dioxide

Interactive responses of Norway spruce (two clones) to different levels of CO_2, O_3, H_2O and K nutrition were studied during one growing season (Barnes et al. 1995). Half of the plants had been preconditioned for 2 years for K deficiency. At normal levels of CO_2, 1-year-old needles of one clone showed a 30% reduction in net photosynthesis following O_3 treatment (80 nl l^{-1}). In the case of high CO_2 (750 $\mu l\ l^{-1}$) and high CO_2 plus high O_3 (80 nl l^{-1}) stomatal conductance was restricted while high O_3 alone did not affect this parameter. The second clone was less sensitive to O_3 although in both clones chlorophyll was found to be reduced after 6 months (Steiger 1990).

The new flush which developed during exposure had reduced photosynthetic capacity in both the -K and +CO_2 treatments. Only high O_3 plus high CO_2 exposure resulted in reductions of photosynthesis and pigments together with a loss of chloroplast fine structure, and there was no indication that high CO_2 protected against O_3 damage (Barnes et al. 1995). A detailed analysis of the photosynthetic apparatus identified O_3 sensitivities in the D-1 protein content as well as in electron transport (Lütz et al. 1992). Ozone strongly enhanced the D-1 content in previous-year needles, and potassium deficiency enhanced levels of D-1 to the same extent. This means that without a knowledge of the nutritional status of the plant, the relationship between the amounts of D-1 protein and O_3 exposure is unknown (Lütz et al. 1992).

The individual and combined effects of O_3 (80 nl l^{-1}), CO_2 (400 and 600 $\mu l\ l^{-1}$), additional N supply and water stress were assessed in 5-year-old spruce trees using a factorial design (Lippert et al. 1996b). The O_3 treatment significantly reduced the biomass of the 1-year-old needles only, and interactions between CO_2 and N supply were observed. Gas exchange studies showed that reductions had occurred in photosynthesis and respiration due to O_3. This held also for the O_3/CO_2 and O_3/drought interactions. The rates of carboxylation were reduced by O_3 as well as by high CO_2. Ozone exposure also diminished water use efficiency while CO_2 showed no effect on stomatal conductance. As in previous experiments, nutrient imbalances (mainly N) changed growth and metabolism more efficiently than exposure to O_3. Photosynthesis of the current-year needles was much less affected by O_3 stress than that of the 1-year-old needles. Late in the year, about 10% of the trees showed banding patterns only on the previous-year needles as a consequence of O_3 treatment (see Sect. 6.6). High N supply partially reduced this effect (Lippert et al. 1996b).

6.3.3 Effects of Mixtures of Air Pollutants

Air pollution in the field typically consists of mixtures of trace gases such as O_3, SO_2 and NO_2. Therefore, several experiments in the GSF phytotron were performed to study possible effects of pollutant mixtures on spruce trees. Concentrations of air pollutants ranged from control and near-ambient field levels up to high, phytotoxic levels. Cytological studies showed that high concentrations (about five times elevated) of mixtures of $O_3 + NO_2$ or $O_3 + NO_2 + SO_2$ (exposure over 16 days, 8 h day^{-1}) yielded a slightly increased starch content of mesophyll cells but the overall appearance of the cells and especially chloroplasts was not changed significantly (Schiffgens-Gruber and Lütz 1992). By comparison, single gas treatments caused significant changes in both membrane formation and stacking in chloroplasts and also influenced starch accumulation.

Mixed pollutant exposure of Norway spruce over 3 months which simulated autumnal conditions strongly enhanced phosphoenolpyruvate carboxylase (PEPC) activity, especially following NO_2 + O_3 treatment (Hohlfeld et al. 1991). This study supports the later findings of Wild and Schmitt (1995) that PEPC activity in damaged trees in the field can be used as an indicator for air pollutants. The rates of protein synthesis in the light as well as the dark were increased by SO_2 in combination with O_3 and/or NO_2 (Schmitz et al. 1993). Schmitz et al. also have described a similar result with older spruce trees in the field when healthy and damaged trees were compared. Photosynthesis was assayed by determining D-1 protein content, CO_2 fixation and electron transport with isolated thylakoids (Steiger 1990; Lütz et al. 1992). It was not possible to find a severe breakdown of photosynthetic functions after 3 months of exposure despite the high concentrations of pollutants in some of the mixtures.

6.4 Biochemical Responses to Ozone

6.4.1 Response Levels

Effects of O_3 on crop and tree species have been studied at various levels, including plant growth, visible injury, leaf lifespan, carbon allocation, photosynthesis, dark respiration, stomatal conductance and cellular biochemistry. Several reviews on these topics have appeared (Guderian 1985; Pye 1988; Smith 1990; Lefohn 1992; Runeckles and Chevone 1992; Runeckles and Krupa 1994; Matyssek et al. 1995). Most studies (e.g. Reich 1987) have emphasized growth (biomass production) and photosynthesis as major factors for crop and forest productivity. Conceptually, O_3 effects on primary metabolism are also important for allocation processes, growth and competition between plant species while those on secondary metabolism are also important for phytopathological and stress defence reactions. Unfortunately, physiological studies on O_3 have usually neglected to follow simultaneously the pathways of both primary and secondary metabolism.

Relatively small changes in stomatal conductance often occur in response to low concentrations of O_3 and stomatal closure has usually been found under conditions of elevated O_3 (Darrall 1989). Stomatal opening, probably of a transient nature, has also been reported, e.g. for European beech (Leonardi and Langebartels 1990; Pearson and Mansfield 1993). A high natural variation in leaf conductance exists among crop and tree species, with the range of stomatal conductance covering around one order of magnitude (Lange et al. 1989). Thus, stomatal conductance appears to play a prominent role in plant responses to O_3. Among the major European tree species, Norway spruce has the lowest stomatal conductance which contributes to its high O_3 tolerance in short-term experiments. In contrast, Scots pine is considered to be relatively O_3 sensitive; European beech is considered to have an intermediate sensitivity (Reich 1987). Other workers have classified beech as being O_3 tolerant (Chappelka and Chevone 1992).

Experiments in OTCs (Guderian et al. 1985) or closed chambers (Langebartels et al. 1989, 1990) have shown that near-ambient O_3 levels do not lead to immediate visible injury in Norway spruce. In short-term exposures for up to 6 weeks, more than twice the ambient level of O_3 was needed to produce symptoms of acute injury in this species, while 1.5 times and twice the ambient level of O_3 was sufficient to cause visible damage in European beech (Lippert et al. 1996a) and Scots pine (Langebartels et al. 1990), respectively. O_3 sensitivity depends largely on the provenance or on the particular clone of the tree material employed, so that genotypic and phenotypic variation may impose an important restriction on the extrapolation of chamber results to the field.

Various biochemical pathways have been found to respond to O_3 exposure. The reactions, which occur after several hours to days of subacute O_3 exposure include changes in polyphenol and phytoalexin synthesis, polyamine and ethylene metabolism, accumulation of lytic enzyme activities as well as cell wall changes caused by the production and integration of callose, lignin and extensin (Sandermann et al. 1989a,b; Langebartels et al. 1990; Kangasjärvi et al. 1994; Sandermann 1996). The fastest response occurs in ethylene biosynthesis (after 0.5–1 h) and early response genes are activated after 1–3 h (Langebartels et al. 1991; Schraudner et al. 1992; Eckey-Kaltenbach et al. 1994; Van Camp et al. 1994; Willekens et al. 1994). Detailed time courses of the early O_3 responses and necrotic development in tobacco (Schraudner et al. 1994, 1996), parsley (Eckey-Kaltenbach et al. 1994) and *Arabidopsis* (Conklin and Last 1995) have been published. Certain aspects of molecular plant reactions to O_3 have recently been summarized by Schraudner et al. (1996) and Sandermann (1996).

6.4.2 Photosynthetic Pigments

Chlorophylls and carotenoids have frequently been studied in trees exposed to various environmental stresses. Samples from damaged spruce stands have been shown to develop distinct patterns of discolouration, mainly yellowing. Pigment studies (e.g. Köstner et al. 1990; Siefermann-Harms 1990) revealed that both chlorophylls and carotenoids are affected, and that changes start at early stages in green needles. Mainly nutrient imbalances and climatic factors (e.g. light intensity), but possibly also air pollutants, seem to be responsible for changes in the pigment spectrum of thylakoids, which in the long run may lower the photosynthetic performance of the trees. Furthermore, chlorophyll and carotenoid contents in spruce needles show marked changes during defined periods of the year (Lange et al. 1989; Lütz 1991) and in dependence of the altitude of the growth site (Lütz 1988, 1991). To test the role of O_3 as a possible detrimental factor for photosynthetic pigments, OTC experiments (Mikkelsen et al. 1995; Siefermann-Harms 1990) and phytotron studies (Senser et al. 1990; Steiger 1990) have been performed.

Spruce needle pigments from phytotron experiments with 4-year-old trees grown in control air and adult trees from Wank mountain (Bavarian Alps, 780 m a.s.l.) showed a similar composition of individual pigments.

This result indicated that the quality of environmental simulation was high and that pigments from young trees in phytotron studies match those of adult trees in the field. This is further supported by photosynthetic measurements (Steiger 1990).

Possible reactions of photosynthetic pigments due to abiotic stress factors are as follows:

(1) A large decrease in the levels of all chlorophylls and carotenoids, resulting in yellowing and browning of needles, together with a large reduction in photosynthesis and subsequent needle loss. This type of reaction has never been observed in phytotron studies, neither after a 12-month-treatment with 80 nl l^{-1} (Lippert et al. 1996b) nor in short-term studies (up to 5 weeks) with 100–800 nl l^{-1} O_3 (Sandermann et al. 1990). It is, however, known to occur in the field and was considered to result largely from N and Mg deficiency (Köstner et al. 1990; Siefermann-Harms 1990).

(2) A decreased stability of the light-harvesting complex (LHC), resulting in light sensitivity of needles, yellowing and reduced photosynthesis. Some reduction in the content of chlorophylls and LHC carotenoids may occur. This type of reaction has often been observed in phytotron experiments, with reductions in chlorophyll and carotenoid content up to 40% of the control levels, and depends on the O_3 dose, water supply, needle age and season of the year. Pigments of the LHC complex are preferentially reduced, while β-carotene, a component of the photosynthetic reaction centres, is only slightly affected. Recent studies by Siefermann-Harms (1996) also point out that the LHC may be destabilized by Mg deficiency and O_3 exposure under high light conditions resulting in yellowing of needles. The sum of the pigments of the xanthophyll cycle is elevated under various O_3 exposure profiles (Siefermann-Harms 1987; Pfündel and Bilger 1994), and an increase in the violaxanthin/antheraxanthin ratio is often found which indicates impaired enzyme activity.

(3) Light-dependent changes in the relative composition of xanthophyll cycle pigments (increase in total content and decreased violaxanthin to antheraxanthin ratios) were found in Norway spruce saplings treated with 80 nl l^{-1} O_3 for one growing season. With normal water and N-supply there were no changes in chlorophyll or carotenoid content nor any visible injury (C. Lütz, unpubl.). However, plants with a higher N-supply had reduced (>30%) levels of xanthophyll cycle pigments. Similar reactions were found in OTC experiments (Mikkelsen et al. 1995). Pigment content increases and epoxidation reactions of the cycle are slowed down during the early period of

exposure while reductions in pigment content of up to 30% and effects on de-epoxidation steps occur after several months (Steiger 1990). This may be a consequence of the reduction in ascorbic acid levels during O_3 stress in the leaf tissue as reducing power is required to maintain the xanthophyll cycle (Pfündel and Bilger 1994). A non-effective cycle may then increase photoinhibitory reactions, probably by non-dissipated radiation or by impaired conformation of the LHC (Horton et al. 1991; Pfündel and Bilger 1994; Siefermann-Harms 1987). Therefore, xanthophyll pigments and especially the formation of zeaxanthin can be used as an early indicator of stress in the photosynthetic apparatus provided that the environmental conditions of the individual plants are known (e.g. in closed chamber experiments). However, its diagnostic value in field studies is questionable because of the influences of numerous stress factors on xanthophyll pigment turnover.

6.4.3 Polyphenolic Metabolites

Plant polyphenolic metabolites are members of a widespread family largely derived from the amino acid phenylalanine via cinnamic acid (Fig. 6.4). Multiple roles have been assigned to plant polyphenols ranging from attractants, repellents and phytoalexins to radical scavengers and UV-B screening pigments (Harborne 1993). High constitutive levels of polyphenolic compounds are typical for coniferous species where total soluble compounds may reach cellular levels of up to 0.5 *M*. In addition to soluble compounds located mainly in the vacuole and to a lower extent in the apoplastic fluid, cell-wall-bound polyphenols (e.g. the complex lignins) can also occur in high amounts. Early studies with herbaceous plants revealed an O_3-induced accumulation of anthocyanins and caffeic acid in *Rumex* and bean, respectively (Koukol and Dugger 1967; Howell 1970), while in soybean, O_3 induced oestrogenic isoflavonoids in parallel to necrotic lesion development (Keen and Taylor 1975). Ozone-induced stress metabolites of tree species are summarized in Table 6.1.

Several of the inducible and constitutive compounds (e.g. flavonoids and polyamine conjugates) are known to be effective radical scavengers (Bors et al. 1989, 1990, 1996) and may play a role in the removal of O_3-derived radicals. An O_3 induction of stilbenes was observed following short-term exposure of Scots pine seedlings, as well as in medium- and long-term experiments with young pine trees (Sandermann et al. 1989a,b; Heller et al. 1995). Stilbenes are phytoalexins of pine sapwood and needles but are constitutive components of pine heartwood. The content of the stilbenes pinosylvin and pinosylvin 3-methyl ether (Fig. 6.4) increased both with time and O_3 dose from levels near the detection limit up to several hundred-fold (Rosemann et al. 1991). An 8-h exposure also significantly stimulated the activities of the biosynthetic enzymes phenylalanine ammonia lyase (PAL, Fig. 6.4) and stilbene synthase (STS). The activity of chalcone synthase (CHS), which uses the same substrates as STS, increased somewhat later (Rosemann et al. 1991). In contrast to PAL and CHS, STS activity was not present in needles of control plants.

The flavan catechin (Fig. 6.4), an antioxidant compound, was markedly increased in both O_3-exposed Norway spruce (Heller et al. 1990) and Scots pine (Langebartels et al. 1990). In the latter, the accumulation was linear with time and O_3 concentration throughout 39 days of exposure with a threshold value between the ambient level and twice the ambient level of O_3. Interestingly, elevated levels of this compound were still present after cultivation of Scots pine trees in the field from November until September of the following year (Fig. 6.5).

In contrast, stilbene accumulation exhibited a pronounced (but transient) peak in the first days of exposure and 50% reduced levels during the following weeks. The disappearance of soluble stilbenes (Lange et al. 1994), tyramine conjugates (Schraudner et al. 1994) and other O_3-induced phenolic compounds seems to be due to immobilization in the cell wall probably by binding to lignins. Pine seedlings treated with 0.15 $\mu l\ l^{-1}$ O_3 initially exhibited only minor reactions but stilbene levels increased after 6 weeks of exposure and elevated levels were still found in the spring of the year following the exposure (day 184 to 340, Fig. 6.5). In contrast to stilbenes and catechin, other phenolic compounds detected by HPLC did not show significant changes in response to O_3 treatment. Acylated flavonoid derivatives of Scots pine were induced by UV-B radiation (Jungblut et al. 1995; Schnitzler et al. 1996) but not by O_3. Thus it seems that a defined subset of phenolic compounds responds to the pollutant. This subset is different to the one reacting to UV-B radiation but may overlap with the subset affected in plant–pathogen interactions. Stilbenes and catechin are normally present in only low amounts in the foliage so that

▶

Fig. 6.4. Biochemical pathways of stilbenes (*left*), catechin (*middle*) and lignin (*right*): ozone-inducible processes in conifers. *CAD* Cinnamyl alcohol dehydrogenase; *CCR* Cinnamoyl-CoA reductase; *CHI* Chalcone isomerase; *CHS* Chalcone synthase; *4CL* 4-Coumarate:CoA ligase; *DFR* Dihydroflavonol 4-reductase; *F3′*H Flavonoid 3′-hydroxylase; *FHT* Flavanone 3-hydroxylase; *LAR* Leucoanthocyanidin 4-reductase; *PAL* Phenylalanine ammonia lyase; *PMT* Pinosylvin-3-*O*-methyl transferase; *STS* Stilbene synthase (pinosylvin forming)

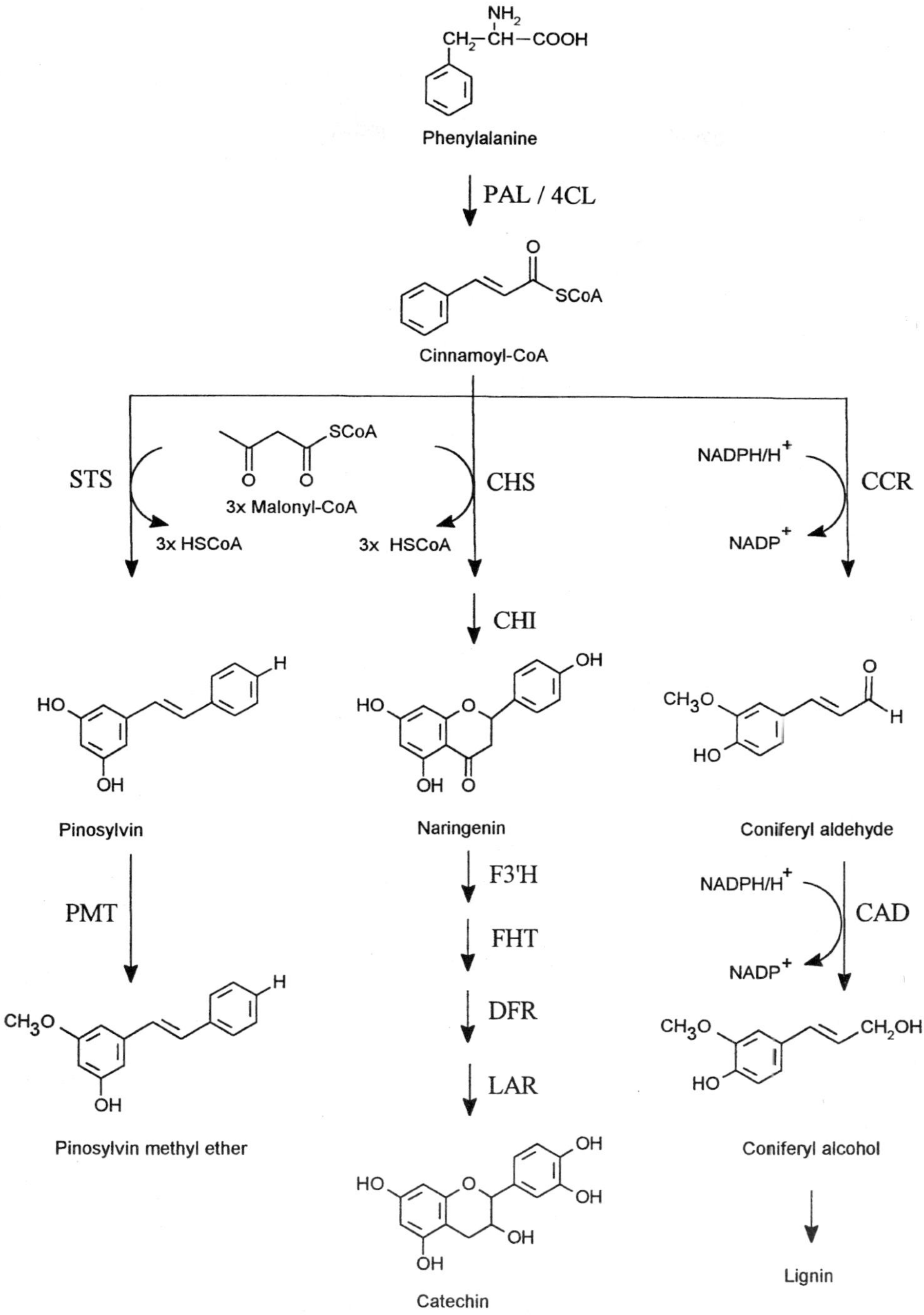

Phenylalanine
PAL / 4CL
Cinnamoyl-CoA
STS
3x Malonyl-CoA
CHS
NADPH/H+
CCR
3x HSCoA
3x HSCoA
NADP+
CHI
Pinosylvin
Naringenin
Coniferyl aldehyde
PMT
F3'H
FHT
DFR
LAR
NADPH/H+
CAD
NADP+
Pinosylvin methyl ether
Coniferyl alcohol
Catechin
Lignin

Table 6.1. Ozone-induced stress metabolites in trees

Compound	Tree species	Reference
Total phenols	Pine	Tingey et al. (1976b)
Kaempferol 3-*O*-glucoside	Spruce	Kicinski et al. (1988)
Pinosylvin, pinosylvin methyl ether	Pine	Rosemann et al. (1991)
Catechin	Spruce	Strack et al. (1987)
		Heller et al. (1990)
	Pine	Langebartels et al. (1990)
3,3′,4,4′-Tetramethoxybiphenyl	Beech	Zielke and Sonnenbichler (1990)
Putrescine, putrescine conjugates	Spruce	Langebartels et al. (1990)
Tyramine conjugates	Beech	Schraudner (1992)
Ethylene, ACC	Pine	Tingey et al. (1976a)
	Spruce	Wolfenden et al. (1988)
		Van den Driessche and Langebartels (1994)
	Beech	Sandermann et al. (1994)

their induction may have pronounced effects on cell metabolism and on interactions with pathogens.

6.4.4 Polyamines and Ethylene

The biosynthesis of polyamines markedly responds to air pollutant exposure as shown for crops and trees (Langebartels et al. 1991; Sandermann et al. 1994; see also Chap. 10 by Heath and Taylor, this Vol.). It was established that putrescine and tyramine are present as free amines and as hydroxycinnamic acid conjugates in the foliage of spruce, pine and beech. Putrescine and putrescine conjugates accumulated in Norway spruce treated with O_3 and acid mist (Sandermann et al. 1989a,b; Dohmen et al. 1990). Tyramine conjugates with cinnamic, ferulic and sinapic acids were induced in O_3-treated beech saplings (Schraudner 1992). Studies with herbaceous plants have shown that polyamines exert several effects which counteract the effects of oxidative stress. They have been implicated in the inhibition of lipid peroxidation of membranes, the activation of membrane-bound ATPases and the reduction of ethylene formation (Heath 1988; Flores et al. 1989; Zarembinski and Theologis 1994). The hydroxycinnamoyl derivatives of the polyamines, but not the free polyamines, are highly efficient scavengers of oxyradicals with rate constants similar to those of known radical scavengers such as ascorbic acid (Bors et al. 1989). Thus, it seems that polyamines and polyphenols (Bors et al. 1990, 1996) form part of the defence system of trees against oxidative stress.

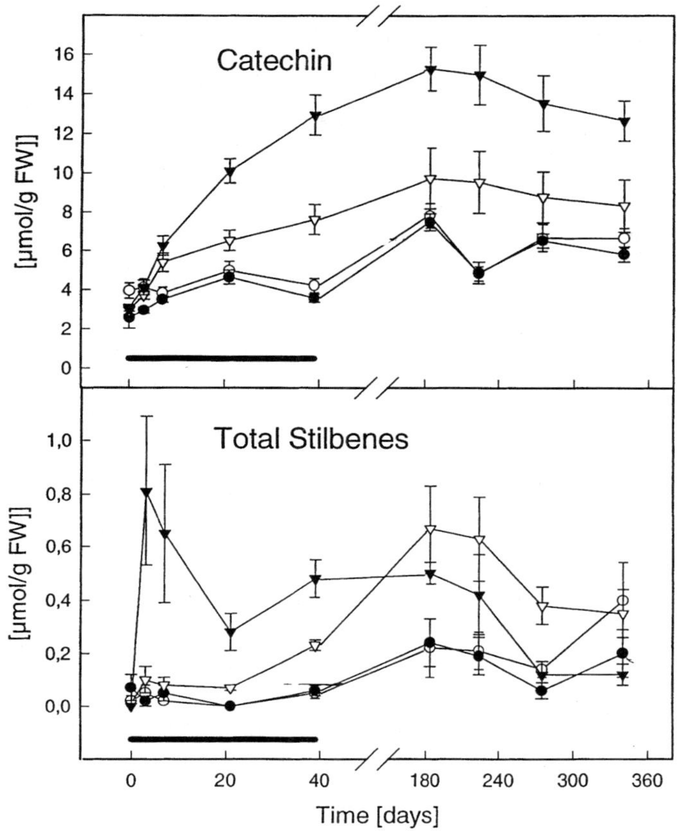

Fig. 6.5. Time course of catechin and stilbene (pinosylvin and pinosylvin 3-methyl ether) levels in Scots pine exposed to four levels of ozone (10 - 300 nl l^{-1}, 8 h per day). After treatment for 39 days the plants were post-cultivated in the field until November of the year following exposure. (Langebartels et al. 1996) ○, 10 nl l^{-1}; ●, 75 nl l^{-1}; ▽, 150 nl l^{-1}; ▼, 300 nl l^{-1} O_3

The amounts of ethylene and of its precursor, 1-aminocyclopropane-1-carboxylic acid (ACC), increase upon O_3 exposure of crops and trees (Tingey et al. 1976a; Sandermann et al. 1994). The level of stress ethylene formation is correlated with visible O_3 injury (Tingey et al. 1976a; Mehlhorn and Wellburn 1987; Langebartels et al. 1991). When ethylene reacts with O_3 it produces a number of highly reactive, water-soluble free oxyradicals and aldehydes which may contribute to plant damage (Elstner et al. 1985; Mehlhorn and Wellburn 1987). Ethylene is also known to promote senescence processes (Abeles et al. 1992; Kende 1993) and to mediate alterations in gas exchange known to occur under adverse conditions (Gunderson and Taylor 1991). Ethylene and polyamines share the same precursor, *S*-adenosyl-methionine (SAM), and mutually inhibit their biosynthesis (Zarembinski and Theologis 1994). O_3-induced ethylene may therefore amplify O_3 effects while O_3-induced polyamines may mitigate them. A metabolic switch towards ethylene or polyamine biosynthesis has been postulated as a major factor in O_3 sensitivity and O_3 tolerance of the tobacco varieties Bel W3 and Bel B, respectively (Langebartels et al. 1991).

It was recently shown that ethylene biosynthesis of O_3-treated potato and tomato plants proceeds via the normal ACC pathway, namely transcriptional activation of the three consecutive enzymes, SAM synthetase, ACC synthase and ACC oxidase (Yin et al. 1994; Betz et al. 1995; Schlagnhaufer et al. 1995). In addition to increases in ethylene emission, levels of free and conjugated ACC were increased in O_3-treated Norway spruce and European beech saplings (Sandermann et al. 1994; Langebartels et al. 1996); however, in contrast to the transient nature of ethylene and ACC peaks which may last only hours in herbaceous plants, the responses in trees were shown to be surprisingly persistent. Daily O_3 treatments led to daily peaks of ACC in spruce seedlings analogous to those previously described for crop plants (Langebartels et al. 1996). The ethylene peaks started each day from higher basal levels leading to a linear ACC accumulation for more than 3 weeks. Conjugated ACC responded in a second phase and showed increased levels over the total experimental period of 6 weeks. In 5-year-old European beech, the levels of free and conjugated ACC were elevated for more than 3 months with daily O_3 exposure (Sandermann et al. 1994). In contrast, combined O_3 and drought treatment of Norway spruce seedlings led to reduced damage and ethylene formation compared with well-watered plants (Van den Driessche and Langebartels 1994). Ethylene emissions and free ACC levels remained elevated in this study for prolonged periods of time, while conjugated ACC was not responsive. In line with these observations, Chen and Wellburn (1989) showed that acid mist treatment resulted in enhanced ethylene emissions as well as in elevated ACC contents for several months in red spruce and Norway spruce. The responses were significantly different from controls even 10 weeks after the misting had been stopped. The levels of conjugated ACC again did not respond to the treatment suggesting that ACC conjugation is not a strict consequence of enhanced ACC formation.

Thus from studies carried out under controlled conditions it seems that free ACC and polyamine content, (and in some cases also conjugated ACC levels) of needles and leaves are remarkably persistent albeit non-specific biomarkers for O_3. These compounds have also been found to correlate with visible damage of adult Norway spruce trees in the field (Wolfenden et al. 1988; Tenter and Wild 1991; Yang et al. 1994). In the latter study

(Yang et al. 1994), the biosynthetic enzyme ACC synthase was also found to be more active in damaged than in healthy needles.

6.4.5 Ozone-Related Proteins

Pathogen infection or externally applied ethylene induces a coordinately regulated group of proteins in herbaceous plants, the so-called pathogenesis-related proteins (PR proteins; Kombrink and Somssich 1995). PR proteins have been classified into eleven groups, PR-1 to PR-11, based on their mobility in native gel electrophoresis (Van Loon et al. 1994). PR-2 proteins are β-1,3-glucanases; PR-3, PR-8 and PR-11 are chitinases, and PR-5, proteinase inhibitors (Bowles 1990; Simmons 1994; Van Loon et al. 1994). β-1,3-Glucanases and chitinases catalyze the hydrolysis of the main components of fungal cell walls, β-1,3/1,6-glucans and chitin. They exhibit antimicrobial properties, particularly when used in combination, and may affect the spread of invading pathogens, thereby playing an important role in plant defence (Boller 1993).

Recent biochemical and molecular biological studies of plant responses to O_3 have indicated increased activities of several PR proteins (e.g. β-1,3-glucanase, chitinase and PR-1) in tobacco (Ernst et al. 1992; Schraudner et al. 1992; Yalpani et al. 1994), European beech (Langebartels et al. 1990) and Norway spruce (Kärenlampi et al. 1994; Table 6.2). These responses were found in tobacco only upon exposure to O_3 or O_3 plus NO_2 and not with SO_2 or NO_2 alone (Schraudner et al. 1994). Predominantly basic isoforms of the glycosidases are known to be induced by ethylene (Ecker 1995), and these were also found to be O_3-responsive. Therefore, O_3 may act in part via stress ethylene (Schraudner et al. 1992).

It was concluded that O_3 is a major abiotic elicitor of plant defence reactions and that these responses are elicited presumably because O_3 activates the ethylene or reactive oxygen species (ROS) signalling pathways (Sandermann 1996; Schraudner et al. 1996). In this respect, O_3 effects in plants resemble those of the hypersensitive response (HR) in which the spread of pathogens is restricted to the site of entry by localized necrotic lesion development.

Furthermore, part of the O_3 responses seems to be similar to processes of systemically acquired resistance (SAR). SAR provides protection throughout the plant against a broad spectrum of pathogens. The SAR genes include those encoding acidic isoforms of the PR proteins (Ryals et al. 1995). The latter are also induced upon prolonged exposure to O_3 (Ernst et al. 1992; Thalmair et al. 1996).

Table 6.2. Ozone-induced stress proteins and genes in trees

Enzyme	Tree species	Induction level[a]	Reference
Cinnamyl alcohol dehydrogenase	Spruce	T/E	Galliano et al. (1993a,b)
	Pine	T/E	Zinser (1996)
Short-chain alcohol dehydrogenase	Spruce	T	Bauer et al. (1993)
Stilbene synthase	Pine	E	Rosemann et al. (1991)
		T	Zinser (1996)
Phenylalanine ammonia lyase	Pine	E	Rosemann et al. (1991)
Chalcone synthase	Pine	E	Rosemann et al. (1991)
3-Hydroxymethylglutaryl-CoA synthase	Pine	T	Wegener (1995)
Extensin	Pine/Spruce	T	Schneiderbauer et al. (1995)
	Beech	T	
β-1,3-Glucanase	Beech	E	Langebartels et al. (1990)
	Spruce	P/E	Kärenlampi et al. (1994)
Chitinase	Beech	E	Langebartels et al. (1990)
	Spruce	P	Kärenlampi et al. (1994)
PR 1a, b, c	Spruce	P	Kärenlampi et al. (1994)
Polyubiquitin	Pine	T	Wegener (1995)
Ozone-related proteins, unknown function	Spruce	P	Schmitt and Sandermann (1990)
		P	Kärenlampi et al. (1994)
	Pine	T	Großkopf et al. (1994)

[a]Transcriptional (T), translational (P) and enzyme activity (E) level.

Table 6.2 summarizes the O_3-induced stress proteins and genes in trees and gives information on the level of their observed induction (e.g. transcript, protein or enzyme activity). An O_3 induction of the cell wall protein, extensin, at the transcript level has been described for Scots pine, Norway spruce and European beech seedlings (Schneiderbauer et al. 1995). Extensins (hydroxyproline-rich glycoproteins) are known to be structural components of the glycoprotein network of the cell wall and to be involved in plant defence against wounding or pathogen attack (Kieliszewski and Lamport 1994; Showalter 1993). Extensin is also a major component of an elicitor-induced spruce stress lignin (Lange et al. 1995). In Scots pine, transcript amounts for stilbene synthase and polyubiquitin were also found to be induced by O_3 (Wegener 1995; Zinser 1996; Table 6.2). In addition to these identified proteins and genes, several other O_3-related proteins (O_3R-proteins) of unknown function were found in Norway spruce (Schmitt and Sandermann 1990; Kärenlampi et al. 1994) and Scots pine (Groβkopf et al. 1994, Table 6.2).

6.4.6 Lignins

It was found that several of the O_3-induced responses are cell wall- or apoplast-directed. They include the formation of secondary compounds and their incorporation into cell wall components (see Sect. 6.4.3) as well as the induction of cell wall proteins and lignin biosynthesis (Sandermann 1996). In addition to being important components of xylem vessels and wood, lignins are also part of the phytopathological defence system of plants. Lignin is present around stomatal guard cells and epidermal cells of conifer needles. It is formed through oxidative polymerization of cinnamyl alcohols. Conifer wood lignin consists mainly (98–99%) of guaiacyl units originating from coniferyl alcohol which is formed from coniferyl aldehyde by a key enzyme in lignin biosynthesis, cinnamyl alcohol dehydrogenase (CAD; Fig. 6.4).

When Norway spruce trees were treated with sub-acute doses of O_3, the needles exhibited a dose-dependent increase in CAD activity (Sandermann et al. 1989a,b; Galliano et al. 1993b; Table 6.2). CAD activity was also induced by a fungal elicitor in spruce suspension cultures, and the enzyme was purified to homogeneity from this material (Galliano et al. 1993b). A full-length cDNA clone was isolated by screening a cDNA library with a specific antibody. The identity of the CAD clone was verified by DNA sequence alignment with amino acid sequences of tryptic peptides from the purified enzyme and by prokaryotic expression of the immunoreactive and active enzyme (Galliano et al. 1993a). Interestingly, it was concluded that the O_3- and elicitor-responsive CAD is identical to wood-forming CAD from spruce cambial tissue on the basis of the following evidence : (1) it had identical catalytic properties and electrophoretic mobility (i.e. molecular weight) and cross-reacted with the same antibody (Western blotting); (2) only one CAD gene was present in the spruce genome (Southern blotting). The full-length cDNA clone isolated from spruce cell cultures (Galliano et al. 1993a) therefore appeared to represent the first isolated gene of wood formation. Ozone treatment of spruce and pine seedlings led to a marked increase in CAD transcript levels (Table 6.2).

Using the recently developed method of thioacidolysis, the composition of elicitor-induced stress lignin was compared to that of other types of lignin (Lange et al. 1995). Stress lignin closely resembled early developmental lignins such as those of compression wood and tracheid middle lamellae. Stress lignin was clearly different from conifer wood lignin since it had a high content of p-hydroxyphenyl units and of β–β-linkages (Lange et al. 1995). The structure of O_3-induced lignin is as yet unknown.

6.5 Premature Senescence in Beech

European beech is the most important broad-leaved tree species in central Europe, but its biochemical and physiological responses to O_3 were rarely examined until recently (Pearson and Mansfield 1993, 1994; Lippert et al. 1996a). Some of the earlier experiments lasted only several weeks, and constant daily O_3 exposures were usually applied (Krause and Prinz 1989). In other studies (Klumpp et al. 1988), the interactions of O_3 with other air pollutants were examined.

Leonardi et al. (1990) observed a transient increase in stomatal conductance (cf. Pearson and Mansfield 1993), a persistent decrease in assimilation as well as premature leaf fall in response to near-ambient O_3 levels. This experiment was conducted under autumnal conditions and lasted for only 2 months. In cooperation with the Institute for Applied Biology in Schönenbuch, Switzerland, a dose–response experiment was undertaken, where beech seedlings were exposed to four different O_3 levels (cf. Fig. 6.3). The experiment lasted for a whole growing season from April until the end of September (bud break to leaf fall). The O_3 responses of beech saplings in open-top and controlled chamber exposures were compared.

Photosynthetic measurements in August revealed the light reactions to be less affected than the dark reactions (Lippert et al. 1996a). Assimilation rates, apparent carboxylation efficiency and maximum photosynthetic capacity were reduced by 1.5 times and twice the ambient O_3 level. These results are in good agreement with gas exchange measurements reported by Krause and Prinz (1989), Taylor and Dobson (1989), Leonardi and Langebartels (1990) and Leonardi et al. (1990). The results of the Critical Level calculation according to the UN-ECE (1994) revealed that an external AOT40 O_3 dose of 25 to 50 $\mu l\ l^{-1}$ h reduced both carboxylation efficiency and maximum photosynthetic capacity by 10% (CL 10; Lippert et al. 1996a). In June when the first measurements were conducted, AOT_{40} had accumulated to about 40 $\mu l\ l^{-1}$ h in the treatment with twice the ambient O_3 level without a negative effect. In August, however, both parameters were heavily affected when the trees had experienced an O_3 exposure of about 80 $\mu l\ l^{-1}$ h. Consequently, a reasonable CL 10 for European beech would be between 40 and 80 $\mu l\ l^{-1}$ h (Lippert et al. 1996a).

Symptom development on beech leaves in the treatments with 1.5 times and twice the ambient O_3 level began in mid-July following a period of elevated O_3 values (Sandermann et al. 1994; Lippert et al. 1996a). Fig 6.6 shows the time course of symptom development. Ozone symptoms as found in mid-August are depicted in Fig. 6.7 (see Appendix). Visible

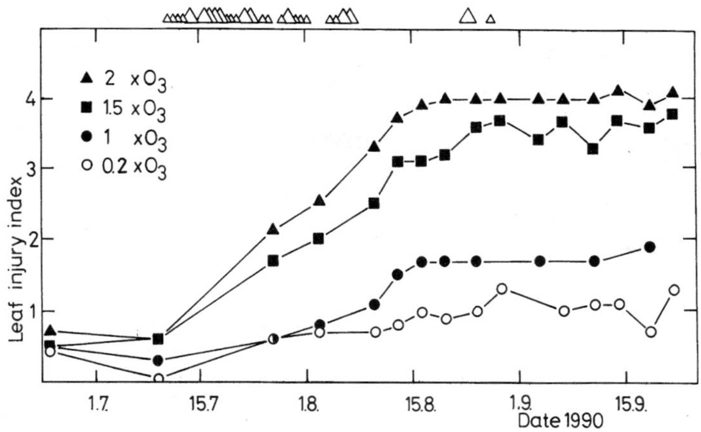

Fig. 6.6. Leaf injury in European beech saplings. Three-year old plants were exposed to four ozone levels (*0.2* to *2x* ambient, cf. Fig. 6.2) from April to October. Days with maximum concentrations exceeding 60 or 80 nl l^{-1} are indicated by small or large triangles, respectively, at the top of the graph

symptoms did not occur in the O_3 treatment with 0.2 times the ambient O_3 level and were few in ambient O_3 levels, but pronounced differences were observed between ambient and 1.5-fold elevated O_3 levels. Maximum leaf injury was reached in mid-August with 1.5 times and twice the ambient O_3 level. The symptoms consisted of spot-like brown necrotic lesions that could initially only be detected when leaves were held over an illuminated surface or when dried. It appeared that the high pigment levels of the leaves overlayed necrotic spots at the onset of lesion development. Typical O_3 injury (Hartmann et al. 1988; Klumpp et al. 1988) occurred in intercoastal areas of the upper leaf surface and spread over the total leaf area in the course of the experiment (Fig. 6.6).

The levels of the ethylene precursor ACC, in free and conjugated form, were persistently increased in beech leaves between July and September. Leaf fall in beech treated with 1.5 times and twice the ambient level of O_3 was induced 2–3 weeks earlier than in the ambient and sub-ambient treatments. Similar observations had been made by Leonardi et al. (1990). These authors used O_3 levels between 10 and 110 nl l^{-1} in combination with acid mist. Mainly O_3 and to a lesser extent acid mist accelerated the leaf fall of European beech. Pearson and Mansfield (1994) additionally reported that exposure to O_3 in the previous year reduced the amount of new growth in beech as a result of reduced internodal expansion.

In the GSF study, biochemical changes, O_3 symptom development and accelerated leaf fall were found in both 3- and 5-year-old saplings grown on different soils. The O_3-induced necrotic symptoms and the accelerated leaf fall are difficult to detect in the field. Symptoms may be concealed by the high contents of leaf pigments and may not become visible before autumn when pigment content is naturally reduced. Furthermore, it is not easy to relate leaf fall in autumn to previous periods of elevated O_3 since no "control" trees are available in the field.

Although ambient O_3 levels did not lead to impaired plant performance in European beech, the range between ambient and phytotoxic O_3 levels appeared to be small for this tree species and for sensitive herbaceous plants. Indeed, similarities between deciduous trees and herbaceous plants have already been recognized with respect to O_3 sensitivities of photosynthesis and biomass production (Reich 1987). Figure 6.8 shows a comparison of monthly ambient O_3 values for Schönenbuch and various stations in Germany for the period April to September 1990. It can be seen that the mean values for all Umweltbundesamt (UBA) Stations in Germany are lower than the simulated values, but that O_3 concentrations of montainous sites in southern Germany (Schauinsland, Brotjacklriegel) definitely were within the range of between ambient and 1.5 times the ambient O_3 levels at the Schönenbuch station. The AOT40 O_3 dose of the 1.5 times ambient O_3 which had accumulated from bud break until mid August was about 80 $\mu l\ l^{-1}$ h and was only slightly higher than that at Schauinsland (78 $\mu l\ l^{-1}$ h; Grennfelt and Beck 1994).

The relative O_3 sensitivity of beech apparent from the GSF data contrasts with the tolerance of beech described in other reports (Chappelka and Chevone 1992). Furthermore, damage inventories on high-O_3 exposure field sites in Germany [e.g. at the Wank Mountain, Bavarian Alps (see Chap. 5, Rennenberg et al., this Vol.) or at the Schauinsland, Black Forest] has not shown high incidents of damage. This contradiction may point to the existence of beech ecotypes or to marked adaptation processes.

6.6 "Memory" Effects for Ozone in Conifers

Biochemical responses to and visible effects of O_3 tend to develop simultaneously in both herbaceous plants and deciduous trees (see Sect. 6.5). For example, the O_3 induction of oestrogenic isoflavonoids in soybean plants was accompanied by necrotic development (Keen and Taylor 1975) and ethylene emission was closely linked to necrotic development in most plant species tested (Tingey et al. 1976a). Conifers were exceptional in this study since *Pinus mugo* and *Pinus ponderosa* gave, under certain conditions, an ethylene response without necrotic

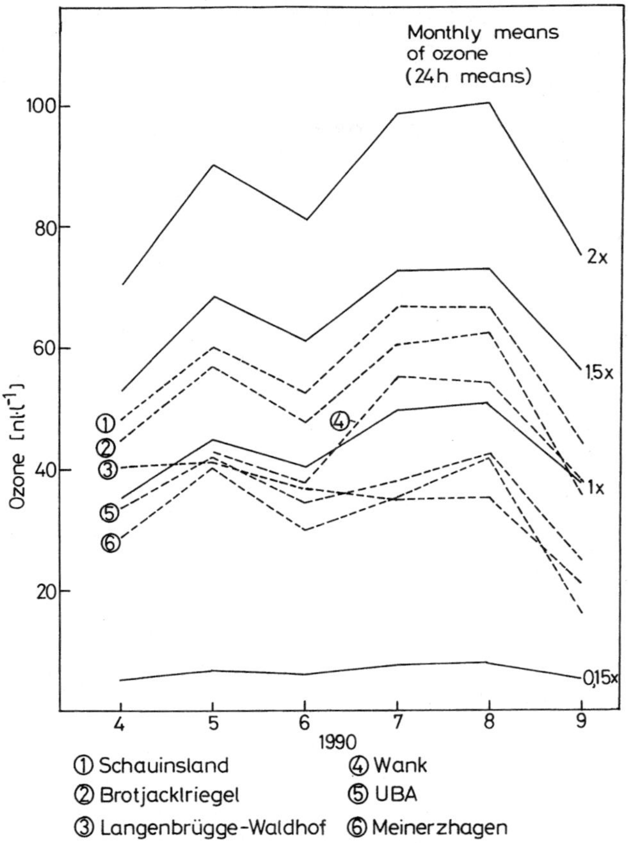

Fig. 6.8. Monthly mean concentrations of ozone in 1990 at various locations in Germany (*broken lines*). Ambient levels at Schönenbuch, Switzerland (*lx*) are compared to those at rural sites at higher elevations (approx. 1000 m.a.s.l.), Schauinsland (*1*), Brotjacklriegel (*2*), Wank (*4*), and sites at similar elevations (approx. 500 m), Langenbrügge-Waldhof (3) and Meinerzhagen (*6*).Mean concentrations of all Umweltbundesamt sites (*5*). In addition, decreased (0.15x) and increased (*1.5x* and *2x*) ozone levels at Schönenbuch are indicated (*solid lines*). (Umweltbundesamt 1990; J.P. Schnitzler and W. Heller, unpubl; Lippert et al. 1996a)

development (Tingey et al. 1976a). On the other hand, certain responses such as the O_3 induction of arginine decarboxylase and polyamines in tobacco Bel B are antagonistic with necrotic development (Langebartels et al. 1991).

When Norway spruce and Scots pine saplings were exposed to two- to four-fold ambient summer O_3 levels for several weeks, only the 1-year-old needles of Scots pine showed a dose-dependent reduction in chlorophyll content at the end of the exposure period (Langebartels et al. 1989, 1990,

1996; Steiger 1990). The plants were then postcultivated in the nursery for two subsequent vegetation periods. In spruce, visible injury appeared in the following year when the new flush emerged. All needle age classes which had been exposed to O_3 levels of 200 nl l^{-1} and above (equivalent to SUMO 100 $\mu l\ l^{-1}$ h) in the preceding year showed dose-dependent pigment changes. Levels of chlorophylls a and b as well as total carotenoids, mainly β-carotene and neoxanthin, were reduced while levels of the xanthophyll cycle pigments increased with medium O_3 exposures but remained at control levels at high doses (Steiger 1990). The visible symptoms consisted of an evenly distributed chlorosis and, after high O_3 levels (300 nl l^{-1} and above), of banding patterns (Fig. 6.7B). The new flush of spruce remained unaffected with respect to growth, photosynthesis and pigment patterns. Symptom development in these experiments was complete in mid-summer and no further alterations were observed in the following two growing seasons.

A similar late symptom development was found in a study in which clonal plants of Norway spruce were treated with 80 nl l^{-1} O_3 for a whole growing season (Langebartels et al. 1996; Lippert et al. 1996b). Again, symptom development occurred together with the onset of the new flush in the spring of the year following the O_3 exposure. In contrast to older needle age classes, the new flush which emerged in ambient O_3 levels in May did not show chlorosis or banding symptoms. Thus, it seems that ambient O_3 level do not affect Norway spruce while twice-ambient O_3 levels may suffice to cause visible damage. Similarities with montane yellowing (see Sect. 6.1) were (1) occurrence during the development of the new flush (Lange et al. 1989); (2) light dependence of the symptoms; and (3) restriction to older needle age classes. However, the banding pattern observed at high O_3 levels has so far not been observed in Norway spruce trees in the field. Late O_3 effects were also recognized in root tip meristems. Ozone raised the aberrations of root tip metaphase chromosomes after exposure to 80 nl l^{-1} for up to one growing season (Müller et al. 1994, 1996). It is not yet understood how this O_3 reaction which was detectable even 21 months after the treatment developed in the roots.

In the case of Scots pine, the shoot length of the new flush of the year following O_3 treatment was significantly reduced as a consequence of exposure to twice the ambient level of O_3 and above (Langebartels et al. 1996). The youngest O_3-treated needles had a reduced chlorophyll content and subsequently lost a major part of the needles starting in August of the year following exposure. These effects were dependent on the O_3 dose given the year before and started with twice-ambient O_3. Again, the youngest needles which developed under field conditions did not show any sign of visible injury while 1-year-old needles were affected as late as 1 year after the exposure period (Fig. 6.7C; Langebartels et al. 1996).

These much delayed symptoms in conifers which followed detectable initial biochemical stress reactions have been termed the "memory effect" for O_3 (Sandermann et al. 1989a, 1990; Langebartels et al. 1990; Sandermann 1996). Similarly delayed symptoms ("carry-over" or "after-effects") have also been reported for Sitka spruce, although without a description of early reactions, and occurred after three summers of exposure to 70 nl l^{-1} O_3 in the spring of the fourth year (Lucas et al. 1993). The symptoms, however, did not appear on Norway spruce treated in parallel. This again points to the high O_3 tolerance of this species. Independent studies (Guderian et al. 1985; Keller and Häsler 1987; Klumpp et al. 1988; Matyssek et al. 1990) also pointed to delayed cellular changes and long-term effects of O_3 in coniferous trees. However, the after-effects could not be differentiated from acute injury in these multi-season exposures. Guderian et al. (1985) interpreted these symptoms as an accumulation of "latent" effects. As detailed in Langebartels et al. (1996), the O_3 doses (SUMO) that led to memory effects in conifers were all in the same range of 100–150 $\mu l\ l^{-1}\ h^{-1}$ irrespective of the time period of exposure. Apparently, some unidentified factor(s) is accumulated which gives rise to the observed symptoms when a threshold O_3 dose has been reached. As described above, persistent biochemical changes (e.g. catechin, stilbenes, ethylene parameters) could indeed be observed in Scots pine and Norway spruce.

Memory symptoms developed in defined periods of the year, namely in spring for Norway spruce and in early autumn for Scots pine. In mature spruce trees, a pronounced decrease in pigment levels of all needle age classes occurs normally in early spring (Lange et al. 1989). The chlorophyll content of the 1-year-old needles is reduced by 25–50% but recovers at the onset of the new flush. The development of needle chloroplasts also shows a reduction in spring (Senser et al. 1975) while the photosynthetic efficiency of conifer species declines until just before bud break (Frey and Phillips 1977). During bud break and shoot development, there is a dramatic re-allocation of carbohydrates (McLaughlin et al. 1982) and nutrients (Lange et al. 1989) from the existing into the newly developing shoot. Consequently, the period before onset of the new flush is a highly sensitive phase of Norway spruce. Ozone effects may therefore be established in spring when the above dramatic developmental processes take place in the 1-year-old needles.

In Scots pine, symptoms develop during the natural shedding phase in early autumn when normally the 2- or 3-year-old needles are lost. This senescence process is not restricted to the shedded needle age class, but also influences photosynthetic parameters (McLaughlin et al. 1982) and biochemical processes of current-year and 1-year-old needles.

Taken together, visible O_3 effects on spruce and pine may develop with a substantial delay and during "sensitive" periods of the year when chlorophyll and needle loss normally take place. Norway spruce and Scots pine differ in their sensitive periods because of the different number of needle age classes normally remaining on the tree. It is speculated that the loss of 2- to 3-year-old needles in Scots pine influences the remaining crown more significantly than a loss of 7- to 10-year old needles in Norway spruce. Further research should reveal which components of the early biochemical O_3 responses interact with the later developmental stress to produce the memory effect of conifers. Ozone doses needed in controlled long-term studies to produce symptoms on older needles (Guderian et al. 1985; Lucas et al. 1993; Langebartels et al. 1996) are higher than O_3 doses occurring per annual growth period on various forest sites (see Chap. 1, Stockwell et al., this Vol.). It is, however, possible that peak concentrations of O_3 are particularly important for the development of visible symptoms (see Chappelka and Chevone 1992; Reich 1987).

6.7 Conclusions

The recent results described above have established that tree species display a number of early biochemical reactions under O_3 stress. SO_2 and NO_2 usually were much less effective as abiotic elicitors (Schraudner et al. 1994), although in tobacco, SO_2 and UV-B induce the same antioxidative genes as O_3 (Willekens et al. 1994). Most of the O_3-induced reactions can be classified as plant defence reactions. These inducible responses reinforce constitutive defences (e.g. ascorbate, glutathione and superoxide dismutases) which often are not responsive before visible injury occurs (Foyer et al. 1994; Inzé and Van Montagu 1995). Furthermore, the specific responses observed at the transcript, protein and metabolite levels point to controlled induction rather than a mere wounding effect as the molecular basis of O_3 responses.

Due to its high reactivity, it is not likely that O_3 is able to cross the plasmalemma (Heath 1988). It is, however, possible that O_3-derived molecules mimic normal second messenger molecules in inducing the various O_3 responses. Since gene induction has by now been well documented, there must be some form of signal transduction from the plasma membrane to the nucleus. These conclusions bring O_3 research into the context of research on phytohormones and on second messenger molecules such as ethylene, salicylic acid and jasmonic acid (see Chap. 10, Heath and Taylor, this Vol.). Lipid peroxides, calcium gradients, protein kinases and many of the other discussed signal transduction elements may play a role in O_3 effects (Schraudner et al. 1996).

A surprising analogy exists between the newly discovered O_3 responses of tree species and the well-known pathogen defence reactions of herbaceous species (Sandermann et al. 1989a,b; Kangasjärvi et al. 1994; Sandermann 1996; Schraudner et al. 1996). In spite of the progress made in recent years, a precise comparison between O_3-induced biochemical responses and important phytopathological reactions such as the hypersensitive response (HR: Dixon et al. 1994) and systemic acquired resistance (SAR: Kessmann et al. 1994) will only be possible after further research.

There appears to be systematic difference between conifers and herbaceous plants and deciduous trees. Conifers are unique in having a considerable time interval between the O_3-inducted early reactions and later visible symptoms. In our studies, early biochemical reactions and the delayed nature of symptoms have been observed for the first time and are the basis of the scenario described in Fig. 6.9. The abiotic pathway for visible symptoms caused by twice-ambient O_3 proceeds via the memory effect in the case of conifers. The early phytopathological defence reactions are likely to change the disposition for pathogens. Normally, these reactions are induced in a highly regulated manner upon bacterial, fungal or viral attack. Ozone induces these reactions out of their proper temporal and spatial context so that a weakening effect may occur. Controlled phytopathological infection experiments have shown that about half of the studies with O_3 led to an induced resistance, and half of the studies resulted in predisposition and increased disease rates (Manning and von

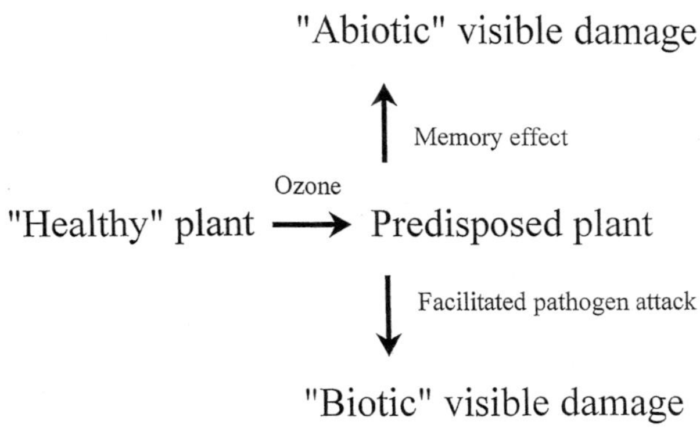

Fig. 6.9. Scenario for the induction of visible symptoms on conifer needles by ozone. The biotic and abiotic mechanisms are explained in the text. (Langebartels et al. 1996)

Tiedemann 1995). No accurate prediction of how O_3 will act in a given tree/pathogen system seems possible at present. On the other hand, the early biochemical O_3 reactions now discovered throw a new light on empirical observations made many years ago in the San Bernardino mountains, California (Guderian et al. 1985; Miller 1993; see Chap. 2, Miller et al., this Vol). Even in the high O_3 area of the San Bernardino mountains, where *Pinus ponderosa* and *P. jeffreyi* had strong visible symptoms, the trees were apparently not killed through the abiotic pathway of Fig. 6.9 but through predisposition for bark beetle and root rot fungal attack. In a model study, O_3 enhanced the root rot disease of Scots pine, mycorrhization being protective (Bonello et al. 1993). Initial studies have revealed that the induction of stilbenes observed for *Pinus sylvestris* (Rosemann et al. 1991) also takes place as an O_3 response in *Pinus ponderosa* (H. Sandermann, W. Heller and unpubl.). In summary, the biochemical results described here and the scenario described in Fig. 6.9 provide a new starting point for field experiments.

Acknowledgements. These studies were financially supported by the Bayerisches Staatsministerium für Landesentwicklung und Umweltfragen, the Bundesministerium für Forschung und Technologie, the Commission of the European Communities, and the Fonds der Chemischen Industrie. They were part of the Münchner Arbeitsgemeinschaft Luftschadstoffe (MAGL) and EUROSILVA research networks. We acknowledge the skilful technical assistance of L. Gößl, M. Stuis, C. Elbert, S. Stich, E. Mattes and E. Kiefer during the past years. Important contributions by P. Bader, G. Seckmeyer, H.K. Seidlitz, D. Strube and the EPOKA technical staff are gratefully acknowledged.

References

Abeles FB, Morgan PW, Saltveit ME (eds) (1992) Ethylene in plant biology. Academic Press, San Diego

Barnes JD, Pfirrmann T, Steiner K, Lütz C, Busch U, Küchenhoff H, Payer HD (1995) Effects of elevated CO_2, O_3 and K deficiency on Norway spruce (*Picea abies* [L.] Karst.). II. Seasonal changes in photosynthesis and non-structural carbohydrate content. Plant Cell Environ 18:1345–1357

Bauer S, Galliano H, Pfeiffer F, Meßner B, Sandermann H, Ernst D (1993) Isolation and characterization of a cDNA clone encoding a novel short-chain alcohol dehydrogenase from Norway spruce (*Picea abies* [L.] Karst.). Plant Physiol 103:1479–1480

Betz C, Tuomainen J, Kangasjärvi J, Ernst D, Yin Z-H, Langebartels C, Sandermann H (1995) Ozone activation of ethylene biosynthesis in tomato. In: Chappelle E, Lichtenthaler HK, Sandermann H (eds) Conference on vegetation stress, Neuherberg, Abstr, p 13

Blank L, Lütz C (eds) (1990) Tree exposure experiment in closed chambers. Environ Pollut 64:189–395

Boller T (1993) Antimicrobial functions of the plant hydrolases, chitinase and ß-1,3-glucanase. In: Fritig B, Legrand M (eds) Developments in plant pathology. Mechanisms of plant defense responses. Kluwer, Dordrecht, pp 391–400

Bonello P, Heller W, Sandermann H (1993) Ozone effects on root-disease susceptibility and defence responses in mycorrhizal and non-mycorrhizal seedlings of Scots pine (*Pinus sylvestris* L.). New Phytol 124:653–663

Bors W, Langebartels C, Michel C, Sandermann H (1989) Polyamines as radical scavengers and protectants against ozone damage. Phytochemistry 28:1589–1595

Bors W, Heller W, Michel C, Saran M (1990) Flavonoids as antioxidants: determination of radical-scavenging efficiencies. Methods Enzymol 186:343–355

Bors W, Heller W, Michel C, Stettmaier K (1996) Flavonoids and polyphenols: chemistry and biology. In: Cadenas E, Packer L (eds) Handbook of antioxidants. Marcel Dekker, New York, pp 409–466

Bosch C, Pfannkuch E, Baum U, Rehfuess KE (1983) Über die Erkrankung der Fichte (*Picea abies*) in den Hochlagen des Bayerischen Waldes. Forstwiss Centralbl (Hamb) 102:167–181

Bowles DJ (1990) Defense-related proteins in higher plants. Annu Rev Biochem 59:873–907

Castillo FJ, Miller PR, Greppin H (1987) Extracellular biochemical markers of photochemical oxidant air pollution damage to Norway spruce. Experientia 43:111–115

Chappelka AH, Chevone BI (1992) Tree response to ozone. In: Lefohn AS (ed) Surface level ozone exposures and their effects on vegetation. Lewis, Chelsea, pp 271–324

Chen Y-M, Wellburn AR (1989) Enhanced ethylene emissions from red and Norway spruce exposed to acidic mists. Plant Physiol 91:357–361

Conklin PL, Last RL (1995) Differential accumulation of antioxidant mRNAs in *Arabidopsis thaliana* exposed to ozone. Plant Physiol 109:203–212

Darrall NM (1989) The effect of air pollutants on physiological processes in plants. Plant Cell Environ 12:1–30

Dixon RA, Harrison MJ, Lamb CJ (1994) Early events in the activation of plant defense responses. Annu Rev Phytopathol 32:479–501

Doehring T, Koefferlein M, Thiel S, Seidlitz HK (1996) Spectral shaping of artificial UV-B irradiation for vegetation stress research. J Plant Physiol 148:115–119

Dohmen GP, Koppers A, Langebartels C (1990) Biochemical response of Norway spruce (*Picea abies* [L.] Karst.) towards 14-month exposure to ozone and acid mist: Effects on amino acid, glutathione and polyamine titers. Environ Pollut 64:375–383

Ecker JR (1995) The ethylene signal transduction pathway in plants. Science 268:667–675

Eckey-Kaltenbach H, Ernst D, Heller W, Sandermann H (1994) Biochemical plant responses to ozone. IV. Cross induction of defensive pathways in parsley (*Petroselinum crispum* L.) plants. Plant Physiol 104:67–74

Elstner EF, Osswald W (1994) Mechanisms of oxygen activation during plant stress. Proc R Soc Edinb 102B:131–154

Elstner EF, Osswald W, Youngman RJ (1985) Basic mechanisms of pigment bleaching and loss of structural resistance in spruce (*Picea abies*) needles: advances in phytomedical diagnostics. Experientia 41:591–597

Ernst D, Schraudner M, Langebartels C, Sandermann H (1992) Ozone-induced changes of mRNA levels of β-1,3-glucanase, chitinase and 'pathogenesis-related' protein 1b in tobacco plants. Plant Mol Biol 20:673–682

Flores HE, Protacio CM, Signs MW (1989) Primary and secondary metabolism of polyamines in plants. Recent Adv Phytochem 23:329–393

Forschungsbeirat Waldschäden/Luftverunreinigungen (ed) (1989) Dritter Bericht, Kernforschungszentrum Karlsruhe, Karlsruhe, ISSN 0931-7805

Foyer C, Lelandais M, Kunert KJ (1994) Photooxidative stress in plants. Physiol Plant 92:696–717

Frey DJ, Phillips IDJ (1977) Photosynthesis of conifers in relation to annual growth cycles and dry matter production. Physiol Plant 40:300–306

Führer G, Dunkl M, Knoppik D, Selinger H, Blank LW, Payer H-D, Lange OL (1990) Effects of low level long-term ozone fumigation and acid mist on photosynthesis and stomata of clonal Norway spruce (*Picea abies* [L.] Karst.). Environ Pollut 64:279–293

Galliano H, Cabané M, Eckerskorn C, Lottspeich F, Sandermann H, Ernst D (1993a) Molecular cloning, sequence analysis and elicitor-/ozone-induced accumulation of cinnamyl alcohol dehydrogenase from Norway spruce (*Picea abies* L.). Plant Mol Biol 23:145–156

Galliano H, Heller W, Sandermann H (1993b) Ozone induction and purification of spruce cinnamyl alcohol dehydrogenase. Phytochemistry 32:557–563

Grennfelt P, Beck JP (1994) Ozone concentrations in Europe in relation to different concepts of the critical levels. In: Fuhrer J, Achermann A (eds) Critical levels for ozone. UN-ECE workshop report, FAC Bern, pp 184–194

Großkopf E, Wegener-Strake A, Sandermann H, Ernst D (1994) Ozone-induced metabolic changes in Scots pine: mRNA isolation and analysis of in vitro translated proteins. Can J For Res 24:2030–2033

Guderian R (ed) (1985) Air pollution by photochemical oxidants. Formation, transport, control and effects on plants. Ecological Studies vol 52. Springer, Berlin Heidelberg New York

Guderian R, Küppers K, Six R (1985) Wirkungen von Ozon, Schwefeldioxid und Stickstoffdioxid auf Fichte und Pappel bei unterschiedlicher Versorgung mit Magnesium und Kalzium sowie auf die Blattflechte *Hypogymnia physodes*. VDI Berichte 560:657–701

Gunderson CA, Taylor GE (1991) Kinetics of inhibition of foliar gas exchange by exogenous ethylene: an ultrasensitive response. New Phytol 110:517–524

Harborne JB (1993) Introduction to ecological biochemistry. Academic Press, London

Hartmann G, Nienhaus F, Butin H (1988) Farbatlas Waldschäden. Diagnose von Baumkrankheiten. Ulmer, Stuttgart

Heath RL (1988) Biochemical mechanisms of pollutant stress. In: Heck WW, Taylor OC, Tingey DT (eds) Assessment of crop loss from air pollutants. Elsevier, London, pp 259–286

Heller W, Rosemann D, Osswald WF, Benz B, Schönwitz R, Lohwasser K, Kloos M, Sandermann H (1990) Biochemical response of Norway spruce (*Picea abies* [L.] Karst.) towards 14-month exposure to ozone and acid mist: Part I - Effects on polyphenol and monoterpene metabolism. Environ Pollut 64:353–366

Heller W, Ernst D, Langebartels C, Sandermann H (1995) Induction of polyphenol biosynthesis in plants during development and environmental stress. In: Brouillard R, Jay M, Scalbert A (eds) Polyphenols 94. INRA, Paris pp 67–78

Hohlfeld H, Lütz C, Strack D (1991) Phosphoenolpyruvate carboxylase activity in Norway spruce needles: effects of air pollutants under controlled conditions. Z Naturforsch 46c:502–505

Horton P, Ruban A, Rees D, Pascal A, Noctor G, Young A (1991) Control of the light harvesting function of chloroplast membranes by aggregation of the LHCII chlorophyll–protein-complex. FEBS Lett 292:1–4

Howell RK (1970) Influence of air pollution on quantities of caffeic acid isolated from leaves of *Phaseolus vulgaris*. Phytopathology 60:1626–1629

Inzé D, Van Montagu M (1995) Oxidative stress in plants. Curr Opin Biotechnol 6:153–158

Jungblut TP, Schnitzler J-P, Heller W, Hertkorn N, Metzger JW, Szymczak W, Sandermann H (1995) Structures of UV-B induced sunscreen pigments of Scots pine (*Pinus sylvestris* L.). Angew Chem Int Ed Engl 34:312–314

Kangasjärvi J, Talvinen J, Utriainen M, Karjalainen R (1994) Plant defence systems induced by ozone. Plant Cell Environ 17:783–794

Kärenlampi SO, Airaksinen K, Miettinen ATE, Kokko HI, Holopainen JK, Kärenlampi LV, Karjalainen RO (1994) Pathogenesis-related proteins in ozone-exposed Norway spruce [*Picea abies* (L.) Karst.]. New Phytol 126:81–89

Keen NT, Taylor OC (1975) Ozone injury in soybeans. Isoflavonoid accumulation is related to necrosis. Plant Physiol 55:731–733

Keller T, Häsler R (1987) Some effects of long-term ozone fumigations on Norway spruce. I. Gas exchange and stomatal response. Trees 1:129–133

Kende H (1993) Ethylene biosynthesis. Annu Rev Plant Physiol Plant Mol Biol 44:283–307

Kessmann H, Staub T, Hofmann C, Maetzke T, Herzog J, Ward E, Uknes S, Ryals J (1994) Induction of systemic acquired disease resistance in plants by chemicals. Annu Rev Phytopathol 32:439–459

Kiciniski HG, Kettrup A, Boos KS, Masuch G (1988) Single and combined effects of continuous and discontinuous O_3 and SO_2 emission on Norway spruce needles. II. Metabolic changes. Int J Environ Anal Chem 32:213–241

Kieliszewski MJ, Lamport DTA (1994) Extensin: repetitive motifs, functional sites, post-translational codes, and phylogeny. Plant J 5:157–172

Klumpp A, Klumpp G, Guderian R (1988) Wuchsleistung und äußere Schädigungsmerkmale bei Buche nach Einwirkung von Ozon, Schwefeldioxid und Stickstoffdioxid Allg Forstztg 26:731–734

Kombrink E, Somssich IE (1995) Defense responses of plants to pathogens. Adv Bot Res 21:1–34

Köstner B, Czygan F, Lange O (1990) An analysis of needle yellowing in healthy and chlorotic Norway spruce (*Picea abies*) in a forest decline area of the Fichtelgebirge (N.E.Bavaria). I Annual time-course changes in chloroplast pigments for five different needle age classes. Trees 4:55–67

Koukol J, Dugger WM (1967) Anthocyanin formation as a response to ozone and smog treatment of *Rumex crispus* L. Plant Physiol 42:1023–1024

Krause GHM, Prinz B (1989) Experimentelle Untersuchungen der LIS zur Aufklärung möglicher Ursachen der neuartigen Waldschäden. LIS-Berichte No 80

Lange BM, Trost M, Heller W, Langebartels C, Sandermann H (1994) Elicitor-induced formation of free and cell-wall-bound stilbenes in cell-suspension cultures of Scots pine (*Pinus sylvestris* L.). Planta 194:143–148

Lange BM, Lapierre C, Sandermann H (1995) Elicitor-induced spruce stress lignin. Structural similarity to early developmental lignins. Plant Physiol 108:1277–1287

Lange OL, Heber U, Schulze E-D, Ziegler H (1989) Atmospheric pollutants and plant metabolism. In: Schulze ED, Lange OL, Oren S (eds) Forest decline and air pollution. A study of spruce (*Picea abies*) on acidic soils. Ecological studies, vol 77. Springer, Berlin Heidelberg New York pp 238–273

Langebartels C, Führer G, Häckl B, Heller W, Kloos M, Payer H-D, Schmitt R, Sandermann H (1989) Dose-dependent biochemical reactions of Norway spruce to ozone fumigation. In: Bucher JB, Bucher-Wallin I (eds) Air pollution and forest decline. Proc 14th IUFRO Meeting, Birmensdorf, pp 446–469

Langebartels C, Heller W, Kerner K, Leonardi S, Rosemann D, Schraudner M, Trost M, Sandermann H (1990) Ozone-induced defense reactions in plants. In: Payer HD, Pfirrmann T, Mathy P (eds) Environmental research with plants in closed chambers. Air pollution research reports, vol 26. CEC, Brussels, pp 358–368

Langebartels C, Kerner K, Leonardi S, Schraudner M, Trost M, Heller W, Sandermann H (1991) Biochemical plant responses to ozone. I. Differential induction of polyamine and ethylene biosynthesis in tobacco. Plant Physiol 95:882–889

Langebartels C, Heller W, Führer G, Lippert M, Simons S, Sandermann H (1996) Memory effects in the action of ozone on conifers (in preparation)

Lefohn AS (ed) (1992) Surface level ozone exposure and their effects on vegetation. Lewis Publ, Chelsea

Leonardi S, Langebartels C (1990) Fall exposure of beech saplings (*Fagus sylvatica* L.) to ozone and stimulated acidic mist: effects on gas exchange and leachability. Water Air Soil Pollut 54:143–153

Leonardi S, Langebartels C, Sandermann H (1990) Fall exposure of beech trees (Fagus sylvatica L.) to ozone and stimulated acidic mist: Immediate and post-treatment effects on whole plant physiology. In: Payer HD, Pfirrmann T, Mathy P (eds) Environmental research with plants in closed chambers. Air pollution research reports, vol 26. CEC, Brussels, pp 369–380

Lippert M, Steiner K, Payer H-D, Simons S, Langebartels C, Payer HD, Sandermannn H (1996a) Assessing the impact of ozone on photosynthesis of European beech (*Fagus sylvatica* L.) in environmental chambers. Trees 10:268–275

Lippert M, Steiner K, Pfirrmann T, Payer H-D (1996b) Interactive effects of elevated O_3 and CO_2 on clonal Norway spruce seedlings. Photosynthetic gas exchange during the exposure and the dynamics of the "memory" effect in the following year. Trees (in press)

Lucas PW, Rantanen L, Mehlhorn H (1993) Needle chlorosis in Sitka spruce following a three-year exposure to low concentrations of ozone: changes in mineral content, pigmentation and ascorbic acid. New Phytol 124:265–275

Lütz C (1988) Photosynthetische Pigmente aus Nadelbäumen unterschiedlicher Höhenstufen des Ötztales. GSF-Bericht 17/88:415–425

Lütz C (1991) Einfluß von Höhenlage, Witterung und Jahreszeit auf das Pigmentmuster von Nadeln ausgewählter Fichten am Wank. GSF-Bericht 29/91:209–224

Lütz C, Heinzmann U, Gülz P-G (1990) Surface structures and epicuticular wax composition of spruce needles after long-term treatment with ozone and acid mist. Environ Pollut 64:313–322

Lütz C, Steiger A, Godde D (1992) Influence of air pollutant and nutrient deficiency on D-1 protein content and photosynthesis in young spruce trees. Physiol Plant 85:611–617

Manning WJ, Krupa SV (1992) Experimental methodology for studying the effects of ozone on crops and trees. In: Lefohn AS (ed) Surface level ozone exposure and their effects on vegetation. Lewis Publ, Chelsea, pp 93–156

Manning WJ, von Tiedemann A (1995) Climate change: potential effects of increased atmospheric carbon dioxide (CO_2), ozone (O_3) and ultraviolet-B (UV-B) radiation on plant diseases. Environ Pollut 88:219–245

Matyssek R, Keller T, Günthardt-Goerg MS (1990) Ozonwirkungen auf den verschiedenen Organisationsebenen in Holzpflanzen. Schweiz Z Forstwes 141:631–651

Matyssek R, Reich PB, Oren R, Winner WE (1995) Response mechanisms of conifers to air pollutants. In: Smith WK, Hinckley TM (eds) Ecophysiology of coniferous forests. Academic Press, San Diego, pp 255–308

McLaughlin SB, McConathy RK, Duvick D, Mann LK (1982) Effects of chronic air pollution stress on photosynthesis, carbon allocation, growth of white pine trees. For Sci 28:60–70

Mehlhorn H, Wellburn AR (1987) Stress ethylene formation determines plant sensitivity to ozone. Nature 327:417–418

Mikkelsen TN, Dodell B, Lütz C (1995) Changes in pigment concentration and composition in Norway spruce induced by long-term exposure to low levels of ozone. Environ Pollut 87:197–205

Miller PR (1993) Mixed conifer forests of the San Bernardino Mountains, California. In: Olson RK, Blinkley D, Böhm M (eds) The response of Western forests to air pollution. Ecological Studies, vol 97. Springer, Berlin Heidelberg New York, pp 461–496

Müller M, Köhler B, Grill D, Guttenberger H, Lütz C (1994) The effects of various soils, different provenances and air pollution on root tip chromosomes in Norway spruce. Trees 9:73–79

Müller M, Köhler B, Tausz M, Grill D, Lütz C (1996) The assessment of ozone stress by recording chromosomal aberrations in root tips of spruce trees (*Picea abies* [L.] Karst.). J Plant Physiol 148:160–165

Payer HD, Blank LW, Gnatz G, Schmolke W, Schramel P, Bosch C (1986) Simultaneous exposure of forest trees to pollutants and climatic stress. Water Air Soil Pollut 31:485–491

Payer HD, Pfirrmann T, Mathy P (eds) (1990) Environmental research with plants in closed chambers. Air pollution research reports of the EC, vol 26. CEC, Brussels

Payer HD, Blodow P, Köfferlein M, Lippert M, Schmolke W, Seckmeyer G, Seidlitz H, Strube D, Thiel S (1994) Controlled environment chambers for experimental studies on plant responses to CO_2 and interactions with pollutants. In: Schulze ED, Mooney HA (eds) Design and execution of experiments on CO_2 enrichment. Commission of the European Communities, Ecosystems Research Report, vol 6. CEC, Brussel, pp 127–145

Pearson M, Mansfield TA (1993) Interacting effects of ozone and water stress on the stomatal resistance of beech (*Fagus sylvatica* L.). New Phytol 123:351–358

Pearson M, Mansfield TA (1994) Effect of exposure to ozone and water stress on the following season's growth of beech (*Fagus sylvativa* L.) New Phytol 126:511–515

Pfündel E, Bilger W (1994) Regulation and possible function of the violaxanthin cycle. Photosynth Res 42:89–109

Pye JM (1988) Impact of ozone on the growth and yield of trees: a review. J Environ Qual 17:347–360

Rehfuess KE, Bosch C (1986) Experimentelle Untersuchungen zur Erkrankung der Fichte (*Picea abies* [L.] Karst.) auf sauren Böden der Hochlagen: Arbeitshypothese und Versuchsplan. Forstwiss Centralbl (Hamb) 105:201–206

Reich PB (1987) Quantifying plant response to ozone: a unifying theory. Tree Physiol 3:63–91

Roberts TM, Brown KA, Blank LW (1988) Methodological aspects of the fumigation of forest trees with gaseous pollutants using closed chambers. In: Mathy P (ed) Air pollution and ecosystems. CEC, Brussels, pp 338–369

Rosemann D, Heller W, Sandermann H (1991) Biochemical plant responses to ozone. II. Induction of stilbene biosynthesis in Scots pine (*Pinus sylvestris* L.) seedlings. Plant Physiol 97:1280–1286

Runeckles VC, Chevone BI (1992) Crop responses to ozone. In: Lefohn AS (ed) Surface level ozone exposures and their effects on vegetation. Lewis Publ, Chelsea, pp 185–266

Runeckles VC, Krupa SV (1994) The impact of UV-B radiation and ozone on terrestrial vegetation. Environ Pollut 83:191–213

Ryals J, Lawton KA, Delaney TP, Friedrich L, Kessmann H, Neuenschwander U, Uknes S, Vernooij B, Weymann K (1995) Signal transduction in systemic acquired resistance. Proc Natl Acad Sci USA 92:4202–4205

Sandermann H (1996) Ozone and plant health. Annu Rev Phytopathol 34:347–366

Sandermann H, Heller W, Langebartels C (1989a) Early biochemical effects of air pollutants: detection and possible significance for forest trees. In: Ulrich B (ed) Proc of the int congr on forest decline research: state of knowledge and perspectives. Friedrichshafen, Kernforschungzentrum Karlsruhe, pp 517–525

Sandermann H, Schmitt R, Heller W, Rosemann D, Langebartels C (1989b) Ozone-induced early biochemical reactions in conifers. In: Longhurst JWS (ed) Acid deposition. Sources, effects and controls. British Library, London, pp 243–254

Sandermann H, Langebartels C, Heller W (1990) Ozonstreß bei Pflanzen. Frühe und "Memory"-Effekte von Ozon bei Nadelbäumen. Z Umweltchem Ökotox 2:14–15

Sandermann H, Ernst D, Heller W, Langebartels C (1994) Biochemical markers for stress detection and ecophysiology. In: Schulze ED, Mooney HA (eds) Design and execution of experiments on CO_2 enrichment. Commission of the European Communities, Ecosystems research report, vol 6. CEC, Brussels. pp 45–51

Schiffgens-Gruber A, Lütz C (1992) Ultrastructure of mesophyll cell chloroplasts of spruce needles exposed to O_3, SO_2 and NO_2 alone and in combination. Environ Exp Bot 32:243–254

Schlagnhaufer DC, Glick RE, Arteca RN, Pell EJ (1995) Molecular cloning of an ozone-induced 1-aminocyclopropane-1-carboxylate synthase cDNA and its relationship with a loss of rbcS in potato (*Solanum tuberosum* L.) plants. Plant Mol Biol 28:93–103

Schmitt R, Sandermann H (1990) Biochemical response of Norway spruce (*Picea abies* [L.] Karst.) towards 14-month exposure to ozone and acid mist: part II – effects on protein biosynthesis. Environ Pollut 64:367–373

Schmitz H, Löffel U, Weidner M (1993) The rate of protein synthesis in needles of Norway spruce (*Picea abies*): light stimulation, regulation through photophosphorylation, stress enhancement. Physiol Plant 87:237–247

Schneiderbauer A, Back E, Sandermann H, Ernst D (1995) Ozone induction of extensin mRNA in Scots pine, Norway spruce and European beech. New Phytol 130:225–230

Schnitzler J-P, Jungblut TP, Heller W, Köfferlein M, Hutzler P, Heinzmann U, Schmelzer E, Ernst D, Langebartels C, Sandermann H (1996) Tissue localization of UV-B-screening pigments and of chalcone synthase mRNA in needles of Scots pine seedlings. New Phytol 132:247–258

Schraudner M (1992) Frühe Reaktionen von Tabak und Buche auf Ozon: Induktion des Tyramin- und Polyaminstoffwechsels. PhD Thesis, Technische Universität München

Schraudner M, Ernst D, Langebartels C, Sandermann H (1992) Biochemical plant responses to ozone. III. Activation of the defense-related proteins ß-1,3-glucanase and chitinase in tobacco leaves. Plant Physiol 99:1321–1328

Schraudner M, Graf U, Langebartels C, Sandermann H (1994) Ambient ozone can induce plant defence reactions in tobacco. Proc R Soc Edinb 102B:55-61

Schraudner M, Langebartels C, Sandermann H (1996) Plant defence systems and ozone. Biochem Soc Transact 24:456-462

Schulze E-D, Lange OL, Oren R (eds) (1989) Forest decline and air pollution. Ecological Studies, vol 77. Springer, Berlin Heidelberg New York

Seckmeyer G, Payer HD (1990) Requirements for artificial irradiation of plants in closed chambers. In: Payer HD, Pfirrmann T, Mathy P (eds) Environmental research with plants in closed chambers. Air pollution research reports of the EC, vol 26. CEC, Brussels, pp 299-308

Seckmeyer G, Payer H-D (1993) A new sunlight simulator for ecological research on plants. J Photochem Photobiol B Biol 21:175-181

Senser M (1990) Influence of soil substrate and ozone plus acid mist on the frost resistance of young Norway spruce. Environ Pollut 64:265-278

Senser M, Schötz F, Beck E (1975) Seasonal changes in structure and function of spruce chloroplasts. Planta 126:1-10

Senser M, Kloos M, Lütz C (1990) Influence of soil substrate and ozone plus acid mist on the pigment content and composition of needles from young Norway spruce trees. Environ Pollut 64:295-312

Showalter AM (1993) Structure and function of plant cell wall proteins. Plant Cell 5:9-23

Siefermann-Harms D (1987) The light harvesting and protective functions of carotenoids in photosynthetic membranes. Physiol Plant 69:561-568

Siefermann-Harms D (1990) The Hohenheim long-term experiment: chlorophyll, carotenoids and the activity of the xanthophyll cycle. Environ Pollut 68:293-303

Siefermann-Harms D (1996) Destabilization of the antenna complex LHC II during needle yellowing of a Mg-deficient spruce tree exposed to ozone pollution – comparison with other types of yellowing. J Plant Physiol 148:195-202

Simmons CR (1994) The physiology and molecular biology of plant 1,3-β-D-glucanases and 1,3;1,4-β-D-glucanases. Crit Rev Plant Sci 13:325-387

Simons S (1993) Biochemische Effekte und Symptomentwicklung bei Buchen (*Fagus sylvatica* L.) und Nadelgehölzen unter realen und proportional erhöhten Ozonkonzentrationen. PhD Thesis, Ludwig-Maximilians University München

Smith WH (1990) Air pollution and forests. Interactions between air contaminants and forest ecosystems. Springer, Berlin Heidelberg New York

Steiger A (1990) Biochemische Grundlagen zur Nadelvergilbung bei Fichten (*Picea abies* (L.) Karst.) unter dem Einfluß von Luftschadstoffen. PhD Thesis, Ludwig-Maximilians Univeristy München

Strack D, Heilemann I, Mömken M, Klinkott ES, Krause GHM, Nowack R, Stannartz B, Wray V (1987) Workshop "Luftverunreinigungen und Waldschaden". Bericht No 1, Essen, pp 329-353

Taylor G, Dobson MC (1989) Photosynthetic characteristics, stomatal responses and water relations of *Fagus sylvatica*: impact of air quality at a site in southern Britain. New Phytol 113:265-273

Tenter M, Wild A (1991) Investigations on the polyamine content of spruce needles relative to the occurrence of novel forest decline. J Plant Physiol 137:647-654

Thalmair M, Bauw G, Thiel S, Doehring T, Langebartels C, Sandermann H (1996) Ozone and ultraviolet B effects on the defense-related proteins β-1,3-glucanase and chitinase. J Plant Physiol 148:222-228

Thiel S, Doehring T, Koefferlein M, Kosak A, Martin P, Seidlitz HK (1996) A phytotron for plant stress research: how far can artificial lighting compare to natural sun light? J Plant Physiol 148:456-463

Tingey DT, Standley C, Field RW (1976a) Stress ethylene evolution: a measure of ozone effects on plants. Atmos Environ 10:969-974

Tingey DT, Wilhour RG, Standley C (1976b) The effect of chronic ozone exposures on the metabolite content of Ponderosa pine seedlings. For Sci 22:234-241

UN-ECE (1994) Critical levels for ozone. In: Fuhrer J, Achermann B (eds) UN-ECE workshop report. Schriftenreihe FAC 16, Bern

Van Camp, W, Willekens H, Bowler C, Van Montagu M, Inzé D, Reupold-Popp P, Sandermann H, Langebartels C (1994) Elevated levels of superoxide dismutase protect transgenic plants against ozone damage. Bio/Technology 12:165-168

Van den Driessche R, Langebartels C (1994) Foliar symptoms, ethylene biosynthesis and water use of young Norway spruce (*Picea abies* [L.] Karst.) exposed to drought and ozone. Water Air Soil Pollut 78:153-168

Van Loon LC, Pierpoint WS, Boller T, Conejero V (1994) Recommendations for naming plant pathogenesis-related proteins. Plant Mol Biol Rep 12:245-264

Wegener A (1995) Ozon-induzierte Änderungen im Transkriptmuster von Kiefernkeimlingen (*Pinus sylvestris* L.): Charakterisierung beteiligter Gene. PhD Thesis, Ludwig-Maximilians University, München

Wild A, Schmitt V (1995) Diagnosis of damage to Norway spruce (*Picea abies*) through biochemical criteria. Physiol Plant 93:375-382

Willekens H, van Camp W, van Montagu M, Inzé D, Langebartels C, Sandermann H (1994) Ozone, sulfur dioxide, and ultraviolet B have similar effects on mRNA accumulation of antioxidant genes in *Nicotiana plumbaginifolia* L. Plant Physiol 106:1007-1014

Wolfenden J, Robinson DC, Cape JN, Paterson IS, Francis BJ, Mehlhorn H, Wellburn AR (1988) Use of carotenoid ratios, ethylene emissions and buffer capacities of the early diagnostics of forest decline. New Phytol 109:85-95

Yalpani N, Enyedi AJ, Leon J, Raskin I (1994) Ultraviolet light and ozone stimulate accumulation of salicyclic acid, pathogenesis-related proteins and virus resistance in tobacco. Planta 193:372-376

Yin Z-H, Langebartels C, Sandermann H (1994) Specific induction of ethylene biosynthesis in tobacco plants by the air pollutant, ozone. Proc R Soc Edinb 102B:127-130

Yang C, Wilksch W, Wild A (1994) 1-Aminocyclopropane-1-carboxylic acid, its malonyl conjugate and 1-aminocyclopropane-1-carboxylate synthase activity in needles of damaged and undamaged Norway spruce trees. J Plant Physiol 143:389-395

Zarembinski TI, Theologis A (1994) Ethylene biosynthesis and action: a case of conservation. Plant Mol Biol 26:1579-1597

Zielke H, Sonnenbichler J (1990) Natural occurrence of 3,3′,4,4′-tetramethoxy-1,1′-biphenyl in leaves of stressed European beech. Naturwissenschaften 77:384-385

Zinser C (1996) Induktion der Gene der Zimtalkohol-Dehydrogenase und der Stilbensynthase durch Ozon und UV-B in der Kiefer (*Pinus sylvestris* L.). PhD Thesis, Ludwig-Maximilians-Universität München

7 Controlled O_3 Exposures and Field Observations of O_3 Effects in the UK

A.R. WELLBURN, J.D. BARNES, P.W. LUCAS, A.R. MCLEOD
and T.A. MANSFIELD

7.1 Introduction

Damage caused by air pollution to conifers in the UK was first recorded by Cohen and Ruston (1912), and it has long been known that one of the most serious effects of air pollution on conifers is increased susceptibility to cold injury and drought (Munch 1933; Lux 1965) including winter desiccation. Consequently, studies of the effects of atmospheric O_3 on tree health using controlled O_3 exposures in the UK have been directed towards evaluating these relationships. The O_3 pollution climate of the UK and several surveys of tree health are discussed along with the results of long-term exposures including data obtained from the large-scale fumigation experiment at Liphook. Detailed results from UK chamber studies of O_3-induced changes in surface waxes, winter hardening, nutrient leaching, photosynthesis, carbon allocation and responses to biotic factors in conifers are followed by a description of the studies of the effects of interactions of drought with atmospheric O_3 on deciduous trees.

7.2 The O_3 Climate of the UK

In response to various recommendations, the UK Department of the Environment expanded its network of O_3 monitoring stations to encompass 17 rural sites varying in their topography and meteorology. Data from this coordinated monitoring network, and from several other sites where routine measurements of atmospheric O_3 are made, have enabled a meaningful assessment to be made of the likely impact of tropospheric O_3 on vegetation, human health and building materials in the UK.

Peak O_3 concentrations in Britain are associated with periods of warm, dry weather typical of stationary or slow-moving anticyclones over northern Europe which provide easterly air flows with fairly low wind

Ecological Studies, Vol. 127
Sandermann et al. (eds) Forest Decline and Ozone

speeds (UK-PORG 1987, 1993). Under such conditions the level of O_3 in ambient air frequently exceeds internationally recognised air quality guidelines. Ozone budgets derived by following air columns between well-mixed phases on successive days during photochemical episodes have revealed a net O_3 generation capability under UK conditions of about 45 nl l^{-1} h^{-1} in April and about 60 nl l^{-1} h^{-1} in mid-summer under similar conditions (Weston et al. 1989). During such photochemical episodes the ground level concentration of O_3 in areas remote from sources of industrial pollution may attain average levels of up to 100 nl l^{-1} between April and September. Under exceptional circumstances even higher concentrations have been recorded, as in the summer of 1976 when peak hourly concentrations of 258 nl l^{-1} were monitored in southern England (UK-PORG 1987). Although climatic conditions in the UK generally restrict photochemical episodes to spring and summer, long-term monitoring has revealed that O_3 can also attain phytotoxic levels for short periods during the winter months (UK-PORG 1993). However, the biological significance of these sporadic episodes remains to be established.

Because of the relatively short atmospheric lifetime of O_3, and its precursors (NO, NO_2 and hydrocarbons), the uneven geographic distribution of its sources, and variations in solar radiation and temperature, there are considerable spatial and temporal variations in the ground level concentrations of O_3 in both rural and urban areas. In rural areas, this is largely a consequence of spatial and temporal changes in the degree to which individual sites are linked to the main reservoir of O_3 in the boundary layer (Cape et al. 1994). Ground level O_3 concentrations in valleys or coastal plains show a clear diurnal cycle, whilst windy upland areas characteristic of land used for traditional forestry purposes are strongly coupled to the free troposphere and show no such diurnal pattern in O_3 concentrations (Fig 7.1). This variability of O_3 cycles leads to exposures which vary with altitude (Derwent and Kay 1987).

Utilizing the available monitoring data and knowledge of the spatial and temporal variability in the ground level concentration of O_3, Fowler and co-workers have recently generated altitude-corrected maps showing the average summer O_3 exposure for every square kilometre of the UK (Cape et al. 1994). This provides the first realistic picture of the summertime exposure to O_3 across the whole of the UK, and draws attention to the pronounced south-east to north-west gradient in ground level O_3 concentrations. Photochemical O_3 episodes increase in frequency from the north-west of Scotland, with 5-30 h in excess of 60 nl l^{-1}, to the south-east of England with 200 h in excess of 60 nl l^{-1} (Fowler et al. 1993). Indeed, much of southern Britain experiences O_3 episodes almost as frequently as parts of central Europe where exposure to 200 h year^{-1} of O_3

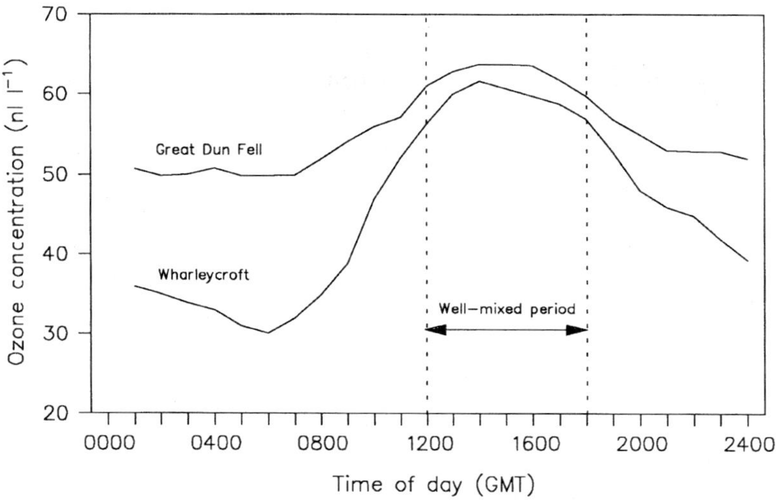

Fig. 7.1. Mean ozone concentrations during photochemical episodes at neighbouring sites varying in altitude; Great Dun Fell (847 m) and Wharleycroft (200 m). (UK-PORG 1993)

levels above 60 nl l^{-1} is common and where there is concern over the deteriorating health of a number of major tree species (Roberts et al. 1989; Schulze and Freer-Smith 1991).

One approach that has been adopted to identify those areas where vegetation may be at the greatest risk of O_3 damage is the concept of critical levels. This approach relies on the identification of limits of exposure to pollutants that will protect vegetation from any measurable adverse effects, and is reviewed extensively elsewhere (see Davison and Barnes 1992; Ashmore and Wilson 1993; UN-ECE 1993). Recently, attempts have been made to define the critical level of O_3 for different types of vegetation, taking into account occasional exposure to O_3 episodes and the period of exposure during daylight hours, by calculating the critical level from the accumulated exposure of O_3, concentrations in excess of a 40 nl l^{-1} threshold (AOT 40; UN-ECE 1993). For the protection of forest trees, this critical level has been provisionally set at an AOT 40 of 10 μl l^{-1} h^{-1} calculated over 24 h during the 6 months of the growing season when the tree is perceived to be at its most sensitive (UN-ECE 1993). Using the AOT 40 as a guideline it becomes possible for the first time to identify those areas of the UK where trees may be considered at risk of O_3 damage. Exceedance maps like that shown in Fig. 7.2 (see Appendix) based on the provisional critical level for the protection of forest trees, reveal that trees in central and southern England as well as Wales may be under the greatest threat from O_3. While the apparent exceedance of the critical level in parts of south Wales remains to be verified, it is possible that lack of monitoring stations in this region results in a skewed distribution of the summer mean height-corrected O_3 concentrations.

As more information becomes available, the critical level of O_3 for the protection of coniferous and broad-leaved trees will undoubtedly be modified. Consequently, any exercise based on current scientific understanding that aims to identify regions of the UK where the critical level is exceeded is fraught with difficulties, and maps like that presented in Fig 7.2 (see Appendix) are probably best regarded as providing no more than a tentative guideline to the likelihood of injury.

Present uncertainties stem partly from the fact that the majority of evidence for the role of O_3 in reducing tree vitality has been drawn from experimental studies rather than field observations, as the O_3 concentrations to which trees are exposed in the field are rarely high enough to induce characteristic visible symptoms. Inevitably, experimental work has focussed on the responses of seedlings or young trees to O_3 over relatively short periods (a maximum of a few years) in open-top chambers or solardomes. In the UK, conifers are grown for up to 60 years before felling, compared with up to 100 years in other parts of Europe, so any feasible experimental exposure to controlled levels of air pollutants corresponds to only a very small part of the life cycle. The dangers in extrapolating from relatively short-term fumigations, often at unrealistic pollutant concentrations, to effects of ambient O_3 on the vitality of mature forest trees are self evident. Not only are there difficulties with the choice of the experimental models and the time-scales over which many studies have been conducted, but there are also concerns over the way in which controlled O_3 exposures have been performed. Controlled experiments rarely if ever reproduce the chemical conditions which are always associated with O_3 episodes (Cape et al. 1994), yet even relatively small changes in microclimatic conditions (e.g. the exclusion of UV light by growing plants behind glass or under artificial illumination) may radically affect the subtle chemical balance of the atmosphere. The possible significance of such shifts in gas-phase chemistry are emphasized by data showing that removal of nitric oxide (NO) influenced the responses of Sitka spruce [*Picea sitchensis* (Bong.) Carr] to O_3 (Neighbour et al. 1990). The uncertainty is also fuelled by the many important gaps that remain in our fundamental knowledge of the impact of O_3 on forest trees – particularly the timing of O_3 exposure in relation to the tree's phenological state, the importance of the plants' prior O_3 history, the impact of O_3 in combination with other pollutants and the way in which O_3 may influence forest pests and diseases.

7.3 Surveys of Tree Health

The UK Forestry Commission (FC; now the Forestry Authority), triggered by reports of forest decline in central Europe and north-eastern America, formally began to assess the health of trees in the UK in 1984. However, as in other European countries, the early surveys were not comparable with later more rigorous assessments. The 1984 FC survey examined three tree species: Sitka spruce, Norway spruce [*Picea abies* (L.) Karst.] and Scots pine (*Pinus sylvestris* L.). Sampling was based on a factorial design, with approximately 100 plots located in plantations throughout the UK (Binns et al. 1985), and the study was repeated in 1985 with minor modifications (Binns et al. 1986). In both years, two indices believed to reflect tree health were assessed: crown density and crown discoloration. On the basis of these indices, neither survey identified a widespread problem although there was clear evidence of specific problems in local regions. This contrasted markedly with the results of an independent survey (Rose and Neville 1985), which concluded that typical symptoms of acid rain damage to trees were widespread in Britain. The different conclusions were attributed to differences in the interpretation of data relating to crown density and discoloration, but the contrasting data sets attracted considerable media interest. During the subsequent debate concerns were expressed over certain aspects of the FC surveys (see Rose 1985; Innes 1988). These prompted increased efforts to standardize the observations made by different surveyors in subsequent years (Innes et al. 1986) and to bring assessments of crown conditions in line with those adopted in Switzerland (Bosshard 1986).

By 1989, over 30 different measures of crown condition were being recorded for each species in more than 350 plots around the UK. This has enabled a more detailed assessment of crown condition to be made than is possible in other standard European surveys (Innes and Boswell 1989, 1990b; Innes 1993). There were two major consequences of recalibration. Firstly, the recalibration to Swiss standards resulted in a recorded deterioration in the condition of trees between 1985 and 1987, but there is no evidence that this was anything other than an artefact resulting from the changes in survey standards. Secondly, recalibration introduced a degree of uncertainty in comparisons of data before and after 1987. Consequently, subsequent analyses have generally been confined to data obtained from 1987 onwards to ensure a safe interpretation (Innes 1993). In addition, several broad-leaved species, sessile oak (*Quercus petraea* (Mattuschka) Lieblein) and English oak (*Quercus robur* L.), and hybrids of the two, as well as beech (*Fagus sylvatica* L.) were added to the FC surveys in 1987. Initially, sample sizes were small, but there has been a gradual expansion so that by 1991 the number of plots for the five main species investigated was approximately equal. There have also been several other assessments of tree health, generally concentrating on trees growing in locations different from those of the FC surveys. The data for the major tree species in the UK are discussed briefly below (Innes 1993; UK-TERG 1993).

7.3.1 Sitka Spruce

Sitka spruce [*Picea sitchensis* (Bong.) Carr.] is the most important commercial forest tree in the UK accounting for 28% of the forested area (Locke 1987). The FC is the only body that has undertaken surveys on the condition of this species, and their data relating to changes in crown density between 1987 and 1993 in 67 plots of Sitka spruce distributed throughout the UK are presented in Table 7.1. Surveys indicate that the majority of trees have crown density reductions of between 10 and 49%, with the most frequent type of defoliation involving the uniform loss of needles throughout the crown (Innes 1993). There are considerable fluctuations in crown density from year to year which make it impossible to identify any temporal trends in crown density at this stage (UK-TERG 1993). Although the reasons for the large year-to-year variations in crown condition are far from fully understood, there is a general consensus that a combination of mild winter weather and outbreaks of the green spruce aphid (*Elatobium abietimum* Walker) are major contributing factors (Redfern et al. 1987). No significant relationships have been found between crown condition and the patterns of pollutant exposure experienced in Britain (UK-TERG 1993), and the condition of individual stands

Table 7.1. Mean percentage of tree crown densities for the period 1987-1993. Class *0* 0-10% reduction in crown density; class *1* represents 11-20% reduction; class *2* represents 21-30% reduction and so on, in relation to a fully-foliated tree (n= total number of sampled trees)

Species	Crown density class						
	0	1	2	3	4	5	6-9
Sitka spruce (n=1064)	11.7	25.1	27.3	19.3	9.4	3.7	3
Norway spruce (n=1431)	16.9	26.9	26.7	18.4	7.6	3.4	2.7
Scots pine (n=1348)	8	23.6	33.4	21.4	7.8	2.6	2.7
Oak (n=724 to 1990 then 1740)	6.3	16.5	26.3	29	14	5.1	3
Beech (n=642 to 1990 then 864)	8.7	24.3	30.4	21.9	8.9	2.5	1.1

is very variable, with adjacent stands showing differences in crown condition as large as those observed for stands separated by long distances (Innes and Boswell 1989).

7.3.2 Norway Spruce

Norway spruce accounts for only 6% of the forested area of the UK (Locke 1987), so the species is of much less commercial importance than in many other parts of central Europe. However, particular attention has been paid to the condition of this species since the widespread decline of Norway spruce in mountainous areas of central Europe has attracted considerable concern (see Roberts et al. 1989; Schulze and Freer-Smith 1991). FC data relating to the condition of Norway spruce in 76 plots distributed throughout the UK are summarized in Table 7.1. Crown densities remained virtually unchanged between 1987 and 1990 although the majority of trees showed crown density reductions of up to 39%. However, between 1990 and 1991, there was a sharp decline in the condition of Norway spruce reflected in the high proportion (45%) of trees that had higher transpiration scores, although this disappeared between 1992 and 1993. The reason for this rapid decline and recovery in health is unknown but a similar trend is also apparent from surveys conducted in other northern and central European countries. The main factors affecting the condition of Norway spruce during this period are believed to be the summer droughts of 1989 and 1990 and a series of mild winters, or a combination of the two. Ozone levels were, however, higher than usual during this period, so photochemical oxidant pollution cannot be excluded as a contributory factor (Innes and Boswell 1991a,b).

In contrast to the most frequent pattern of defoliation seen in the UK between 1987 and 1990 (i.e. the uniform loss of needles throughout the crown), between 1990 and 1992 there was a sharp increase in the number of trees developing gaps in the upper crown, a condition often called subtop dying (Schröter and Aldinger 1985). This disorder is associated with the most severe type of spruce decline in central Europe, but other characteristic symptoms (e.g. nutrient imbalances. Mg deficiency, upper surface chlorosis of older needles) have not been observed in the UK (Innes 1993). This is thought to be because Mg deposition in precipitation in the UK and in other countries along the maritime seaboard of northern Europe is much greater than that in central Europe and may be sufficient to supply the trees' demand (Roberts et al. 1989). This conclusion is also supported by reports that UK forest stands generally do not respond to Mg fertilization (Binns and Grayson 1967; Binns et al. 1980).

7.3.3 Scots Pine

Scots pine is the second most common tree species in the UK, constituting 13% of the forested area (Locke 1987). Although it is one of the few coniferous species native to the UK, current plantations are predominantly derived from foreign stock (Lines 1987). There have been two surveys of Scots pine condition in the UK, the most extensive being undertaken as part of the FC's annual monitoring programme. The other by researchers at Imperial College was concentrated on the condition of isolated individuals growing on lowland heaths in southern England. The FC's assessment of changes in crown density in 81 plots distributed throughout the UK during the period 1987-1992 is presented in Table 7.1. In general, the condition of trees deteriorated over the survey period, although the latest assessments of tree health suggest that the situation may have stabilized (Redfern et al. 1994). There was a particularly sharp decline in health between 1990 and 1991, but the factors underlying the deterioration in crown condition remain to be elucidated. The summer droughts of 1989 and 1990, combined with high levels of O_3, storm damage and a series of late spring frosts may have played a major role. As with Sitka and Norway spruce, some of the crown thinning can be attributed to the rate of growth of individual trees, the fastest growing trees generally having thinner crowns than slower growing ones. Since the first assessments of crown condition in 1987, a small proportion of trees with extremely thin crowns, and even some mortality, has been observed during the survey period. Crown density appears to be particularly poor in the south of Scotland, where it is associated with the combined effects of the needle-cast fungus (*Lophodermium seditiosum* Minter), scleroderris canker [*Brunchorstia pinea* (P Karsten) Höhnel] and the pine shoot beetle (*Tomicus piniperda* L). However, the reason why these should affect stands so severely in this region is unknown.

The Imperial College survey showed that 66% of trees had a reduction in crown density of greater than 25%. However, no significant relationships were found between crown density and site-factors, climate or pollution levels, a broadly similar position to that identified by the more extensive FC survey of tree health.

7.3.4 Oak

Oak is a common component of broad-leaved woodland in the UK and constitutes 9% of the forested area (Locke 1987). There are two main species, sessile oak [*Quercus petraea* (Mattuschka) Lieblein] and pedunculate or English oak (*Quercus robur* L.). These are known to hybridize freely and no distictions have been made in surveys, although it is recognized that the two species differ in their ecological requirements and, when planted together, one may decline while the other remains healthy (Innes 1993). There have been two independent assessments of the condition of oak in the UK, the most extensive undertaken as part of the FC's annual survey, the other by Greenpeace which concentrated on sites of conservation interest in the south of England (Tickle 1988).

The results of the FC surveys between 1987 and 1993 in 77 plots around the UK are presented in Table 7.1. In 1990, 50% of the trees surveyed were reported to exhibit some form of dieback (Innes and Boswell 1990a), and between 1990 and 1992, tree condition deteriorated still further with the result that, by the end of 1992, the condition of oak was poorer than that of any of the other species surveyed in the UK (Redfern et al. 1993). The current condition of oak in the UK is a cause for some concern, but the factors underlying the poor health of this species are far from fully understood. It is believed that summer droughts combined with a series of mild winters, outbreaks of oak mildew (*Microsphaera alphiloids*) and defoliation by insect pests may be responsible. However, a new phenomenon known as oak decline (Greig 1992), which is sometimes lethal, has been shown to be affecting trees in certain parts of southern and eastern England. At present, not much is known about this decline.

The Greenpeace survey (Tickle 1988) reported that 52% of the trees in the south of England displayed losses in crown density of 25%, or more. Although comparisons with the FC's 1988-1989 survey are difficult, as both woodland and non-woodland trees were included and the survey was restricted to the south of England, it is interesting to note that the full FC survey of oak found a very similar number of trees (48%) with losses in crown density of 25% or more.

7.3.5 Beech

Beech (*Fagus sylvatica* L.) is not an important commercial forestry species in the UK, and constitutes only 4% of the forested area (Locke 1987). However, it has been given special attention because of widespread reports of a decline in the health of the species in Europe, and because of its importance for conservation and amenity purposes in the UK. The main FC survey has relatively few beech sites (864 trees scattered throughout the UK in 60 plots) as the condition of the species has received a great deal of attention in other studies. Although crown density data (see Table 7.1) revealed considerable year-to-year fluctuations in tree condition between 1987 and 1990, no clear trends in vitality during this period were evident. Some difficulties with the widely adopted system devised by Roloff (1985a,b) to assess the crown condition of beech have been noted by the FC, and latterly a more detailed classification system described by Westman (1989) has been adopted (see Innes 1993). This system is primarily concerned with the development of gaps within the canopy. The results obtained for trees assessed by the Westman method suggest that a large proportion of trees have gaps in the crown (UK-TERG 1993). Both methods of assessment show a sharp deterioration in tree condition between 1990 and 1991. Although there has been a slight, but encouraging, recovery of trees since 1991, the overall condition of beech in the UK remains poor (Redfern et al. 1993). FC data also show that the proportion of trees with thin crowns increases with increasing openness of the stand, a phenomenon that has also been observed in central Europe (Keller and Imhof 1987).

Beech has also been the subject of a special 2-year study conducted by the Forestry Commission, with the aim of identifying whether trees in the UK showed any symptoms of the widespread deterioration in health noted in Germany (Lonsdale 1986a,b). When biotic and site-related factors were taken into account, there was little evidence of widespread changes in crown density, architecture, or colour that could be judged unusual. However, subsequent detailed analyses of shoot extension at 15 sites along a transect from south-west to eastern England revealed that trees at seven sites had entered a period of declining growth at varying times between 1971 and 1979 (Lonsdale et al. 1989). Prolonged droughts were shown to have had a major adverse effect on shoot growth. However, modelling exercises revealed that even given the impacts of a series of prolonged droughts the overall performance of the trees was significantly poorer than would be expected. The additional factors affecting tree growth were not explained and, although there was a strong suspicion that exposure to phytotoxic concentrations of O_3 may have contributed to the observed decline in tree performance, there were insufficient data on O_3 concentrations to enable correlations to be established.

Researchers at Imperial College have also examined the health of beech. Trees were surveyed at 72 sites of conservation interest in southern Britain between 1987 and 1990. Of the 1728 trees assessed in 1987–1988, 25% were in poor health, 23% had a loss of crown density of more than 25%, and a similar proportion of trees were placed in categories 2 and 3 using the Roloff classification system. However, the incidence of moderate to severe chlorosis was generally low at 5%. In subsequent years, the condition of 17–24 trees at 16 sites was reassessed. These data reveal a steady decline in the health of surveyed trees between 1987 and 1990

(Power 1992). Differences in site conditions (disturbance, tree age, openness of the stand, and the soil pH and drainage) could explain much, but not all, of the differences in tree health between and within adjacent sites, and there was evidence that ambient levels of acidifying air pollutants may have contributed to the changes in tree health (Ling et al. 1993). Although there was the suspicion that O_3 may have also contributed to the observed decline in tree vitality, this could not be tested because of a lack of monitoring data.

Subsequent dendrochronological analyses applied to a limited number of trees at six sites revealed a sharp decline in twig growth associated with the severe drought in southern England in 1976; healthy trees quickly regained their pre-1976 growth rates but unhealthy trees at two of the sites failed to recover (Power 1994). At most sites, a significant proportion of the variance in twig growth could be explained by combinations of climatic variables. However, O_3 was implicated in the best-fit model for unhealthy trees at two sites. More recent attempts to link modelled pollution data with survey data have revealed some evidence to support the hypothesis that O_3 may be a factor contributing to the decline in health of beech in southern England (Mather and Savill 1994). However, compared with the Imperial College survey, a greater number of trees (39%) were found by Greenpeace to have a loss of crown density of 25% or more, and the percentage of trees in Roloff categories 2 and 3 (48%) was also higher than that found by the Imperial College group (Tickle 1988).

7.3.6 Ash

Ash (*Fraxinus excelsior* L.) is not a particularly important forest species and accounts for only 4% of the forested area of the UK (Locke 1987). However, it has been widely planted in hedgerows, and it forms an increasingly important component of the British landscape following the loss of elms throughout the UK. Despite this, the species has not featured in any specific survey, other than a very small number of trees assessed by the FC as part of the UN-ECE survey. A condition known as ash dieback, which was first reported around 40 years ago, is known to affect hedgerow trees in the east of the UK (Hull and Gibbs 1991). The factors underlying this condition remain poorly understood, but it has been proposed that agricultural practices, atmospheric pollutants, pathogenic infections and insect infestations may contribute to it (Pawsey 1983). The importance of O_3 in the decline of hedgerow ash remains to be established, but the lack of a correlation between the incidence of the condition and the geographic distribution of O_3 suggests that the pollutant may have only a minor role to play in ash decline.

7.3.7 Yew

Yew (*Taxus baccata* L.) is not a commercially important forestry species. Its main importance is as an ornamental tree for amenity planting, although it does occur naturally in certain parts of the UK. The only data relating to the condition of yew were collected by Greenpeace (Tickle 1988). The survey was conducted at 40 sites (320 trees) in southern England; 74% of the trees were found to exhibit a reduction in crown density of 25% or more, and the presence of adventitious shoots and dying sub-branches were reported to be common. Many of the surveyed trees were located in churchyards, or on roadsides, where local influences cannot be assessed so the value of the data is difficult to judge.

7.3.8 UN-ECE Surveys

In addition to its own surveys, the FC also monitors the crown condition of trees at 71 sites in the UK on behalf of the EU. At each of the sites, mainly located in southern England and Scotland, the condition of 24 trees was assessed (Innes and Boswell 1990b). Most (63%) were conifers and the majority (76%) less than 60 years old. Despite difficulties in interpretation because of the small sample sizes involved (Innes and Boswell 1991a), the nature and distribution of sample plots (Innes 1988) and slight differences in protocols in different countries (Landmann 1990), the results of recent UN-ECE tree surveys must be a cause for some concern as they show the condition of UK trees to be markedly poorer than in any other European country (UN-ECE 1992). The 1991 survey revealed 52% of the trees observed in the UK to be moderately or severely defoliated, compared with 25% in Germany, 21% in Switzerland, and 7% in France and Austria. Moreover, UN-ECE surveys reveal a progressive deterioration in tree health in the UK, in terms of defoliation, since 1987 when the international surveys commenced (see UK-TERG 1993).

7.3.9 The Early Diagnosis Survey

In an effort to quantify the impact of air pollution on forest trees, researchers from the Institute of Terrestrial Ecology and Lancaster University carried out a survey of 12 forest sites across north-west Europe from northern Scotland to southern Germany. At each site, samples of Norway spruce, Scots pine and beech trees were collected and subjected to a wide range of diagnostic tests (Cape et al. 1988, 1989; Wolfenden et al. 1988). Stepwise discriminant analysis of the results for 2-year-old Norway spruce needles indicated that 14 of these tests, five of them relating to soil components, were capable of making a significant contribution (> 5%) to the discrimination between sites. When these were combined in a simple average cluster analysis, a discrimination between geographically adjacent sites was achieved, with a separation along the transect from northern Scotland to southern Germany (Cape et al. 1988). Canonical discriminant analysis of the same data then showed that the first canonical variable correlated with mean rainfall, pH and O_3 level. The ranking of sites by this variable was very similar to the clustering produced by the average linkage method (Cape et al. 1988). Three of the tests in these studies (differences in dry weight/fresh weight ratio between current and 2-year-old needles, ethylene emissions and violaxanthin/antheraxanthin ratios) showed a significant relationship to tree damage in areas affected by forest decline which were independent of site effects and again suggested that atmopheric O_3 may contribute to this problem (Mehlhorn et al. 1988). Unfortunately, large-scale surveys across Europe like this one have not been continued.

7.4 Long-Term Ozone Exposures

7.4.1 Filtration Experiments

Between 1988 and 1993, the FC undertook a series of experiments to assess the effects of ambient air pollution on the growth and physiology of four major tree species in the UK, Sitka spruce, Norway spruce, Scots pine and beech. Open-top chambers (OTCs) were situated at Headley (Hampshire), Chatsworth (Derbyshire) and Glendevon (Fife), the sites being chosen to provide a range of pollution climates (Willson et al. 1987). At each site, eight chambers were ventilated with ambient air and eight with filtered air, and there was an equal number of outside plots enabling the effects of ambient air quality and the effects of the chambers to be identified. Because it was not possible to grow every species in each chamber, there were effectively four replicate chambers per treatment and species at each site (Lee et al. 1990). Sitka spruce was not grown at Headley and beech was not grown at Glendevon because the climate is unsuitable. Only data up to 1991 are currently available. At Headley, where the predominant air pollutant was O_3 (Durrant et al. 1992), the height growth of Norway spruce and beech was found to be reduced in ambient compared with filtered air for 3 years in succession, and there were marked differences in stem and needle dry weights in Norway spruce and Scots pine (Durrant et al. 1992). Specific studies on beech and Sitka spruce have also revealed effects of ambient air on root morphology with the production of longer and thinner roots in OTCs ventilated with unfiltered air at Headley (Taylor et al 1989; Taylor and Davies 1990). In contrast, no significant effects of ambient air quality were found on height, stem diameter, or fresh or dry weights of any of the species at Glendevon. At Chatsworth, results for the three coniferous species were similar to those reported above for Glendevon, but slightly increased needle dry weights and improved height growth were found in filtered compared with ambient air. Beech, on the other hand, exhibited a marked improvement in growth in the filtered air treatment at Chatsworth. After 3 years, height growth showed a 25% improvement in filtered air, stem diameter increased by 15% and stem dry weights increased by 80% (Durant et al. 1992).

Additional studies on poplar conducted in the OTCs at Headley by Taylor and Frost (1992) have revealed significant effects of ambient air pollution on leaf production and growth. These authors reported that leaf growth was significantly depressed early in the season in response to exposure to ambient air, but this effect was lost later in the year when leaf growth accelerated in expanding leaves at the top and middle of the canopy. During this later period, polluted air simultaneously enhanced the senescence of leaves lower down the plant. Subsequent experimental work has provided strong support for the view that these two effects are linked, with nutrients (especially N) being retranslocated from prematurely senescing leaves to those growing higher up the canopy (Freer-Smith 1993)

7.4.2 Fumigation Experiments

The effects of long-term exposure to O_3 on the growth of Sitka and Norway spruce seedlings in the UK have been examined at Lancaster. Both species are frequently planted on upland sites in Britain, particularly Sitka spruce, and during the summer such areas may experience a larger mean concentration and longer exposure to elevated O_3 concentrations (> 60 nl l^{-1}) than low altitude sites (UK-PORG 1993). The exposure regime used in these experiments therefore sought to represent the type of O_3 exposure which these species might experience in the field during the summer months.

Two year-old seedlings of both spruce species were grown in four solardome chambers, two of which received O_3 at a 7-h mean concentration of 70 nl l^{-1}, 5 days per week (AOT40, 25 nl l^{-1} h^{-1}). The remaining two chambers acted as controls and received filtered air. On average, the trees were exposed to O_3 for 13 weeks for each of three successive summers. In the case of Sitka spruce, fumigation was from 1986 until 1988, while for Norway spruce fumigation was from 1987 until 1989. Throughout each growing season the height and stem diameter of each tree were measured at regular intervals. For the Sitka spruce, growth measurements were also continued during 1989 following removal of the trees from the exposure chambers. The growth data for each year and for each species were analysed separately using a repeated measurements statistical model (Diggle 1990) to describe mean growth response, i.e. logarithmic volume growth, [$\log(D^2H)$].

For the Norway spruce, exposure to O_3 had no statistically significant effect on growth in any year. For the Sitka spruce, however, a significant treatment effect was observed during 1988 and 1989 (Fig. 7.3). Moreover, approximately 30% of the O_3-treated Sitka exhibited chlorotic mottle on the older needles, symptoms more typical of injury caused by acute rather than chronic exposure to O_3. Subsequent analysis of this data revealed no evidence of an association between reduced growth and symptoms of chlorosis. At Lancaster, however, similar studies with Sitka spruce for a longer period do suggest that the cumulative uptake of O_3 over three summers of exposure can produce senescence-like visible symptoms of

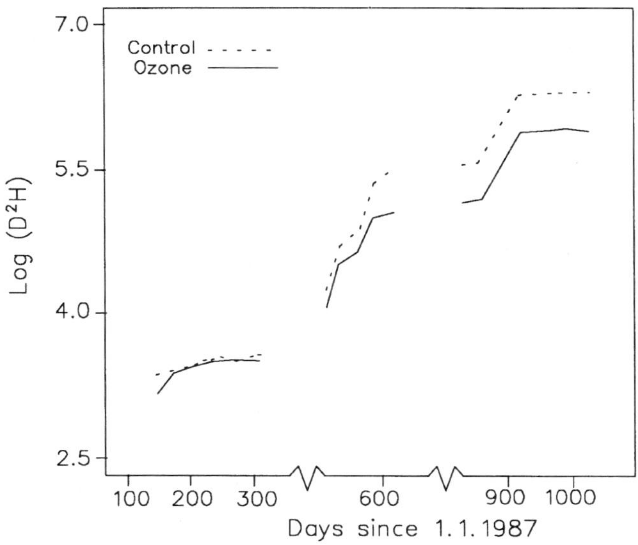

Fig. 7.3. The effect of a realistic exposure to O_3 (7 h-mean of 70nl l^{-1} for 5 days week^{-1}) on the volume growth ($\log[D^2H]$) of two-year old Sitka spruce seedlings grown in solardomes and fumigated for approximately 13 weeks during consecutive summers between 1986 and 1988. Measurements were continued for 1 year after the termination of the O_3 treatment

chlorosis in early summer (Lucas et al. 1993; Fig. 7.4, see Appendix). Unlike deciduous species, however, the needles did not begin to abscise until the year after expression of the symptoms.

These results clearly illustrate the value of multi-season exposure experiments for perennial species, since cessation of the study after 1 year would have produced very different conclusions from those reached after 3 years. In addition to being species dependent, the effects appear to be both cumulative and irreversible since for the Sitka spruce a significant decrease in growth was still apparent 1 year later in the absence of the pollutant. Although little is known about the temporal pattern of O_3 distribution that most affects trees, some studies suggest that acute peak concentrations are more important than longer-term chronic exposures (reviewed by Lefohn 1992). However, the work reported here suggests that for conifers which retain their needles for several years, repeated exposure to lower concentrations of O_3 should also be considered as an important modifier of tree growth.

7.4.3 The Liphook Field Exposure Experiment

The Liphook Forest Fumigation project was a 7-year study of the effects of SO_2 and O_3 on coniferous trees using the technique of open-air fumigation (McLeod et al. 1992; McLeod 1995; McLeod and Skeffington 1995). More than 4000 young trees of Scots pine and Norway and Sitka spruce were exposed to these two pollutants both singly and in combination for nearly 4 years at a site in south-east England where episodes of O_3 pollution occur fairly frequently. The advantages of this exposure method were that it minimizes changes in ambient microclimatic conditions and eliminates chamber effects, and the large plot sizes (50 m in diameter) permit the simultaneous study of many ecosystem components. The major aim of this project was to determine whether realistic concentrations of SO_2 produced symptoms of forest damage and, if so, whether such damage could be expected in the UK.

Two-year-old seedlings were planted in 1985 in seven plots (Fig. 7.5, see Appendix) and their growth was assessed for 2 years before fumigation with SO_2 began in 1987 and with O_3 in 1988. Further details on the plots and exposure concentrations (ambient annual mean = 25 nl l^{-1} and AOT40 = 10.73 μl l^{-1} h; elevated annual mean = 30 nl l^{-1} and AOT40 = 30 μl l^{-1} h), as well as the full range of studies undertaken at the Liphook site, have been given by McLeod (1995) and McLeod and Skeffington (1995). Holland et al. (1995) found that additional O_3 had no effect on the growth of any species either alone or in combination with SO_2, and typical chlorotic mottle or banding caused in some conifers by O_3 (Skelly 1987) was not seen (Shaw et al 1993b). Similarly, Holland and Mueller (1992) found that both Norway and Sitka spruce were affected by frost damage just after bud burst in the high SO_2 treatment plots but could find no effect of the O_3 treatment on spring frost damage. Shaw and McLeod (1995) also reported no effects of the O_3 treatment on the concentration of foliar nutrients and Cape et al. (1995) found no effects of the O_3 treatment on needle wettability or electrolyte leakage although they were able to detect effects of the SO_2 treatment in Norway spruce.

The composition of throughfall below Norway and Sitka spruce was studied by Skeffington and Sutherland (1995). They observed no effects of the above-ambient O_3 treatment on the rate at which nutrients were leached from foliage, although SO_2 treatments did increase foliar leaching. They did, however, observe a trend for SO_2 deposition to be enhanced by simultaneous O_3 fumigation during 1988. McLeod and Skeffington (1995) suggested that this was unlikely to be due to O_3 because, although it may enhance SO_2 oxidation in solution (Chameides 1987), this is limited by its low solubility and a rate of reaction that decreases with decreasing pH (Finlayson-Pitts and Pitts 1986). However, the more reactive oxidant H_2O_2 is known to be produced from O_3 in solution and in the presence of organic molecules (Heikes 1984).

The N_2O_5 produced by air-fed O_3 generators may also have confounded studies of O_3 effects (McLeod et al. 1989). At Liphook, Holland et al. 1995) attributed an increase in the lammas growth of Norway spruce in 1988 in the O_3-treated plot to the N_2O_5 content because the effect was not observed in subsequent years after a scrubber was fitted. Similarly, Shaw and Johnston (1993) observed an elevated microbial activity in fallen pine needles from the O_3-treated plot in 1988 when N_2O_5 was present but not in subsequent years.

Darrall (1992a) examined net photosynthesis and transpiration of detached current and 1-year-old needles of Scots pine in 1988 and 1989 and found no effects of the additional O_3, although net photosynthesis and quantum yield in current year foliage was reduced by the combined high SO_2 + O_3 treatment. Further measurements made in 1990 revealed no effects of either O_3 or SO_2 treatments on net photosynthesis, transpiration, quantum yield, light compensation point, dark respiration, CO_2-saturated photosynthesis, carboxylation efficiency or stomatal conductance in Scots pine (Darrall 1992b). She also found no treatment effects in 1989 and 1990 on net photosynthesis and transpiration of attached shoots of Norway spruce, and no effects on these parameters were found in 1991 (Darrall 1992b), the year after fumigation had ceased. Photosynthesis of detached shoots of Norway spruce was also examined by Wedler et al. (1995), who found a reduction in net photosynthesis in the combined high SO_2 + O_3 treatment in 1988 but not in 1989.

Peace et al. (1995) investigated a range of substrates and enzymes involved in carbohydrate metabolism of Scots pine and Norway spruce between February and July 1989. They found that activity of sucrose phosphate synthase (SPS) was decreased in needles of Scots pine in all the O_3 and SO_2 treatment plots during the summer months of May and July, in particular in those receiving the O_3 only treatment. In Norway spruce, there was also evidence for reduced SPS activity following O_3 treatment. They suggested that this would lead to a reduced rate of sucrose synthesis because this enzyme is a key regulatory point in sucrose or starch allocation. This conclusion is supported by reduced amounts of total soluble sugars (including sucrose, glucose, fructose and their phosphates) in Scots pine needles exposed to both O_3 and SO_2 treatments during the summer, and to a lesser extent in Norway spruce needles (Peace et al. 1995).

Endophytic, phyllosphere and pathogenic fungi were studied at Liphook by Magan et al. (1995). Although the pathogenic fungus *Lophodermium piceae* was occasionally isolated from needles, there was no relationship with treatments. This study, however, did find evidence that the endophyte *Rhizospaera kalkoffii* colonized the needles of Sitka spruce at a higher level in the O_3-treated plot but there was no evidence that O_3 (or SO_2) affected the occurrence of fungal pathogens. Pearce and McLeod (1995) also found no effect of any treatment on resistance to wood invasion or on the amount of stilbene antifungal compounds in the bark of either Norway or Sitka spruce. However, Heller et al. (1991) detected an increase in the level of secondary metabolites having antimicrobial properties in the needles of Scots pine at Liphook with both O_3 and SO_2 treatments.

Studies of insect performance at Liphook supported past observations that shoot aphids feeding on foliage exposed to SO_2 have a higher growth rate (McLeod and Shaw 1990). However, parallel studies by Salt and Whittaker (1995) examining the root aphids of Norway spruce at Liphook reported that the occurrence of *Pachypappa tremulae* and *Prociphilus xylostei* was decreased by O_3 treatment and the high SO_2 treatments with and without O_3. Despite many reports that air pollution influences mycorrhizae (Jansen 1988), Shaw et al. (1992, 1993a) found no effects of O_3 or SO_2 during extensive surveys of mycorrhizae of Scots pine and Norway spruce at Liphook using both fruit body counts and root harvests for morphotype analysis.

7.5 Ozone-Induced Changes in Epicuticular Waxes

Epicuticular waxes, which cover the surface of leaves and needles, constitute the main interface between a plant and its aerial environment. Because of their vulnerable position, and the relative ease by which structural changes in the epicuticular waxes can be visualized, the wax layer is often reported as the first target for the effects of air pollutants. In conifers, the wax layer comprises a delicate and intricate network of fine tubular crystals which fill and surround the stomatal antechambers (Jeffree et al, 1971). However, in response to a variety of environmental factors (see Günthardt-Goerg et al 1994), the physico-chemical nature of the waxes changes as the needles age (Günthardt-Goerg 1986; Crossley and Fowler 1986). This ageing process is characterized by aggregation and fusion of the tubular wax crystals, a process believed to represent the transition of the predominant wax constituent, nonacosan-10-ol (m/z M^+ 496;M-15 481) from a metastable crystalline state to an amorphous form. Conifers growing in polluted regions have exhibited premature wax degradation (Fowler et al. 1980; Cape and Fowler 1981; Cape 1983; Crossley and Fowler 1986). In the UK, this effect is believed to reflect the known effects of SO_2 and aqueous acidity on the production and subsequent ageing of the epicuticular waxes (Cape 1994), but there is growing evidence that ambient levels of O_3 may also contribute toward the physico-chemical changes in leaf surface waxes (Barnes and Brown 1990; Barnes et al. 1990b,c; Kerfourn and Garrec 1992; Percy et al. 1992; Percy et al. 1994)

Interest in the impact of O_3 on conifer needle waxes was triggered by correlations between degraded wax structure and needle loss in declining stands of Norway spruce in the Bavarian Forest, implying that air pollutants may contribute to forest decline via effects on the surface waxes (Sauter and Voss 1986). At the time, it was known that the heterogeneous mixture of saturated hydrocarbons ($C_{12} > C_{64}$), which comprise the epicuticular wax layer in spruce and other species of Pinaceae form a chemically inert surface layer which is unlikely to react directly with atmospheric pollutants such as O_3 (Hoigne 1988). Consequently, the view that O_3 may contribute to forest decline via effects on the surface waxes was treated with a certain degree of scepticism, which seemed to be justified when a series of short-term exposures to relatively high O_3 concentrations administered after the period of needle expansion failed to show any changes in the structure of the epicuticular waxes on the surface of spruce needles (Günthardt-Goerg and Keller 1987; Lütz et al. 1990; Cape 1994). However, in recent years our understanding of the impact of O_3 on leaf surface waxes has improved considerably, and with it a growing awareness that the timing of the O_3 exposure may be vital in determining the impact of O_3 on the surface waxes.

Barnes and coworkers (Barnes et al. 1990 b,c; Barnes and Brown 1990) drew attention to the fact that long-term exposure to realistic O_3 concentrations resulted in qualitative changes in the waxes on the surface of conifer needles. They were also able to demonstrate that controlled exposure to O_3 in solardomes and OTCs over several seasons resulted in a significant increase in needle wettability (determined by measuring the contact angle of water droplets placed on the surface). These observations provided indirect evidence that O_3 influenced the nature of the waxes exposed on the leaf surface. Recent studies on spruce have confirmed these reports (Percy et al. 1990), and have also shown that long-term O_3 exposure is associated with significant changes in wax composition (Kerfourn and Garrec 1992; Percy et al. 1992, 1993; Günthardt-Goerg et al. 1994).

During these studies it has been shown that qualitative changes in waxes induced by long-term exposure to realistic O_3 concentrations are not always linked with changes in the final amount of wax on the needle surface (Barnes and Brown 1990; Günthardt-Goerg et al. 1994). In plants exposed to O_3, the rate of de novo wax biosynthesis may continue at a slower rate for longer (Percy et al. 1992). In addition to effects on epicuticular waxes, O_3 may increase the thickness of the cuticular membrane through the production of a new amorphous wax layer overlying the reticulate layer (Günthardt-Goerg and Keller 1987; Percy et al. 1990). Chemical analysis of this amorphous layer suggests that it consists of insoluble estolides ($>C_{64}$) constituted from several esterified ω-hydroxy fatty acids (Günthardt-Goerg and Keller 1987). Although the function of this non-reticulate layer is unclear, it may act as an additional mechanism to control gas exchange and leaching of ions in O_3-exposed needles, since a similar layer increasing in thickness as needles age has been found in Norway spruce trees growing at the Alpine timberline (Günthardt-Goerg and Wanner 1982).Three possible mechanisms that may underpin the effects of O_3 on the physicochemical nature of conifer waxes are described below.

Direct Reaction with Wax Constituents. The wax extracted from the surface of conifer needles consists of a heterogeneous mixture of saturated hydrocarbons ($C_{12} > C_{64}$), predominantly secondary alcohols and diols, free *w*-hydroxy fatty acids, free fatty acids and estolides (Riederer 1989), which form a chemically inert layer that is unlikely to react directly with O_3. This conclusion is supported by recent experimental evidence (Percy et al. 1994) which has shown no structural or chemical changes in wax recrystallized from young red spruce trees, or in recrystallized single homologue references representing all predominant odd- and even-chain wax components, in response to controlled fumigation with relatively high concentrations of O_3 for short periods (150 nl l^{-1}, 8 h day^{-1} for 7 days). Furthermore, studies in which O_3 has been administered to conifers after the period of needle expansion have revealed no effects on epicuticular waxes (Günthardt-Goerg and Keller 1987; Lütz et al 1990; Cape 1994) or on the permeability of isolated cuticles (Kerstiens and Lendzian 1989). The available evidence thus suggests that O_3 probably does not act via direct effects on the surface waxes in situ.

Indirect Effects. It has been proposed that O_3 may indirectly affect the surface waxes under the combined influence of the pollutant itself and reactive products formed as a result of interactions with other gases in the boundary layer in either the gaseous or liquid phase (Osswald and Elstner 1987). Ozone is known to react with a number of hydrocarbons (Salter and Hewitt 1992) to yield a number of highly reactive products (including free radicals) that may be able to react directly with the saturated hydrocarbons of leaf surface waxes. Some biogenic molecules (e.g. ethylene, ethane, ethanol, terpenes and isoprene) are also very soluble in lipids and may accumulate from the atmosphere resulting in changes in the physical properties and structure of semi-crystalline waxes, a situation that may be exacerbated in the presence of O_3. Plants themselves may be major sources of such molecules, usually released through the stomata, which may explain why the greatest effects of O_3 on the structure of conifer needle waxes are often reported to be in and around the stomatal antechambers. Although, O_3–hydrocarbon interaction afford an elegant explanation for the observed effects of O_3 on conifer needle waxes, little convincing experimental data exist at this time to demonstrate the importance of this type of reaction.

Effects on Wax Biosynthesis. There is increasing evidence that the timing of the O_3 exposure in relation to the phenology of the tree is critical in determining the impact of O_3 on the physico-chemical nature of the surface waxes. In conifers, wax secretion onto the needle surface begins prior to bud burst and continues until the needle is fully expanded (Juniper 1960; Günthardt-Goerg et al. 1994). Consequently, effects of O_3 on the microsomal-based epidermal elongase–decarboxylase or elongase–reductase enzyme systems responsible for wax biosynthesis may only be revealed if the O_3 exposure occurs during this critical period. When O_3 was administered to conifers after needle expansion, no significant effects of the pollutant on the epicuticular wax layer were found (Cape 1994;

Percy et al. 1994). This implies that the main impact of O_3 on the epicuticular waxes occurs via effects on wax production (Percy, et al. 1990; Kernfourn and Garrec 1992; Percy et al. 1992, 1994; Günthardt-Goerg et al. 1994), rather than direct interactions with the needle surface. However, whether such effects result from a direct effect of O_3 on the microsomal-based enzyme systems responsible for wax biosynthesis, or an indirect effect of the pollutant on secondary metabolism, remains to be established.

7.6 Effects of O_3 on Winter Hardiness and Frost Injury

A physiological and biochemical understanding of winter hardiness is now emerging (Wellburn 1995) and there is considerable evidence that summer exposure to elevated concentrations of O_3 may reduce the winter hardiness of conifers. The first report of an interaction between summer O_3 and winter hardiness was made by UK researchers working for the Central Electricity Generating Board research laboratories at Leatherhead (now closed). They noted that some clonal trees of Norway spruce, left over after a summer-long O_3 fumigation in solardomes similar to those described by Lucas et al. (1987) showed signs of visible winter injury in the spring of 1986 following exposure to an early out-of-season frost. The damage, however, was restricted to the older needles (i.e. the 1984 flush) and only occurred on those trees that had previously been exposed to O_3 in 1985 when 3 years old (Brown et al. 1987). Unfortunately, at the same time, these researchers found that their O_3 chambers may have been contaminated with small amounts of N_2O_5 because their O_3 generator used a stream of air rather than O_2 (Brown and Roberts 1988). Nevertheless, a series of studies both in the UK and the USA confirmed their findings that summer O_3 decreases the ability of conifers to harden properly.

The following year, Barnes and Davison (1988) exposed the same eight clones of 3-year-old Norway spruce from Bavaria to 120 nl l^{-1} O_3 for 70 days using controlled environment cabinets, starting in November 1986. Excised shoots were frozen to temperatures between –6 and –19 °C in January 1987. In four clones, severe freezing damage, scored as percent visible injury, subsequently developed on the older needles of plants (i.e. 1985 flush) that had been previously exposed to O_3. Initially, this damage was suggested to provide evidence that O_3 predisposed plants to freezing injury (Barnes and Davison 1988). However, although the visible injury observed was similar to that reported by Brown et al. (1987), it was different from frost damage observed in the field. Moreover, this experiment was conducted late in the year when needless would have been expected already to have acquired considerable frost resistance. In this case, it is more likely that this type of injury was due to O_3-induced desiccation (see Barnes et al. 1995a).

Using 2-year-old Sitka spruce seedlings exposed to either charcoal filtered air (<5 nl l^{-1} O_3) or different O_3 regimes (70-170 nl l^{-1}) for 7 h day^{-1} for 3 months (June–August 1986) in solardomes, Lucas et al. (1988) were able to demonstrate reduced winter hardiness on the first of two occasions (November 1986) when shoots were scored for visible damage after an artificial freezing treatment. The majority of their plants were frost hardy below −20 °C at the second analysis in December 1986, but at this stage, variation between individuals obscured any evidence that O_3-treated shoots were less hardy than those grown in clean air. Consequently, they concluded that Sitka spruce exposed to O_3 in summer would be more sensitive to early autumn frosts.

Fincher et al. (1989) exposed young red spruce trees to O_3 in OTCs and followed subsequent winter injury. If only those trees showing visible injury were considered, there was a significant effect of summer (1987) O_3 on the number of injured shoots observed in spring 1988 following winter frosts. However, the summer O_3 exposure had no effect on subsequent winter injury if all the experimental trees (damaged and undamaged) were ranked according to visible injury. This same situation was observed when this experiment was continued for a further year with a similar summer O_3 exposure regime (Fincher and Alscher 1992). This contrasts with the observations of Waite et al. (1994) who found no effect of summer O_3 on winter hardening in red spruce. Indeed, some of their seedlings which had received charcoal-filtered air were significantly less cold tolerant than those which had received elevated O_3.

In the GSF facility near Munich, six different clones of Norway spruce planted in two different soils were exposed to control levels of O_3 (25 nl l^{-1}) and acid mists (pH 5.6) or treatments with O_3 (50 nl l^{-1} with episodic excursions up to 170 nl l^{-1}) and acid mist (pH 3). Following an artificial freezing regime, representative of an alpine site showing forest decline, the shoots were scored for visible injury (Senser and Payer 1989; Senser 1990). Although the severity of the injury was related to both soil type and the clone of spruce tested, no significant effects of exposure to high O_3 plus acid mist were detected.

Following these initial experiments, the problems associated with visual assessments of winter injury after programmed cooling and thawing were solved when Burr (1987) and Murray et al. (1990) developed quantitative assays of frost damage based on changes in conductivity as a measure of the rate of electrolyte leakage. These techniques were rapidly adopted, and a number of research groups using this procedure soon confirmed that atmospheric O_3 increased sensitivity to winter injury. In addition, Chappelka et al. (1990) found visible injury after a late season frost in 1989 on loblolly pine (*Pinus taeda* L.) seedlings that had previously been exposed to O_3 in the summer of 1988 in OTCs. Furthermore, damage was much greater on individual trees from an O_3-sensitive family (GAKR 15-91), and visible symptoms were confirmed using the electrolyte leakage procedure. Edwards et al. (1990), also using loblolly pine seedlings in OTCs, were able to confirm the observations of Chappelka et al. (1990) and showed that summer exposure to O_3 resulted in delayed winter hardening. However, this latter group were unable to find evidence of parallel visible symptoms after thawing. This distinction may be important and may partially explain some of the discrepancies regarding O_3 and winter hardening in the literature. Certainly, there are clonal difference in winter hardiness following exposure of Norway spruce to O_3 at concentrations as low as 40 nl 1^{-1} (Skre and Mortensen 1990).

Meanwhile, Neighbour et al (1990) fumigated red spruce seedlings for two seasons (1987 and 88) in solardomes with either charcoal-filtered air (i.e. plus NO, up to 14 nl 1^{-1}) and O_3 in the first season and either charcoal-filtered air or charcoal and Purafil-filtered air (i.e. 40% less NO) and the equivalent O_3 regimes (all 70 nl 1^{-1} O_3) with and without Purafil^{TM} filtration the following season. Following programmed cooling of needles in the winters of 1987–1988 and 1988–1989, there was a significant effect of summer O_3 plus NO exposures on mid-winter electrolyte leakage, but no effect of O_3 with reduced NO were found after the second year of exposure. From this they concluded that NO takes part in the mechanism by which summer O_3 affects winter hardening.

There is also evidence that summer O_3 may affect individual components of winter hardening. For example, the increases in raffinose (and stachyose) and other soluble sugars that occur in parallel with increasing cold hardiness are delayed in those conifers that have previously been exposed to summer O_3 (Alscher et al. 1989; Barnes et al. 1990a). In the case of fatty acids, however, the full significance of any effect of summer O_3 on winter hardening is obscured if only total fatty acids are analyzed and the fatty acids are not first hydrolyzed from separated lipid classes. This is the case, unfortunately, with several studies of the effects of summer O_3 exposure on the fatty acid composition in conifers (e.g. Fangmeier et al. 1990; Kyburz et al. 1991). The molar percentage of octadecatetraenoate (18:4) in MGDG was found to increase and the molar ratios of two forms of linoleate, $\Delta^{5,9}$18:2 and $\Delta^{9,12}$ 18:2 in MGDG were found to decrease during frost hardening in 5-year-old Norway spruce trees growing in charcoal-filtered air, but not in similar trees exposed to episodes of summer O_3 for three consecutive summers (Wolfenden and Wellburn 1991). This suggested that the pollutant interferes with the biosynthesis of 18:4 fatty acids by inhibiting Δ^5 -desaturation, although significant changes in the degree of unsaturation of MGDG could not then be detected. More recently, however , Wellburn et al. (1994) have found significant decreases in the degree of unsaturation of both C_{16} and C_{18} fatty acids, the molar percentage of $\Delta^{5,9,12,15}$ 18:4, and the molar ratio of $\Delta^{5,9}$ 18:2 to $\Delta^{9,12}$ 18:2 in MGDG at the time of maximum winter hardiness (December) in Norway spruce of similar age which had been exposed to O_3 over four consecutive summers in the OTCs at Bush, Scotland. Prior to this and using the same set of plants, Cape et al. (1990) had already shown that 20% shoot death occurs at -23.5 °C by October in those needles exposed to summer O_3 compared with a similar mortality at -28.5 °C for needles from trees growing in charcoal-filtered air. The changes in plastidic MGDG found by Wellburn et al. (1994) could be traced to a significant effect of summer O_3 on the Δ^5- and Δ^{12}-desaturases acting upon phosphatidylcholine (PC) in the endoplasmic reticulum. The replacement of the Δ^6-subset of C_{18} fatty acids by an equivalent Δ^5-series was confirmed by gas chromatography and mass spectrometry. Molecular modelling also showed that the Δ^5-forms, which resembled the Δ^9-isomers, are very different in shape to the Δ^6-series which were not found and this may account, in part, for the ability of Norway spruce needles to recover from exposure to extremely low winter temperatures. Wellburn (1995), using differences in Δ^5 patterns, has also been able to demonstrate a strong genetic component in lipid changes; the needle lipids of Norway spruce provenances growing for many years in the UK change during winter hardening as though they were still on trees in the original sites throughout Europe from which the seeds of these trees were initially collected.

7.7 Effects of Ozone on Nutrient Leaching

The leaching of solutes from foliage by water is a well-documented phenomenon and several factors are known to influence this process (Tukey 1971). These include the pH of the bathing medium, frequency, duration and intensity of precipitation, and most importantly, the proportion of the surface in contact with a given volume of water (i.e. the wettability of the needle surface; McCune and Boyce 1992). In conifers and xerophytes, the dangers of mineral loss by leaching are reduced to a minimum by the hydrophobic nature of the needle surface and the presence of a relatively thick and impermeable cuticle. However, O_3 is known to influence the nature of the needle surface and to damage cell membranes. This situation will favour the increased exchange of H^+ (deposited on the surface in rainfall) for K^+, Ca^{2+}, and Mg^{2+} on sites in the cuticle, epidermal cell walls or apoplastic space. Excessive leaching of nutrients from the foliage of trees under the combined influence of O_3 and acidic deposition (rain, mist, fog and dew) was proposed as a major factor contributing to the nutrient deficiencies characteristic of the most serious types of spruce decline in central Europe (Prinz et al. 1982; Arndt et al. 1982). Consequently, the influence of O_3, and acid mist, on the rate at which nutrients are leached from foliage has been the subject of extensive laboratory investigation in recent years.

Experiments conducted at Leatherhead during the mid 1980s examined, in detail, the effects of O_3 and acid mist on the rate at which nutrients were leached from the crown/needles of Norway spruce (reviewed in Roberts et al 1988). These investigations showed, as did other researchers around the same time (Krause et al. 1985), that exposure of Norway spruce to O_3 concentrations > 100 nl l^{-1} accompanied by acid mist applications of pH 3-3.5 for 60 days leached a significant fraction (<6%) of the Mg^{2+} from needles. However, these authors also found an increase in the foliar concentration of a number of elements, including N, an effect also observed by Krause (1987, 1988). This led Brown and Roberts (1988) to question the mechanisms underlying the impact of O_3. Was the N accumulated O_3-fumigated plants a result of an effect of pollutant on needle nitrogen metabolism, or was this an artefact associated with the method of O_3 generation? During the course of their studies, they convincingly demonstrated that the reported effects of O_3 on foliar leaching were confounded by the presence of higher oxides of nitrogen (e.g. N_2O_5) and nitric acid vapour, produced as inevitable by-products of air-operated electric discharge O_3 generators. When these artefacts were removed, O_3 was shown to accelerate the rate at which cations were leached from foliage by <3% (Roberts et al. 1989). This conclusion is supported by more recent data gathered during the course of multi-season experiments of juvenile Norway spruce trees in which uncontaminated O_3 was used (see Pfirrmann et al. 1990, 1996a). Moreover, the majority of studies report little effect of O_3 at ambient, or higher concentrations, on the nutrient composition of foliage (see Roberts et al. 1989; Pfirrmann et al 1996a).

7.8 Effects of Ozone on Photosynthesis and Carbon Allocation

7.8.1 Photosynthetic Capacity

Photosynthesis is seen by some as the metabolic process most affected by air pollutants including O_3 (see Pye 1988); but, as Darrall (1989) and others have concluded, other processes (e.g. shoot/root allocation, nitrogen fixation and stress ethylene emissions) are more sensitive. Indeed, the O_3 exposures that have commonly been adopted for photosynthetic studies are somewhat higher than those occurring under natural conditions and have often failed to reproduce the seasonal and diurnal variations in O_3, or the duration of exposure, occurring in the field. Furthermore, few photosynthetic measurements have been made on a specified needle-year-age-class. However, improvements in the design of leaf cuvettes and diagnostic gas exchange techniques have enabled advances to be made in our understanding of the way in which photosynthesis of needles of different ages is affected by O_3.

The photosynthetic rate of older needles has generally been found to be depressed by O_3 at ambient (Wallin et al. 1990), or slightly higher concentrations (see Barnes et al. 1990a, 1995b). These effects have been primarily attributed to decreases in carboxylation efficiency rather than changes in the degree of stomatal limitation (Wallin and Skärby, 1992; Wallin et al. 1992a,b), conclusion supported by shifts (+0.75-1.5%) in the $\delta^{13}C$ isotope ratio of leaves/needles in response to long-term exposure to realistic O_3 concentrations in controlled experiments (Martin et al. 1988; Matyssek et al. 1992; Pfirrmann et al. 1996b).

There is, however, a growing number of reports indicating that O_3 may have a contrasting effect on different needle year ages in evergreen species (reviewed in Barnes et al. 1995b). Although the photosynthetic rate of older needles may be depressed, recent measurements suggest that the photosynthetic rate of the new season's growth may be stimulated by O_3 (see Dobson 1990; Eamus et al. 1990; Wallin et al. 1990; Eamus and Murray 1991; Pier et al. 1992; Beyers et al. 1992). This not only emphasizes the importance of measuring gas exchange in more than one needle age class but also leads to questions regarding possible mechanisms. Although a number of theories have been proposed, Barnes et al. (1995b) consider the most likely explanation to be that O_3 exposure results in premature needle ageing (i.e. O_3 accelerates the entire life cycle of the shoot, so that needles mature earlier and age faster). Irrespective of the underlying mechanism, the O_3-induced stimulation of photosynthesis in the current-year foliage may compensate (at least to some extent) for the detrimental effects of O_3 on older needles.

7.8.2 Carbon Allocation and Utilization

In contrast to the well-documented effects of O_3 on photosynthesis, comparatively little attention has been paid to the impact of O_3 on carbohydrate allocation and utilization in trees. In general, the dark respiration (R_D) rate of leaves and needles is reported to be increased by O_3 exposure (see Darrall 1989; Barnes et al. 1990c; Dizengremel and Petrini 1994). This response is believed to represent the increased costs associated with the detoxification of O_3 with its solution products and the repair of damaged membranes and proteins (Amthor 1988). In view of the effects of O_3 on R_D it is probably not surprising that several authors have found a decrease in the soluble carbohydrate content of O_3-treated leaves and needles (Barnes et al. 1990a, 1995b). Starch, on the other hand, has been reported to accumulate in the needle plastids of O_3-treated conifer needles (Ebel et al. 1990; Schiffgens-Gruber and Lütz 1992). Studies recently conducted by Wellburn and Wellburn (1994) using solardome fumigation chambers to examine the influence of high episodic levels of O_3 on Aleppo pine (*Pinus halepensis* Mill.), an important Mediterranean landscape tree, have shown a marked seasonality in starch accumulation in response to long-term O_3 exposure. A similar effect has been reported by Günthardt-Goreg et al. (1993) for O_3-exposed birch (*Betula pendula* Roth), where starch was observed to accumulate along the fine leaf vein network in summer, but was entirely retranslocated in the autumn prior to senescence. The observed reduction in export from O_3-exposed leaves and needles, damage to phloem sieve cells and the increased partitioning of carbohydrate to starch at the expense of soluble carbohydrates (predominantly sucrose) is entirely consistent with a restriction in the translocation process. This conclusion is supported by observations made on darkened O_3-treated Aleppo pine needles using the very old technique of Sachs from the middle of the last century (Wellburn and Wellburn 1994). These authors excluded light from some trees in the fumigation systems for 2 days in mid-summer, and found that a large proportion of the starch which accumulated in high O_3-exposed pine needles was not mobilized. This contrasts with starch in needles exposed to lower O_3 concentrations in which carbohydrate levels fell dramatically following darkening. These observations support the view that O_3 impairs the processes associated with phloem loading (Okano et al. 1984; Spence et al. 1990; Günthardt-Goerg et al. 1993); however, there are contrasting reports of O_3-treatment on foliar carbohydrate levels in conifers (Paynter et al. 1991; Friend and Tomlinson 1992; Barnes et al. 1995b), and further research is needed to elucidate the mechanism(s) underlying the effects of the pollutant on carbon allocation.

Several studies in which coniferous and broad-leaved trees have been fed $^{14}CO_2$ during, or immediately following, exposure to O_3 have indicated that the pollutant decreases the amount of assimilate moving from leaves to roots (Spence et al. 1990; Friend and Tomlinson 1992). This is reflected in a strong effect of O_3 on the partitioning of dry matter above and below ground, with root growth more severely affected than shoot growth (Davidson et al. 1992; Wellburn and Wellburn 1994) – an effect likely to predispose trees to water stress.

7.9 Interactive Effects of Ozone

7.9.1 Biotic Factors

Although there is increasing awareness that O_3, like a number of other air pollutants, may influence the degree of attack by fungal pathogens and invertebrate pests (see Kvist 1986; Dowding 1988; McNeill and Whittaker 1990; Whittaker and Warrington 1990; Riemer and Whittaker 1992; Bell 1993; Brown 1994), comparatively little attention has been paid to herbivores of woody plants. A number of studies have examined the effects of controlled O_3 fumigation on chewing insects (see Brown 1994) but few have been performed on woody species. In general, O_3-exposure of the host plant has been shown to increase palatability, increase consumption and enhance insect performance. Studies on the performance of phloemfeeding insects on O_3-treated crops and trees are far more difficult to interpret. Both stimulations and reductions in performance have been observed for the same aphid/host combination, the direction and magnitude of response being influenced by a number of factors, such as concentration, duration and episodicity of the pollutant exposure, and the prevailing temperature (Brown et al. 1993; Walmer-Kristiansen et al. 1993). There is, however, general agreement that effects of O_3 on aphid performance result indirectly from changes in host plant chemistry rather than any direct effects on the aphids themselves (Brown et al. 1992). In agreement with this conclusion, Braun and Flückiger (1989) showed that population growth of *Phyllaphis fagi* on beech was stimulated by exposure to ambient levels of O_3 (seasonal mean 40 nl l^{-1}) in semi-open top chambers, and that the effect was linked to a significant increase in the amino acid/sugar ratio of phloem exudate in trees ventilated with ambient air. Other studies, on the response of the spruce shoot aphid (*Cinara pilicornis*) to 100 nl l^{-1} O_3, revealed both increased and decreased rates of aphid growth on 2-year-old pre-fumigated Sitka spruce trees, whilst no significant effects of O_3 were found on *Schizolachnus pineti* feeding on Scots pine (Brown et al. 1993). Similarly, a series of short-term (5-day) experiments by Warrington, Moore and Whittaker (reported in McNeill and Whittaker 1990) failed to detect any significant effects of O_3 (70 nl l^{-1}) on *Eucallipterus tiliae* feeding on lime (*Tilia x europea*).

The response of plant – aphid systems to O_3 is very complex. However, there is cause to be concerned since there is sufficient evidence that O_3, as well as other air pollutants, might affect plant productivity through alteration of plant – insect interactions (see McNeill and Whittaker 1990). The potential for ecologically and economically significant effects on forest trees is real, but as yet it is too early to make predictions as to the extent of these effects or their significance at current ambient levels of O_3.

7.9.2 Water Relations of Deciduous Trees

The extent to which O_3 may influence the water relations of deciduous trees growing under natural conditions in the UK cannot as yet be determined. All of the relevant research on deciduous trees which has been conducted in Britain has been done with seedlings grown in some form of ventilated chamber. Such systems can never replicate conditions in the natural environment. Nevertheless, results from controlled exposure experiments indicate how important physiological mechanisms, which are responsible for the control of plant water status, may be influenced by O_3. Such information is of value in interpreting some of the variety of effects observed in the field.

Two basic physiological mechanisms are available to plants in order to conserve water during a drought. Firstly, reduction of transpiration may be achieved through a decrease in stomatal conductance. Secondly, new assimilates may be allocated to roots to support their growth at the expense of the above-ground parts of the plant. This has the dual function of increasing the capacity of new roots to explore the deeper layers of soil for further supplies of water and restricting transpiration by reducing leaf expansion, and hence leaf area.

In most plants, the strategy for avoiding drought stress involves limiting the rate of water loss, but both of the above mechanisms also result in a decrease in CO_2 assimilation and thus biomass production. It is therefore not unexpected that most research into interactions between O_3

and drought have concentrated on the above mechanisms, particularly changes in stomatal conductance, in view of the importance of stomata as the primary route of O_3 uptake.

Due to the obvious practical difficulties involved, it is not surprising that few studies have attempted to investigate the effects of O_3 on the distribution of carbon in ways that change the capacity of the tree to respond to soil moisture deficits. Using filtered and unfiltered OTCs at a site in southern England, Taylor and Davies (1990) studied the impact of ambient air quality on the growth and development of the roots of 3-year-old beech seedlings during 1988. In contrast to many previous studies, root length as well as root biomass was used as an indicator of development and function. This was based on findings by others (Christie 1975; Paul and Ayres 1986) that the acquisition of water and nutrients was more closely related to total root length and volume than to biomass. The trees were grown in 1-m lengths of plastic tubing which could be split longitudinally so that periodic non-destructive measurements of root growth could be made. Approximately 6 months after planting, water was withheld from half of the seedlings for 10 days. All the seedlings were harvested in early September. In contrast to many previous studies, where root shoot ratios had decreased following exposure to O_3 (see Mansfield 1988), there was no evidence that the allocation of dry matter between roots and shoots was significantly altered in this experiment. However, both root extension and specific root length (root length per unit root dry weight) were significantly greater for beech grown in the unfiltered chambers, which suggests that the root system was not only longer but was also made up of thinner roots. Although water was withheld from the trees for only a short period, both root length and root biomass were reduced, irrespective of whether they were grown in filtered or unfiltered air. No interactive effects of air quality and soil water deficits were therefore observed and these results do not appear to support the hypothesis that trees exposed to ambient air pollutants in southern England are made more susceptible to drought because of effects on roots.

The absence of any interaction between air quality and drought may, however, have been due to the comparatively infrequent occurrence of O_3 episodes during the period of the experiment. During May and June, there were only 23 days in which the concentration of O_3 was in excess of 60 nl l^{-1} and only 5 days in which it exceeded 80 nl l^{-1}. From the point of view of comparing these results with what may be occurring in the field, with hindsight it is perhaps unfortunate that this experiment was not carried out one year later (1989) when the frequency of O_3 episodes in southern England was much higher. Some sites in southern England recorded over 200 h in excess of 60 nl l^{-1} in 1989, compared with less than

100 h in 1988 (UK-PORG 1993). It is also unfortunate that there was no opportunity to make measurements of soil water potential or stomatal conductance since these would have provided indications of the rate and severity of soil moisture deficits produced when water was withheld from the beech seedlings, and of the extent to which stomatal conductance had been affected by this treatment.

These two parameters, in addition to leaf water and solute potentials, photosynthesis and transpiration, were measured in an experiment carried out in solardomes. The experiment was designed to investigate the effects of a combined O_3/drought treatment on the growth and physiology of *Prunus avium* (wild cherry). Unlike the previous study, the plants were exposed to controlled concentrations of O_3 in the range of 60–120 nl l^{-1} superimposed on a background concentration of 30 nl l^{-1}. These concentrations were used to produce a realistic fumigation regime, with the plants being exposed to the highest concentrations on hot sunny days between 12.00–19.00 h. Provision for the imposition of a realistic drought treatment was accomplished by growing the seedlings in large volumes of soil contained in boxes (450 × 915 × 610 mm) with eight plants per box. Each of the four solardomes (two controls and two O_3 treatments) contained four boxes, from two of which water was withheld for 63 days. The soil in the boxes dried gradually, only reaching the lowest soil water potential (-0.075 MPa) after 45 days of withholding water. Stomatal conductance fell in response to soil moisture deficit with minimum conductance values being recorded 30 days after water was withheld when they were approximately one order of magnitude less than well-watered plants. There were no statistically significant effects of O_3 exposure on stomatal conductance, although a reduction in the rate of photosynthesis was observed in the drought treatments.

With one exception, for the majority of the parameters measured, including growth, drought was of overriding importance, and any subtle effects which O_3 may have had on the water relations of the cherry were not detected in this experiment (Sier 1995). The exception was the higher concentration of leaf abscisic acid (ABA) observed in those plants which had been treated with a combination of episodic O_3 exposure and drought (Fig. 7.6). There have been a few studies on the effects of O_3 on ABA production (Kobriger et al. 1984; Fackler et al. 1986) and to our knowledge none has been carried out over the full course of a drought treatment using realistic episodic concentrations of O_3. The significant feature of the data shown in Fig. 7.6 is the timing of the increase in ABA concentration; it occurred approximately 50 days after withholding water and at a time when senescence and abscission of the foliage were starting. It is known that ABA plays a prominent role in the regulation (promotion) of se-

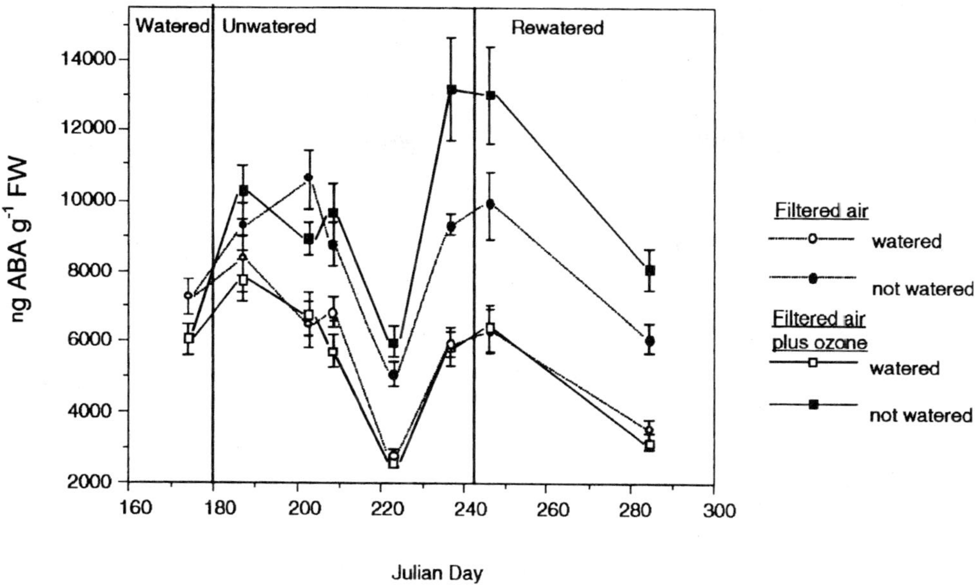

Fig. 7.6. Contents of abscisic acid in leaf discs of well-watered and unwatered *Prunus avium* saplings which were exposed in solardomes to clear air or episodes of O_3 from June to October 1992

nescence (Noodén 1988) and it has been found that levels of endogenous ABA can often rise during early senescence and decrease later (Lindoo and Noodén 1978). Whether the increase in ABA is a direct result of exposure to O_3 or because the O_3 has accelerated the onset of senescence is unknown, although there is good evidence that exposure to O_3 does indeed promote leaf senescence (Keller 1988; Pell 1988; Tjoelker and Luxmoore 1991).

Many of the previous studies in which O_3 has been seen to promote premature leaf senescence in deciduous species have been carried out using O_3 concentrations well in excess of typical concentrations measured at rural sites in England. However, both the work reported here with cherry and that carried out by Wiltshire et al. (1993) on mature apple trees (*Malus domestica*) in which early leaf fall was also observed, used realistic O_3 regimes which implies that this type of damage may occur in the field.

From the examples of UK research considered so far, it appears that the major environmental stress factor influencing the growth and water relations of deciduous trees is not O_3 but soil moisture deficit. If O_3 does have any effect it is likely to act in a subtle manner, for example sensitivity to O_3 could vary depending on the developmental stage of the tree. This was certainly the case for the study on apple by Wiltshire et al. (1993) where the older leaves appeared to be the more susceptible, which suggests that either seasonal accumulated uptake of O_3 was responsible or that leaf aging per se may predispose leaves to O_3 damage.

A further example of the type of subtle interaction which can occur between soil water deficit and O_3 is the work reported by Pearson and Mansfield (1993) in which 3-year-old beech seedlings, grown in solardomes, were exposed to artificially created episodes of O_3 over a 128-day period. The timing of the episodes corresponded to local meteorological conditions which were deemed suitable for the production of ambient photochemical episodes, i.e. high temperatures and incident solar radiation. Each episode lasted from 10.00 to 16.00 h. Throughout the experiment, twice daily measurements of stomatal resistance were made at 08.15 and 11.00 h, on days considered suitable for fumigation with 100 nl l^{-1} O_3. Thirty-nine days after the start of the experiment water was withheld from half the trees. The effects of O_3 on stomatal behaviour were seen in both the well-watered and unwatered trees. For the well-watered group, comparison of stomatal resistance between control and polluted trees on the same days in the absence (08.15 h) and presence (11.00 h) of O_3, showed that the pollution regime caused a two-fold response: first, exposure to successive episodes of O_3 caused a prolonged and lasting increase in stomatal resistance, and second, during pollution episodes an additional increase in resistance was observed, although only transiently. However, with the drought treatment it was found that the stomatal resistance of the O_3-exposed trees did not increase to the same extent as in their comparable controls, which would indicate that exposure to O_3 had disrupted stomatal function in some way.

The indications from this experiment suggest that long-term realistic exposures to O_3 may affect beech in two ways: (1) in well-watered trees by impairing carbon assimilation through increased stomatal resistance, either causing or resulting in reduced photosynthesis; and (2) in droughted trees through an inability to close stomata adequately thus disrupting seasonal control of water status. In the following year (1992) Pearson and Mansfield investigated the subsequent growth of their beech trees after removal from the fumigation chambers (Pearson and Mansfield 1994). In the absence of O_3 and despite uniform growth conditions, the previous drought treatment caused bud break to begin slightly earlier. In addition, rates of shoot growth and the length of the first flush branches were reduced. By the end of the 1992 growing season, total above-ground growth in the trees that had previously been water stressed had been reduced by 36%. In the well-watered plants, shoot growth was reduced during the first week after bud break and exposure to O_3 in the previous year reduced the amount of new growth by 17%, which was largely due to reduced internodal expansion. Ozone in combination with water stress appeared not to have resulted in any additional reduction in new growth, but there was a fall in the number of internodes relative to water-stressed trees grown in filtered air. If effects such as these were to occur in the field they could result in significant changes in canopy architecture since any reduction in the number of internodes would also reduce the number of leaves on a branch and hence crown density as has been observed during recent surveys of beech health in southern England (see above).

It is difficult to extrapolate from measurements of water loss (i.e. stomatal conductance) of individual leaves using porometry to the water loss of a whole tree, due to factors such as leaf age, leaf position and interactions between stomata and the environment. However, methods for measuring whole-tree water use are available, for example, the heat-balance technique for measuring sap flow (Baker and Van Bavel 1987). This procedure was adopted by Wiltshire, Unsworth and Wright (1994) to investigate the effects of O_3 episodes on water use by 4-year-old ash trees (*Fraximus excelsior* L.). The trees were planted directly in the soil and exposed in OTCs to episodes of O_3 (150 nl l^{-1}) for 8 h day^{-1} for 1–4 days in succession when conditions were suitable. Between May and November 1992 there were 27 such occasions. The use of the sap flow technique enabled both daily and seasonal changes in stem flow to be monitored and, although no changes were observed in response to individual O_3 episodes, the longer-term data showed a clear trend with O_3-exposed trees using more water per day than the controls during the early part of the growing season (June-July). Later in the season (August-September), the reverse was the case. It seems likely that these changes were caused by increases in stomatal resistance during the later stages of the experiment. If so, these results appear to agree well with those of Pearson and Mansfield (1993) and suggest that O_3 is a factor involved in the decline of ash in England (Hull and Gibbs 1991).

7.10 Conclusions

Phytotoxic O_3 concentrations are experienced in the UK, especially in Wessex and in Kent, and can occur in the winter as well as the summer. They also vary with altitude and higher land currently used for forestry does not show a clear diurnal cycle of exposure. Maps based on height and on the AOT40 have been included to provide a tentative guide to the possibility of injury across the UK. Tree health surveys have shown reductions in crown densities for most species, with oak and beech in particular showing cause for concern, and that the general condition of trees in the UK is poorer than elsewhere in Europe. Long-term experimental exposures have shown a variety of detrimental effects, which are age dependent, but a large-scale field exposure system at Liphook failed to find detrimental effects on growth of O_3 at means of 5 nl l^{-1} above ambient levels. Controlled exposure experiments, however, suggest that the effects of soil moisure deficit and O_3 in combination appear to be complex, often acting in a subtle manner, and may have long-term implications for the growth and health of deciduous trees. In conifers, O_3-induced restrictions on phloem loading have marked effects on carbon allocation, with roots more severely affected, causing trees to be more susceptible to subsequent droughts. O_3-exposure also changes the physico-chemical nature of needle waxes but no evidence has been found for increased rates of cation leakage. Episodes of O_3 in summer, however, have a significant effect on rates of winter hardening in autumn. In Norway spruce, this had been traced to diminished desaturation of membrane lipids and reduced activity of an unusual Δ^5 fatty acid desaturase.

Acknowledgements. We are grateful to the UK Department of the Environment, Forestry Authority and Natural Environment Research Council, National Power (UK) Ltd., and the Commission of the European Union (DG XII) for funding some of the studies described in this chapter.

References

Alscher RG, Amundson RG, Cumming JR, Fellows S, Fincher J, Rubin G, Van Leuken P, Weinstein LH (1989) Seasonal changes in the pigments, carbohydrates and growth of red spruce as affected by ozone. New Phytol 113:211–223

Amthor JS (1988) Growth and maintenance of respiration in leaves of bean (*Phaseolus vulgaris* L.) exposed to O_3 in open-top chambers in the field. New Phytol 110:319–325

Arndt U, Seufert G, Nobel W (1982) Die Beteiligung von Ozon an der Komplexkrankheit der Tanne (*Abies alba* Mill.) - Eine prüfenswerte Hypothese. Luft 42:243–247

Ashmore MR (1993) Critical levels and agriculture in Europe. In: Jäger HJ, Unsworth M, Temmerman L De, Mathy P (eds) Effects of air pollution on agricultural crops in Europe. Air Pollution Research Report 46. CEC, Brussels, pp 105–130

Ashmore MR, Wilson RB (1993) Critical levels for ozone. Proc UNECE Critical Levels Workshop, Egham, Creative Press, Reading

Baker JM, Van Bavel CHM (1987) Measurements of mass flow of water in the stems of herbaceous plants. Plant Cell Environ 10:777–782

Barnes JD, Brown KA (1990) The influence of ozone and acid mist on the amount and wettability of the surface waxes in Norway spruce (*Picea abies* [L.] Karst). New Phytol 114:531–535

Barnes JD, Davison AW (1988) The influence of ozone on the winter hardiness of Norway spruce [*Picea abies* (L.) Karst]. New Phytol 108:159–166

Barnes JD, Eamus D, Brown KA (1990a) The influence of ozone, acid mist and soil nutrient status on Norway spruce. II. Photosynthesis, dark respiration and soluble carbohydrate status of trees in late autumn. New Phytol 115:149–156

Barnes JD, Eamus D, Brown KA, Davison AW (1990b) The influence of ozone and acid mist on the needle surface and water relations of *Picea abies* [L.] Karst. In: Payer H-D, Pfirrmann T, Mathy P (eds) Environmental research with plants in closed chambers. Air Pollution Research Report 26, CEC Brussels, pp 310–321

Barnes JD, Eamus D, Davison AW, Ro-Poulsen H, Mortensen L (1990c) Persistent effects of ozone on needle water loss and wettability in Norway spruce. Environ Pollut 63:345–363

Barnes JD, Hull MR, Davison AW (1995a) Impacts of air pollutants and rising CO_2 in winter-time. In: Yunus M, Iqbal M (eds) Plant growth and air pollution. J Wiley, London

Barnes JD, Pfirrmann T, Steiner K, Lütz C, Busch U, Küchenhoff H, Payer H-D (1995b) Effects of elevated CO_2, O_3 and K deficiency on Norway spruce (*Picea abies* [L.] Karst. II. Seasonal changes in photosynthesis and non-structural carbohydrate content. Plant Cell Environ (submitted)

Bauer H, Larcher W, Walker RB (1975) Influence of temperature stress on CO_2-gas exchange. In: Cooper JP (ed) Photosynthesis and productivity in different environments. Cambridge University Press, Cambridge, pp 557–586

Bell JNB (1993) Biotic and abiotic interactions with air pollutants. In: Jäger HJ, Unsworth M, Temmerman L De, Mathy P (eds) Effects of air pollution on agricultural crops in Europe. Air Pollution Research Report 46. CEC, Brussels, pp 383–409

Beyers JL, Reichers GH, Temple PJ (1992) Effects of long-term ozone exposure and drought on the photosynthetic capacity of ponderosa pine (*Pinus ponderosa* Laws). New Phytol 122:81–90

Binns WO, Grayson AJ (1967) Fertilisation of established crops: Prospects in Britain. Scott For 21:81–98

Binns WO, Mayhead GJ, Mackenzie JM (1980) Nutrient deficiencies of conifers in British forests. Forestry Commission Leaflet 76. HMSO, London

Binns WO, Redfern DB, Rennolls K, Betts AJA (1985) Forest health and air pollution: 1984 survey. Forestry Commission Research and Development Paper 142. Forestry Commission, Edinburgh

Binns WO, Redfern DB, Boswell RC, Betts AJA (1986) Forest health and air pollution: 1985 survey. Forestry Commission Research and Development Paper 147. Forestry Commission, Edinburgh

Bosshard W (1986) Kronenbilder. Eidgenössiche Anstalt für das forstliche Versuchswesen, Birmensdorf

Braun S, Flückiger W (1989) Effect of ambient ozone and acid mist on aphid development. Environ Pollut 56:177–187

Brown KA, Roberts TM (1988) Effects of ozone on foliar leaching in Norway spruce (*Picea abies* L. Karst): confounding factors due to NO_x production during ozone generation. Environ Pollut 55:55–73

Brown KA, Roberts TM, Blank LW (1987) Interaction between ozone and cold sensitivity in Norway spruce: a factor contributing to the forest decline in central Europe? New Phytol 105:149–155

Brown VC (1994) Insect herbivores and gaseous air pollutants: current knowledge and predictions. In: Stork NE, Harrington R (eds) Insects in a changing environment. Proc 17th Royal Entomological Society Symp. Academic Press, London

Brown VC, McNeill S, Ashmore MR (1992) The effects of ozone fumigation on the performance of the black bean aphid, *Aphis fabae* Scop., feeding on broad beans, *Vicia faba* L. Agric Ecosyst Environ 38:71–78
Brown VC, Ashmore MR, McNeill S (1993) Effects of chronic and acute ozone fumigation on the performance of the black bean aphid, *Aphis fabae*, feeding on *Vicia faba*. In: Jäger HJ, Unsworth M, Temmerman L De, Mathy P (eds) Effects of air pollution on agricultural crops in Europe. Air pollution Research Report 46. CEC, Brussels, pp 491–494
Burr KE (1987) Cold hardiness, root growth capacity, and bud dormancy testing of conifer seedlings. PhD Thesis, Colorado State University, Fort Collins
Cape JN (1983) Contact angle of water droplets on needles of Scots pine (*Pinus sylvestris*) growing in polluted atmospheres. New Phytol 93: 293–299
Cape JN (1994) Evaluation of pollutant critical levels from leaf surface characteristics. In: Percy KE, Cape JN, Jagels R, Simpson C (eds) Air pollutants and the leaf cuticle. NATO ASI Series, vol 36. Springer, Berlin, Heidelberg New York pp 123–138
Cape JN, Fowler D (1981) Changes in the epicuticular wax of *Pinus sylvestris* exposed to polluted air. Silva Fenn 15:457–458
Cape JN, Paterson IS, Wellburn AR, Wolfenden J, Mehlhorn H, Freer-Smith PH, Fink S (1988) Early diagnosis of forest decline: report of a one-year pilot study. Institute of Terrestrial Ecology, Grange-over-Sands
Cape JN, Paterson IS, Wolfenden J (1989) Regional variation in surface properties of Norway spruce and Scots pine needles in relation to forest decline. Environ Pollut 58:325–342
Cape JN, Fowler D, Eamus D, Murray MB, Sheppard LJ, Leith ID (1990) Effects of acid mist and ozone on frost hardiness of Norway spruce seedlings. In: Payer HD, Pfirrmann T, Mathy P (eds) Environmental research with plants in closed chambers. Air Pollution Research Report No 26. Commission of the European Communities, DG XII, Brussels, pp 331–335
Cape JN, Smith RI, Fowler D (1994) The influence of ozone chemistry and meteorology on plant exposure to photo-oxidants. Proc R Soc Edinb 102B:11–31
Cape JN, Sheppard LJ, Binnie J (1995) Leaf surface properties of Norway spruce needles exposed to sulphur dioxide and ozone at Liphook. Plant Cell Environ 18:285–289
Chameides WL (1987) Acid dew and the role of chemistry in the dry deposition of reactive gases to wet surfaces. J Geophys Res 92:11895–11908
Chappelka AH, Kush JS, Meldahl RS, Lockaby BG (1990) An ozone-low temperature interaction in loblolly pine (*Pinus taeda* L.) New Phytol 114:721–726
Christie EK (1975) Physiological responses of semi-arid grasses II. The pattern of root growth in relation to external phosphorus concentration. Aust J Agric Res: 26:437–446
Cohen JB, Ruston AG (1912) Smoke: a study of town air. Edward Arnold, London
Crossley A, Fowler D (1986) The weathering of Scots pine epicuticular wax in polluted and clean air. New Phytol 103:207–218
Crowe JH, Crowe LM (1986) Water and carbohydrate interactions with membranes. Studies with infrared and differential scanning calorimetric methods. Methods Enzymol 127:696–703
Darrall NM (1989) The effect of air pollutants on physiological processes in plants. Plant Cell Environ 12:1–30
Darrall NM (1992a) Measurements of photosynthesis and associated gas exchange parameters in Scots pine exposed to elevated levels of SO_2 and O_3 using the Liphook fumigation system: I 1988–1990 Results, National Power Report TEC/L/0566/R92, Swindon
Darrall NM (1992b) Measurements of photosynthesis and associated gas exchange parameters in Scots pine and Norway spruce exposed to elevated levels of SO_2 and O_3 using

the Liphook fumigation system: II 1989–1991 Results, National Power Rep TEC/L/0567/R92, Swindon

Davidson SR, Ashmore MR, Garretty C (1992) Effects of ozone and water deficit on the growth and physiology of *Fagus sylvatica*. For Ecol Manage 51:187–194

Davison AW, Barnes JD (1992) Patterns of air pollution. The use of the critical loads concept as a basis for abatement strategy. In: Newsom MD (ed) Managing the human impact on the natural environment: patterns and processes. Bellhaven, New York, pp 109–129

Derwent RG, Kay PJA (1987) Modelling and assessments group, Environmental and Medical Sciences Division, Harwell Laboratory. Air Pollution Research Report 6. EUR 11432 EN

Diggle PJ (1990) Time series: a biostatistical introduction. Clarendon Press, Oxford

Dizengremel P, Petrini M (1994) Effects of air pollutants on the pathways of carbohydrate breakdown. In: Alscher RG, Wellburn AR (eds) Plant responses to the gaseous environment: molecular, metabolic and physiological aspects. Chapman & Hall, London, pp 255–278

Dobson MC (1990) Fluxes to, and effects of O_3 and SO_2 on *Picea abies, P. sitchensis* and *Fagus sylvatica*. PhD Thesis, University of Ulster, Ulster

Dowding P (1988) Air pollutant effects on plant pathogens. In: Schulte-Hostede S, Darrall NM, Blank LW, Wellburn AR (eds) Air pollution and plant metabolism. Elsevier, London pp 329–355

Durrant DWH, Waddell DA, Benham SE, Houston TJ (1992) Air quality and tree growth: results of the open-top chamber experiments 1991. Forestry Commission Research Information Note 221. Forestry Commission, Edinburgh

Eamus D, Murray M (1991) Photosynthetic and stomatal conductance responses of Norway spruce and beech to ozone, acid mist and frost – a conceptual model Environ Pollut 67:23–45

Eamus D, Barnes JD, Mortensen L, Ro-Poulsen H, Davison AW (1990) Persistent stimulation of CO_2 assimilation and stomatal conductance by summer ozone fumigation in Norway spruce. Environ Pollut 63:365–379

Ebel B, Rosenkranz J, Schiffgens-Gruber A, Lütz C (1990) Cytological observations on spruce needles after prolonged treatment with ozone and acid mist. Environ Pollut 64:323–335

Edwards GS, Pier PA, Kelly JM (1990) Influence of ozone and soil magnesium status on the cold hardiness of loblolly pine (*Pinus taeda* L.) seedlings. New Phytol 115:157–164

Fackler U, Huber W, Hock B (1986) Influence of mineral nutrition, ozone and acid fog on abscisic acid and indole acetic acid in needles of *Picea abies* (L) Karst. Forstwiss Centralbl (Hamb) 105:254–257

Fangmeier A, Kress LW, Lepper P, Heck, WW (1990) Ozone effects on the fatty acid composition of loblolly pine needles (*Pinus taeda* L.). New Phytol 115:639–647

Fincher J, Alscher RG (1992) The effect of long-term ozone exposure on injury in seedlings of red spruce (*Picea rubens* Sarg) New Phytol 120:49–59

Fincher J, Cumming JR, Alscher RG, Rubin G, Weinstein L (1989) Long-term ozone exposure effects winter hardiness of red spruce (*Picea rubens* Sarg.) seedlings. New Phytol 113:85–96

Finlayson-Pitts BJ, Pitts JN (1986) Atmospheric chemistry: fundamentals and experimental techniques. John Wiley, New York, pp 645–692

Fowler D, Cape JN, Nicholson IA, Kinnaird JW, Paterson IS (1980) The influence of polluted atmosphere on cuticle degradation in Scots pine. In: Drablos D, Tollan R (eds) Ecological impact of acid precipitation. SNSF Project, Oslo, pp 146–150

Fowler D, Cape JN, Smith R, Leith I, Erisman J-W (1993) The pollution climate of Europe. In: Jäger HJ, Unsworth M, Temmerman L De, Mathy P (eds) Effects of air pollution on agricultural crops in Europe. Air Pollution Research Report 46. CEC, Brussels, pp 43–64

Freer-Smith PH (1993) The interacting effects of drought and pollution on whole-plant physiology IN: Jäger HJ, Unsworth M, Temmerman L De, Mathy P (eds) Effects of air pollution on agricultural crops in europe. Air Pollution Research Report 46. CEC, Brussels, pp 307–320

Friend AL, Tomlinson PT (1992) Mild ozone exposure alters ^{14}C dynamics in foliage of *Pinus taeda* L. Tree Physiol 11:215–227

Greig BJW (1992) Occurence of decline and dieback of oak in Great Britain. Forestry Commission Research Information Note 214. Forestry Commission, Edinburgh

Günthardt-Goerg MS (1986) Epicuticular wax of needles of *Pinus cembra, Pinus sylvestris* and *Picea abies.* Eur J For Pathol 16:400–418

Günthardt-Goerg MS (1994) The effect of the environment on the structure, quantity and composition of spruce needle wax. In: Percy KE, Cape JN, Jagels R, Simpson C (eds) Air pollutants and the leaf cuticle vol 36. NATO ASI Series, Springer, Berlin Heidelberg New York, pp 165–174

Günthardt-Goerg MS, Keller T (1987) Some effects of long-term ozone fumigation on Norway spruce II. Epicuticular wax and stomata. Trees 1:145–150

Günthardt-Goerg MS, Wanner H (1982) Die Menge des cuticulären Wachses auf Nadeln von *Pinus cembra* L. und *Picea abies* (L.) Karsten in Abhängigkeit von Nadelalter und Standort. Flora 172:125–137

Günthardt-Goerg MS, Matyssek R, Scheidegger C, Keller T (1993) Differentiation and structural decline in the leaves and bark of birch (*Betula pendula*) under low ozone concentrations. Trees 7:104–114

Günthardt-Goerg MS, Keller T, Matyssek R, Scheidegger C (1994) Environmental effects on Norway spruce needle wax. Eur J For Pathol 24:92–111

Heikes BG (1984) Aqueous H_2O_2 production from O_3 in glass impingers. Atmos Environ 18:1433–1445

Heller W, Rosemann D, Pearce RB, McLeod AR, Sandermann H Jnr (1991) Polyphenol metabolism in Scots pine and Sitka spruce: Effects of ozone and sulfur dioxide in the Liphook open-air fumigation experiment. Conf Abstr, Acidic Deposition Conf, Glasgow, Sept 1990, Royal Society of Edinburgh, Edinburgh, 174 pp

Hoigne J (1988) Bildung und chemische Bedeutung von Photooxidantien in der wässerigen Phase. In: Schulte-Hostede S (ed) Verteilung und Wirkung von Photooxidantien im Alpenraum. GSF-Bericht vol 17. GSF, Munchen, pp 166–175

Holland MR, Mueller PW (1992) Measurements of tree growth and health in the Liphook Forest Fumigation Project: an evaluation of large scale open-air fumigation experiments. In: Schneider T (ed) Acidification research, evaluation and policy applications. Elsevier, Amsterdam pp 357–362

Holland MR, Mueller PW, Rutter AJ, Shaw PJA (1995) Growth of coniferous trees exposed to SO_2 and O_3. Plant Cell Environ 18:227–236

Hull SK, Gibbs JN (1991) Ash dieback – a survey of non-woodland trees. For Comm Bull 93. HMSO, London

Innes JL (1988) Forest health surveys – a critique. Environ Pollut 54:1–15

Innes JL (1993) Forest health: its status and assessment. CAB International, Wallingford, 573 pp

Innes JL, Bosewell RC (1987) Forest health surveys 1987. Part 1. Results. For Comm Bull 74. HMSO, London

Innes JL, Bosewell RC (1989) Monitoring of forest condition in the United Kingdom 1988. For Comm Bull 88. HMSO, London

Innes JL, Boswell RC (1990a) Monitoring of forest condition in Great Britain 1989. For Comm Bull 94. HMSO, London

Innes JL, Boswell RC (1990b) Reliability, presentation and relationships amongst data from inventories of forest condition. Can J For Res 20:790–799

Innes JL, Boswell RC (1991a) Forest monitoring programme: 1991 results. Forestry Commission Research Information Note 209. Forestry Commission, Edinburgh

Innes JL, Boswell RC (1991b) Monitoring of forest condition in Great Britain 1990. For Comm Bull 98. HMSO, London

Innes JL, Boswell RC, Binns WO, Redfern DB (1986) Forest health and air pollution: 1986 survey. Forestry Commission Research and Development Paper 150. Forestry Commission, Edinburgh

Jansen AE (1988) Air pollutant effects on ectomycorrhizal fungi. In: Cape JN, Mathy P (eds) Scientific basis of forest decline symptomalogy. Air Pollution Report 12. CEC, Brussels, pp 181–189

Jeffree CE, Johnson RPC, Jarvis PG (1971) Epicuticular wax in the stomatal antechamber of Sitka spruce and its effect on the diffusion of water vapour and carbon dioxide. Planta 98:1–10

Juniper BE (1960) Growth and development and effect of the environment on the ultrastructure of plant surfaces. J Linn Soc Lond Bot 56:413–419

Keller T (1988) Growth and premature leaf fall in American aspen as bioindicators for ozone. Environ Pollut 52:183–192

Keller W, Imhof P (1987) Zum Einfluss der Durchforstung auf die Waldschäden. 1. Teil: Erste Ergebnisse von Waldschadenuntersuchungen in Buchen-Durchforstungsflächen der EAFV. Schweiz Z. Forstwes 138:39–54

Kerfourn C, Garrec JP (1992) Modifications in the alkane composition of cuticular waxes from spruce needles (*Picea abies*) and ivy leaves (*Hedera helix*) exposed to ozone fumigation and acid fog: Comparison with needles from declining spruce trees. Can J Bot 70:861–869

Kerstiens G, Lendzian KJ (1989) Interactions between ozone and plant cuticles II. Water permeability. New Phytol 112:21–27

Kobriger JM, Tibbitts TW, Brenner ML (1984) Injury, stomatal conductance and abscisic acid levels of pea plants following ozone plus sulphur dioxide exposures at different times of the day. Plant Physiol 76:823–826

Krause GH (1987) Ozone-induced nitrate formation in above ground parts of trees. In: Perry R, Harrison RM, Bell JNB, Lester JN (eds) Acid rain: scientific and technical advances. Selper, London pp 596–601

Krause GH (1988) Ozone-induced nitrate formation in needles and leaves of *Picea abies*, *Fagus sylvatica* and *Quercus robur*. Environ Pollut 52:117–130

Krause GH, Jung KD, Prinz B (1985) Experimentelle Untersuchungen zur Aufklärung der neuartigen Waldschäden in der FRG. VDI-Berichte 560:627–656

Kvist K (1986) Fungal pathogens interacting with air pollutants in agricultural crop production. In: How are the effects of air pollutants on agricultural crops influenced by the interaction with other limiting factors? CEC, Brussels pp 67–78

Kyburz S, Eichenberger W, Keller T, Rennenberg H, Schröder P (1991) Fatty acids of ozone-exposed Scots pine and poplar trees and of damaged field-grown Norway spruce trees. Eur J For Pathol, 21:49–56

Landmann G (1990) Les Symptômes du dépérissement forestier. In: La Sante Des Forets 1990. Ministere de l'agriculture et de la fôret, Paris

Lee HSJ, Willson A, Benham SE, Durrant DWH, Houston T, Waddell DA (1990) The effect of air quality on tree growth. Forestry Commission Research Information Note 182. HMSO, London

Lefohn AS (1992) The characterization of ambient ozone exposures. In: Lefohn AS (ed) Surface level ozone exposures and their effects on vegetation. Lewis, Chelsea, pp 31–92
Lindoo SJ, Noodén LD (1978) Correlations of cytokinins and abscisic acid with monocarpic senescence in soybean. Plant Cell Physiol 19:997–1006
Lines R (1987) Choice of seed origins for the main forest species in Britain. For Comm Bull 66. HMSO, London
Ling KA, Power SA, Ashmore MR (1993) A survey of the health of *Fagus sylvatica* in southern Britain. J Appl Ecol 30:295–306
Locke GML (1987) Census of woodlands and trees 1979-82. For Comm Bull 63. HMSO, London
Lonsdale D (1986a) Beech health study 1985. Forestry Commission Research and Development Paper 164, Forestry Commission, Edinburgh
Lonsdale D (1986b) Beech health study 1986. Forestry Commission Research and Development Paper 149, Forestry Commission, Edinburgh
Lonsdale D, Hickman IT, Mobbs ID, Matthews RW (1989) A quantitative analysis of beech health and pollution across southern Britain. Naturwissenschaffen 76:571–573
Lucas PW, Cottam DA, Mansfield TA (1987) A large-scale fumigation system for investigating interactions between air pollution and cold stress on plants. Environ Pollut 43:15–28
Lucas PW, Cottam DA, Sheppard LJ, Francis B (1988) Growth responses and delayed winter hardening in Sitka spruce following summer exposure to ozone. New Phytol 108:495–504
Lucas PW, Rantanen L, Mehlhorn H (1993) Needle chlorosis in Sitka spruce following a three-year exposure to low concentrations of ozone: changes in mineral content, pigmentation and ascorbic acid. New Phytol 124:265–275
Lütz C, Heinzmann U, Grulz P-G (1990) Surface structures and epicuticular wax composition of spruce needles after long-term treatment with ozone and acid mist. Environ Pollut 64:313–322
Lux H (1965) Ergebnisse von Zuwachsuntersuchungen (Bohnspananalysen) im Rauchschadensgebiet Dübener Heide. Arch Forstwes 14:1103–1121
Magan M, Kirkwood IA, McLeod AR, Smith MK (1995) Effect of open-air fumigation with sulphur dioxide and ozone on phyllosphere and endophytic fungi of conifer needles. Plant Cell Environ 18:291–302
Mansfield TA (1988) Factors determining root: shoot partitioning. In: Cape JN, Mathy P (eds) Scientific basis of forest decline symptomology. COST/CEC, Brussels, pp 171–180
Martin B, Bytnerowicz A, Thorstenson YR (1988) Effects of air pollutants on the composition of stable carbon isotopes, $\delta^{13}C$, of leaves and wood, and on leaf injury. Plant Physiol 88:218–223
Mather R, Savill P (1994) Forest facts. New Sci 143:47
Matyssek R, Günthardt-Goerg MS, Saurer M, Keller T (1992) Seasonal growth, $\delta^{13}C$ in leaves and stem, and phloem structure of birch (*Betula pendula*) under low ozone concentrations. Trees 6:69–72
McCune DC, Boyce RL (1992) precipitation and the transfer of water, nutrients and pollutants in tree canopies. Trends Ecol Evol 7:4–7
McLeod AR (1995) An open-air system for exposure of young forest trees to sulphur dioxide and ozone. Plant Cell Environ 18:215–225
McLeod AR, Shaw PJA (1990) The Liphook fumigation project progress report – June-December 1989. Report No TEC/ELS/0050/M90. National Power Centre, Swindon
McLeod AR, Skeffington RA (1995) The Liphook forest fumigation project: an overview. Plant Cell Environ 18:327–335

McLeod AR, Brown KA, Skeffington RA, Roberts TM (1989) Studies of ozone effects on conifers and the formation of nitrogen during ozone generation. In: Air pollution and forest decline. Eidgenössische Anstalt für das Versuchswesen, Birmensdorf, pp 181–187

McLeod AR, Shaw PJA, Holland MR (1992) The Liphook forest fumigation project studies of sulphur dioxide and ozone effects on coniferous trees. For Ecol Manage 51:121–127

McNeill S, Whittaker JB (1990) Air pollution and tree-dwelling aphids. In: Watt AD, Leather SR, Hunter MD, Kidd NAC (eds) Population dynamics of forest insects. Intercept Andover, pp 195–208

Mehlhorn H, Francis BJ, Wellburn AR (1988) Prediction of the probability of forest decline damage to Norway spruce using three simple site-dependent diagnostic parameters. New Phytol 110:525–534

Munch E (1993) Winterschäden an immergrünen Gehölzen. Ber Dtsch Bot Ges 51:21–26

Murray MB, Cape JN, Fowler D (1990) Quantification of frost damage in plant tissues by rates of electrolyte leakage. New Phytol 113:307–311

Neighbour EA, Pearson M, Mehlhorn H (1990) Purafil-filtration prevents the development of ozone-induced frost injury: a potential role for nitric oxide Atmos Environ 24:711–715

Noodén LD (1988) Abscisic acid, auxin and other regulators of senescence. In: Noodén LD, Leopold AD (eds) Senescence and aging in plants. Academic Press: London, pp 330–368

Okano K, Ito O, Takeba G, Shimizu A, Totsuka T (1984) Alterations of ^{13}C-assimilate partitioning in plants of *Phaseolus vulgaris* exposed to ozone. New Phytol 97:155–163

Osswald WF, Elstner EF (1987) Investigations on spruce decline in the Bavarian Forest. Free Rad Res Comm 3:185–192

Paul ND, Ayres PG (1986) The effects of nutrient deficiency and rust infection on the relationship between root dry weight and length in groundsel (*Senecio vulgaris* L.).Ann Bot 57:353–360

Pawsey RG (1983) Ash dieback survey: summer 1983. Commonwealth Forestry Institute Occasional Paper No 24, Oxford, 22 pp

Paynter VA, Reardon JC, Shelburne VB (1991) Carbohydrate changes in shortleaf pine (*Pinus echinata*) exposed to acid rain and ozone. Can J For Res 21:666–671

Peace EA, Lea PJ, Darrall NM (1995) The effect of SO_2 and O_3 on sucrose metabolism in Scots pine and Norway spruce. Plant Cell Environ 18:277–283

Pearce RB, McLeod AR (1995) Effects of sulphur dioxide and ozone on stilbenes and disease resistance in spruce trees from the Liphook Forest Fumigation Project. Plant Cell Environ 18:303–307

Pearson M, Mansfield TA (1993) Interacting effects of ozone and water stress on the stomatal resistance of beech (*Fagus sylvatica* L.). New Phytol 123:351–358

Pearson M, Mansfield TA (1994) Effects of exposure to ozone and water stress on the following season's growth of beech (*Fagus sylvatica* L.). New Phytol 126:511–515

Pell EJ (1988) Secondary metabolism and air pollutants. In: Schulte-Hostede S, Darrall NM, Blank LW, Wellburn AR (eds) Air pollution and plant metabolism. Elsevier, London, pp 222–237

Percy KE, Krause CR, Jensen KF (1990) Effects of ozone and acid fog on red spruce needle epicuticular wax ultrastructure. Can J For Res 20:117–120

Percy KE, Jensen KF, McQuattie CJ (1992) Effects of ozone and acid fog on red spruce needle epicuticular wax production, chemical composition, cuticular membrane ultrastructure and needle wettability. New Phytol 122:71–80

Percy KE, Jagels R, Marden S, McLaughlin CK, Carlisle J (1993) Quantity, chemistry and wettability of epicuticular waxes on needles of red spruce along a fog-acidity gradient. Can J For Res 23:1472–1479

Percy KE, McQuattie CJ, Rebbeck JA (1994) Effects of air pollutants on epicuticular wax chemical composition. In: Percy KE, Cape JN, Jagels R, Simpson C (eds) Air pollutants and the leaf cuticle. NATO ASI Series, vol 36. Springer, Berlin, Heidelberg New York pp 67–79

Pfirrmann T, Runkel K-H, Schramel P, Eisenmann T (1990) Mineral and nutrient supply, content and leaching in Norway spruce exposed for 14 months to ozone and acid mist. Environ Pollut 64:229–254

Pfirrmann T, Barnes JD, Steiner K, Schramel P, Busch U, Küchenhoff H, Payer H-D (1996a) Effects of elevated CO_2, O_3 and K deficiency on Norway spruce (*Picea abies* [L.] Karst: nutrient supply, content and leaching. Plant Cell Environ (in prep)

Pfirrmann T, Barnes JD, Steiner K, Busch U, Küchenhoff H, Broadmeadow MSJ, Payer H-D (1996b) Effects of elevated CO_2, O_3 and K deficiency on Norway spruce (*Picea abies* [L.] Karst: growth and water use efficiency. Plant Cell Environ (in prep)

Pier PA, Thornton FC, McDuffie C, Hanson PJ (1992) CO_2 exchange rates of red spruce during the second season of exposure to ozone and acidic cloud deposition. Environ Exp Bot 32:115–124

Power SA (1992) Factors influencing the health of *Fagus sylvatica* in southern Britain. PhD Thesis, University of London

Power SA (1994) Temporal trends in twig growth of *Fagus sylvatica* L. and their relationships with environmental factors. Forestry 67:13–30

Prinz B, Krause GH, Stratmann H (1982) Forest damage in the Federal republic of Germany. LIS Report No 28, LIS, Essen, CEGB Translation T14240

Pye JM (1988) Impact of ozone on the growth and yield of trees. A review. J Environ Qual 17:347–360

Redfern DB, Gregory SC, Pratt JE, MacAskill GA (1987) Foliage browning and shoot death in Sitka spruce and other conifers in northern Britain during winter 1983-84. Eur J For Pathol 17:166-180

Redfern DB, Boswell R, Proudfoot J (1993) Forest condition 1992. Forestry Commission Research Information Note 236. Forestry Commission, Edinburgh

Redfern DB, Boswell R, Proudfoot J (1994) Forest condition 1993. Forestry Commission Research Information Note 251. Forestry Commission, Edinburgh

Riederer M (1989) The cuticles of conifers: structure, composition and transport properties In: Schulze E-D, Lange OL, Oren R (eds) Forest decline and air pollution. Ecological Studies 77. Springer, Berlin Heidelberg New York, pp 158–192

Riemer J, Whittaker JB (1992) Air pollution and insect herbivory: observed interactions and possible mechanisms. Insect-Plant Interact 1:73–105

Roberts TM, Skeffington RA, Brown KA, Blank LW (1988) Interactive effects of ozone, acid mist and cold stress on Norway spruce. In: Lavendar DP (ed) Woody plant growth in a changing physical and chemical environment. IUFRO, Vancouver

Roberts TM, Skeffington RA, Blank LW (1989) Causes of Type I spruce decline in Europe. Forestry 62:179–222

Roloff A (1985a) Schadstufen bei der Buche. Forst- Holzwirt 40:131–134

Roloff A (1985b) Untersuchungen zum vorzeitigen Laubfall und zur Diagnose von Trockenschäden in Buchenbeständen. Allg Forst Z 40:157–160

Rose C (1985) Acid rain falls on British woodlands. New Sci 14th Nov:52–56

Rose C, Neville M (1985) Tree dieback survey. Final Report. Friends of the Earth, London

Salt DT, Whittaker JB (1995) Populations of root-feeding aphids in the Liphook forest fumigation experiment. Plant Cell Environ 18:321–325

Salter L, Hewitt CN (1992) Ozone-hydrocarbon interactions in plants. Phytochemistry 31:4045–4050

Sauter JJ, Voss, J-U (1986) SEM observations on the structural degradation of epistomatal waxes in Norway spruce in *Picea abies* L. Karst. and its possible role in 'Fichtensterben'. Eur J For Pathol 16:408–423

Schiffgens-Gruber A, Lütz C (1992) Ultrastructure of mesophyll cell chloroplasts of spruce needles exposed to O_3, SO_2 and NO_2 alone and in combination. Environ Exp Bot 32:243–254

Schröter H, Aldinger E (1985) Beurteilung des Gesundheitzunstandes von Fichte und Tanne nach der Benadelungsdichte. Allg Forst Z 40:1353–1357

Schulze E-D, Freer-Smith PH (1991) An evaluation of forest decline based on field observations focussed on Norway spruce, *Picea abies*. In: Last FT, Walting R (eds) Acidic deposition: its nature and impacts. Proc R Soc Edinb 97B:169–192

Senser M (1990) Influence of soil substrate and ozone plus acid mist on the frost resistance of young Norway spruce. Environ Pollut 64:265–278

Senser M, Payer H-D (1989) The frost hardiness of young conifers exposed to different loads of ozone, sulfur dioxide or nitric oxide in environmental chambers. In: Bucher JB, Bucher-Wallin I (eds) Air pollution and forest decline. IUFRO P2.05, Birmensdorf, pp 523–526

Shaw PJA, Johnston JPN (1993) Effects of SO_2 and O_3 on the chemistry and FDA activity of coniferous leaf litter in an open-air fumigation experiment. Soil Biol Biochem 25:897–908

Shaw PJA, McLeod AR (1995) The effects of SO_2 and O_3 on the foliar nutrient content of Scots pine, Norway spruce and Sitka spruce in an open-air fumigation experiment. Plant Cell Environ 18:237–245

Shaw PJA, Dighton J, Poskitt J, McLeod AR (1992) The effects of sulphur dioxide and ozone on the mycorrhizas of Scots pine and Norway spruce in a field fumigation system. Mycol Res 96:785–791

Shaw PJA, Dighton J, Poskitt J (1993a) Studies on the mycorrhizal community infecting trees in the Liphook forest fumigation experiment. Agric Ecosyst Environ 47:185–191

Shaw PJA, Holland MR, Darrall NM, McLeod AR (1993b) The occurence of SO_2-related foliar symptoms on Scots pine (*Pinus sylvestris* L) in an open-air fumigation experiment. New Phytol 123:143–152

Sier ARJ (1995) The effects of O_3 pollution and soil moisture deficit on the growth and physiology of wild cherry. PhD Thesis, University Lancaster.

Skeffington RA, Sutherland PM (1995) The effect of SO_2 fumigation on deposition and foliar leaching in the Liphook forest fumigation experiment. Plant Cell Environ 18:247–261

Skelly JM (1987) Diagnosing injury to eastern forest trees. USDA Forest Service, Atlanta

Skre O, Mortensen LM (1990) Effects of ozone on frost resistance, growth and carbohydrate status in Norway spruce seedlings (*Picea abies*). Aquilo Ser Bot 29:51–68

Spence RD, Rykiel EJ, Sharpe PJH (1990) Ozone alters carbon allocation in loblolly pine: assessment with carbon-11 labeling. Environ Pollut 64:93–106

Sutherland PM, Skeffington RA (1992) The effects of SO_2 and O_3 on soils at Liphook: equipment installation and results from the first year of fumigation. National Power Report TEC/L/0554-R92, Swindon, UK

Taylor G, Davies WJ (1990) Root growth of *Fagus sylvatica*: impact of air quality and drought at a site in southern Britain. New Phytol 116:457–464

Taylor G, Dobson MC (1989) Photosynthetic characteristics, stomatal responses and water relations of *Fagus sylvatica*: Impact of air quality at a site in southern Britain. New Phytol 113:265–273

Taylor G, Frost DL (1992) Impacts of gaseous air pollution on leaf growth of hybrid poplar. For Ecol Manage 51:151–162

Taylor G, Dobson MC, Freer-Smith PH (1989) Changes in partitioning and increased root lengths of spruce and beech exposed to ambient pollution concentrations in southern Britain. Ann Sci For 46s:573–576

Tickle A (1988) Tree survey of southern England. Greenpeace Environmental Trust, Air Pollution Report 4, London

Tjoelker MG, Luxmoore RJ (1991) Soil nitrogen and chronic ozone stress influence physiology, growth and nutrient status of *Pinus taeda* L. and *Liriodendron tulipifera* L. seedlings. New Phytol 119:69–81

Tukey HB (1971) Leaching of substances from plants. In: Preece TF, Dickinson CH (eds) Ecology of leaf surface mirco-organisms. Academic Press, London, pp 67–80

Uemura M, Steponkus PL (1989) Effect of cold acclimation on the incidence of two forms of freezing injury in protoplasts isolated from rye leaves. Plant Physiol 91:1131–1137

UK-PORG, United Kingdom Photochemical Oxidant Review Group (1987) Ozone in the United Kingdom, Interim Report, Air Quality Division, Department of the Environment. HMSO, London

UK-PORG, United Kingdom Photochemical Oxidant Review Group (1993) Ozone in the United Kingdom, 3rd Report. Air Quality Division, Department of the Environment. HMSO, London

UK-TERG, United Kingdom Terrestrial Effects Review Group (1993) Air pollution and tree health in the United Kingdom. Department of the Environment. HMSO, London

UN-ECE (1992) Forest conditions in Europe, Executive Summary 1992, Report from Programme Co-ordinating Centre, Hamburg

UN-ECE (1993) Critical levels for ozone. Fuhrer J, Achermann J (eds) Swiss Federal Research Station for Agricultural Chemistry and Environmental Hygiene, Liebefeld-Berne

Waite CE, de Hayse DH, Rebbeck J, Schier GA, Johnson AH (1994) The influence of elevated ozone on freezing tolerance of red spruce seedlings. New Phytol 126:327–335

Wallin G, Skärby L (1992) The influence of ozone on the stomatal conductance and non-stomatal limitation of photosynthesis in Norway spruce, *Picea abies*, exposed to soil moisture deficit. Trees 6:128–136

Wallin G, Skärby L, Selldèn G (1990) Long term exposure of Norway spruce, *Picea abies*, to ozone in open top chambers I. Effects on the capacity of net photosynthesis, dark respiration and leaf conductance of shoots of different ages. New Phytol 115:335–344

Wallin G, Skärby L, Selldèn G (1992a) Long term exposure of Norway spruce, *Picea abies*, to ozone in open top chambers III. Effects on the light response of net photosynthesis in shoots of different ages. New Phytol 121:387–394

Wallin G, Ottosson S, Selldèn G (1992b) Long term exposure of Norway spruce, *Picea abies*, to ozone in open top chambers IV. Effects on the stomatal and non-stomatal limitation of photosynthesis and on carboxylation efficiency. New Phytol 121:395–401

Walmer-Kristiansen L, Brown V, Ashmore MR (1993) CEC assessment on biotic interactions. In: Jäger HJ, Unsworth M, Temmerman L De, Mathy P (eds) Effects of air pollution on agricultural crops in Europe. Air Pollution Research Report 46. CEC, Brussels, pp 409–432

Wedler M, Weikert RM, Lippert M (1995) Photosynthetic performance, chloroplast pigments, and mineral contents of Norway spruce [*Picea abies* (L.) Karst] exposed to SO_2 and O_3 in an open-air fumigation system. Plant Cell Environ 18:263–276

Wellburn AR (1995) Genetic and environmental contributions to the winter hardiness of conifers. Bot J Scotl 47:193–209

Wellburn AR, Robinson DC, Thomson A, Leith ID (1994) Influence of episodes of summer O_3 on Δ^5 and Δ^9 fatty acids in autumnal lipids of Norway spruce [*Picea abies* (L.) Karst]. New Phytol 127:355–361

Wellburn FAM, Wellburn AR (1994) Atmospheric ozone affects carbohydrate allocation and winter hardiness of *Pinus halepensis* Mill. J Exp Bot 45:607–614

Westman L (1989) A system for regional inventory of damage to birch. Int Congr on Forest decline research: state of knowledge and perspectives, Friedrichshafeu, 2-6 Oct, 1989, Poster Abstr 1:57–58

Weston KJ, Kay PJA, Fowler D, Martin A, Bower JS (1989) Mass budget studies of photochemical ozone production over the UK. Atmos Environ 23:1349–1360

Whittaker JB, Warrington S (1990) Effects of atmospheric pollutants on interactions between insect pests and their food plants In: Burdon JJ, Leather SR (eds) Pests, pathogens and plant communities. Blackwell, Oxford, pp 97–110

Willson A, Waddell DA, Durrant DWH (1987) Experimental work on air pollution. Forestry Commission Research Information Note 121. HMSO, London

Wiltshire JJJ, Wright CJ, Unsworth MH, Craigon J (1993) The effects of ozone episodes on autumn leaf fall in apple. New Phytol 124:433–437

Wiltshire JJJ, Unsworth MH, Wright CJ (1994) Seasonal changes in water use of ash trees exposed to ozone episodes. New Phytol 127:349–354

Wolfenden J, Wellburn AR (1991) Effects of summer ozone on membrane lipid composition during subsequent frost hardening in Norway spruce [*Picea abies* (L.) Karst]. New Phytol 118:323–329

Wolfenden J, Robinson DC, Cape JN, Paterson IS, Francis BJ, Mehlhorn H, Wellburn AR (1988) Use of carotenoid ratios, ethylene emissions and buffer capacities of the early diagnosis of forest decline. New Phytol 109:85–95

8 Controlled Ozone Exposures and Field Observations in Fennoscandia

G. Selldén, S. Sutinen and L. Skärby

8.1 Land Use in Fennoscandia

Land use in Fennoscandia varies considerably. In Denmark, the largest proportion (60%) of the land is used for agriculture, while forests cover 56 and 66% of the land areas in Sweden and Finland, respectively. Norway is dominated by mountainous regions and only 23% is covered by forests; 3% is farmland. Forests in Fennoscandia are dominated by Norway spruce, *Picea abies*, and Scots pine, *Pinus silvestris*. Norway spruce is somewhat more important in Norway and Sweden, comprising about 45% of the total growing stock, while pine is the dominating tree species in Finland where it comprises 45% of the total growing stock. Deciduous trees such as birch (*Betula pendula and Betula pubescens*), different Salix species, beech (*Fagus silvatica*) and oak (*Quercus*) are less important. However, in all countries there has been a successive shift in attitude towards deciduous trees such as birch. In the 1960s and 1970s, birch was removed from the coniferous forests manually, but also to a large extent by chemical means. Recent results have shown a high productivity of pure forests of *Betula pendula* in Finland and Sweden, and the species is now considered for the reforestation of farm land. Furthermore, results have shown that the productivity of coniferous stands increased when admixed. At present, farmers in Finland are encouraged to use an admixture of about 20% deciduous tree species in coniferous forests(Parvianen and Seppänen 1994).

Land use reflects the importance of the forest industry. In all countries, except Denmark, forest industry is of great economic importance. Since it is vital that a high forest productivity can be sustained, issues regarding forest health have received much interest.

Ecological Studies, Vol. 127
Sandermann et al. (eds) Forest Decline and Ozone

8.2 Ozone Concentrations in Fennoscandia

Ozone levels in Fennoscandia are generally regarded as high enough to cause negative effects on plants, and many types of unidentified injuries to crops and wild plants are regularly reported. For example, necrotic spots on clover (*Trifolium*) species and tomato (*Lycopersicon esculentum*), and brown and violet stipples on potato (*Solanum*). These injuries often look similar to ozone-induced injury although definite proof of this is lacking (Skärby and Jönsson 1985). Several surveys of ozone occurrence using the cultivar Bel W3 of tobacco (*Nicotiana tabacum*) have been performed in Sweden, Denmark and Norway (Skärby et al. 1979; Ro-Poulsen et al 1981; Myhre et al. 1988). These surveys clearly showed that phytotoxic concentrations of ozone occur regularly. However, no visible ozone injury on conifers or deciduous trees has been reported from the field.

Ozone concentrations are measured at several sites belonging to the Environmental Monitoring and Evaluation Programme, EMEP (Fig 8.1).

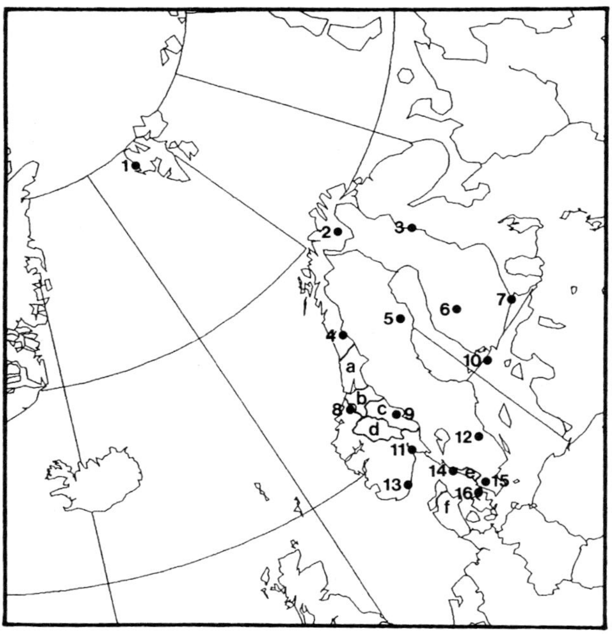

Fig. 8.1. The location of the Nordic stations belonging to the Environmental Monitoring and Evaluation Programme. *1* Zeppelinfjellet; *2* Jergul *3* Oulanka; *4* Trustvatn; *5* Vindeln; *6* Ähtäri; *7* Virolathi; *8* Kaarvatn; *9* Osen; *10* Utö; *11* Jeløya; *12* Norra Kvill; *13* Birkenes; *14* Rörvik; *15* Vavihill *16* Frederiksborg. The following counties are indicated: *a* North-Trøndelag; *b* South-Trøndelag; *c* Hedmark; *d* Oppland, Norway; *e* Halland, Sweden; *f* Jylland, Denmark

From April to September, a 7-h mean ozone concentration of 25 nl l^{-1} is exceeded at most sites most years (Laurila and Lättila 1994). Ozone concentrations in 1991 for a number of sites are presented in Table 8.1. Although the highest 7-h mean ozone concentrations are found in the south, the concentrations in the northern parts are not much lower. However, the pollution climate of the north is different from that of the south in that ozone episodes are fewer in the north. In 1991 the accumulated ozone dose above a threshold of 40 nl l^{-1} (AOT40) at the three most northern sites ranged between 1.2 and 2.3 μl l^{-1}h. At more southern sites, AOT40 ranged between 4.3 and 9.9 μl l^{-1}h. In 1991, 10 μl l^{-1}h, the suggested critical ozone dose for effects on trees(Fuhrer and Acherman 1994), was only exceeded at Utö. However, in 1989 and 1990, an AOT40 of 10 μl l^{-1}h was exceeded at five and two sites, respectively(Laurila and Lättila 1994).

Ozone concentrations are usually measured 3–5 m above the ground. Ozone is readily deposited on surfaces and in order to minimise microscale influences, which could cause large deviations from the regional

Table 8.1. Ozone statistics for selected sites in the Nordic countries in 1991. The locations of the monitoring stations are presented in Fig. 8.1

Station	Latitude	Longitude	Altitude (m)	Yearly mean[a] (nl l^{-1})	7-h mean[b] (nl l^{-1})	AOT30[c] (nl l^{-1}h)	AOT40[d] (nl l^{-1}h)	max[e] (nl l^{-1})	Date
Frederiksborg	55°58′	12°20′	10	22	32	15098	4293	69	14/04
Äthäri	62°33′	24°13′	162	27	38	16968	3980	62	20/06
Utö	59°47′	21°23′	7	35	41	44253	14595	77	20/06
Virolathi	60°31′	27°41′	8	28	39	25472	7111	72	11/06
Oulanka	66°19′	29°25′	310	31	36	18922	3302	63	12/04
Birkenes	58°23′	08°15′	190	29	40	30523	9911	71	24/04
Jergul	69°27′	24°36′	255	29	30	12462	1671	53	13/04
Kårvatn	62°47′	08°53′	210	27	35	17575	4670	63	09/07
Osen	61°15′	11°47′	440	27	37	20408	5320	70	12/04
Zeppelinfjellet	78°54′	11°53′	474	33	31	13894	1231	61	30/04
Jeløya	59°26′	10°36′	3	28	42	26855	7857	77	29/07
Trustvatn	65°50′	13°55′	439	30	32	16372	2837	54	12/04
Rörvik	57°25′	11°56′	10	28	40	29572	8847	80	23/08
Vavihill	56°01′	13°09′	175	25	34	19711	5360	72	14/04
Norra Kvill	57°49′	15°34′	261	30	38	26135	6476	70	29/07
Vindeln	64°15′	19°46′	225	27	34	14338	2558	63	24/08

[a]Mean value of all hourly ozone concentrations during 1 year.
[b]April 1st to September 30th, 09.00–15.59.
[c]Sum of all hourly ozone concentrations >30 nl l^{-1} from 1st April to 30th September.
[d]Sum of all hourly ozone concentrations >40 nl l^{-1} from 1st April to 30th September.
[e]Maximum concentration in 1991.

ozone concentration, the monitoring stations are placed in open terrain. Mechanical turbulence, that is friction between moving air and the earth surface, is one important factor influencing the transport of ozone from the atmosphere to plant surfaces. The rougher the surface, the more mechanical turbulence is produced by a certain wind speed. Compared with field crops, forest ecosystems have rough surfaces. This means that the aerodynamic resistance to ozone transport from air to plant is larger for crops than for forest trees. Ozone concentrations were measured for 27 days at a wind-exposed edge of a 60-year-old 15–20 m tall Norway spruce forest in southwest Sweden. Ozone was also measured at the same time over a barley field, 5 km away at Östad. The ozone gradient over the crop field was considerably steeper than that in the forest (Pleijel et al. 1995). The gradients were even steeper when AOT40 was used. Over the crop field the AOT40 at 1.1 m above the ground was only 50% of that at 9 m above the ground. However, 45 m into the forest, the AOT40 at 13 and 3 m above the ground level were 88 and 71% of the AOT40 at 13 m above the ground 15 m in front of the forest edge, respectively. Thus, the error made when combining data from monitoring stations with dose-response functions in order to predict effects of ozone is less for forest ecosystems than for crop fields.

8.3 Forest Health

8.3.1 Finland

Yearly nationwide assessments of defoliation in Norway spruce, Scots pine and deciduous trees were started in Finland in 1985–1986 (Jukola-Sulonen et al. 1987). In 1993 the assessment covered 404 plots. Altogether 4430 trees, of which 1360 were spruce trees, 2430, pine trees, and 640, deciduous trees, were assessed. In the summer of 1993 a defoliation of >20% was found in 12% of the pine trees, in 38% of the spruce trees, and in 18% of the deciduous trees. However, defoliation has varied from year to year. From 1986 to 1993 the number of trees with a defoliation of >20% had increased by 5% in pine, 12% in spruce, and 11% in deciduous trees.

In general, older trees of all species were more affected than younger trees. Not only age but also latitude affected the degree of defoliation. Northern spruce trees as well as deciduous trees, particularly those over 60 years of age, had lost more foliage than trees from southern Finland. Defoliation of pine trees was not affected by latitude (Lindgren and

Salemaa 1994). In southern Finland, coastal pines were more defoliated than inland pines.

8.3.2 Norway

In 1994, 911 plots were assessed in the nation-wide inventory of forest health in Norway (NIJOS 1995). Altogether, 3950 spruce trees, 2889 pine trees and 1744 birch trees were assessed for crown density and crown colour. The result showed an increase in defoliation in spruce and the decline has been steady since 1989, except for 1992 to 1993. Defoliation in pine has been stable from 1992 to 1994, although distinctly lower compared to the period 1989-1991. However, changes in crown density show geographical differences. Improvements in spruce and pine were observed at some plots in the southern parts. Spruce had worsened in Oppland, Hedmark and Trøndelag, while pine has worsened in west Norway, Hedmark and South-Trøndelag (Fig 8.1). In birch the average crown density decreased by 2.1% from 1993 to 1994.

8.3.3 Sweden

Yearly assessments of defoliation and other tree vitality parameters in Norway spruce and Scots pine were started in Sweden in 1984. In the nation-wide survey, about 17000 trees from 4800 plots are assessed. In 1993, a defoliation of >20% was found in 25 and 8% of all spruce and pine trees, respectively (Wijk et al. 1994). However, compared with the 2 previous years, the inventory in 1993 showed that defoliation has increased only in spruce in southern Sweden. In the whole country, except for northern parts, there is a trend towards increased defoliation in older spruce trees. Changes in pine are less obvious. No clear trend is found in northern parts, while there is a tendency of an increase in defoliation in the southern parts.

Damage on birch was surveyed in southern Sweden in 1990, in central Sweden in 1991 and in northern Sweden in 1992. Birch damage was reinvestigated in southern Sweden in 1993. In 1990, 21.8% of the trees had a defoliation of >25%. In 1993 this proportion had had increased to 29.2%. The proportion of trees with a defoliation >60% had increased from 1.3 to 3.9%, while the proportion of trees with a defoliation of <10% was similar both years.

8.3.4 Denmark

An annual survey of forest damage in Denmark was started in 1989 and in 1991, 1296 trees in 54 plots, in a nation-wide network, were assessed. Altogether, 39% of the conifers and 32% of the deciduous trees had a defoliation of more than 25% while 40 and 37% of the conifers and deciduous trees, respectively, were healthy i.e. defoliation class 0 (Miljø- og Energiministeriet 1995).

Compared with 1989, defoliation had increased in Norway spruce from 9 to 24% in 1994, in Sitka spruce from 22 to 30%, in Scots pine from 30 to 36% and in oak trees from 25 to 33%. In beech, defoliation was 24% in 1989; in 1992 it stabilized and remained constant at about 19%.

8.3.5 Resin-Flow Disease in Norway Spruce

The first reports of an increased resin flow from stems of Norway spruce came at the end of the 1980s. This phenomenon was first observed in Norway spruce in the county of Halland, on the Swedish west coast (Fig 8.1). During the winter 1991/1992 severe damage was observed in one stand. The damage has increased and in the southern parts of the county of Halland, almost all stands with trees over 50 years of age were damaged in spring 1994. This phenomenon has been found in other parts of the country, and increased resin flows from stems of Norway spruce trees are now found in the southern and middle parts of Sweden. Occasional observations of this have also been made in northern Sweden. Similar symptoms have also been observed in Jylland, Denmark (Fig. 8.1), and in Germany (Barklund et al. 1995).

At present, it is debated whether the increased resin flow, as it appears now, is a new phenomenon.

8.4 Exposure Facilities – Open-Top Chambers and Open-Release Systems

Open-top chambers have been widely used to expose trees to elevated concentrations of pollutants. However, the technique was developed for crop plants (Heagle et al. 1973) and when a 5-year open-top chamber experiment was set up in Sweden in 1985, the experience of long-term exposure of trees to elevated concentrations of pollutants in open-top chambers was limited. The experimental site was situated at Rörvik, on the Swedish west coast, 35 km south of Göteborg (57° 40′ N,11° 90′ E), 0.5 km from the coastline. A Norway spruce clone of a Polish provenance was planted in the ground, in a sandy clay loam. The trees were exposed to charcoal-filtered air (CF), non-filtered air (NF) and non-filtered air enriched with ozone (NF+) in open-top chambers (3 m in diameter) for five seasons. A constant amount of ozone was added daily between 11.00 and 18.00, local time. Except for the first year, the aim was to expose the trees from 1 April to 15 November. Two chambers were used in each treatment and two spruce trees were planted in each chamber. For further information on experimental facilities and plant maintenance see Wallin et al. (1990) and Skärby et al. (1995).

Over the 5 years the 7-h seasonal ozone concentrations ranged between 4 and 8, 25 and 35, 42.5 and 51 nl l^{-1} in the CF, NF and NF+ treatments, respectively (Skärby et al. 1995). The seasons started on 15 July, 12 June, 8 May, 24 April and 13 May and ended on 5, 15, 6, 1 and 12 November in 1995, 1986, 1987, 1988 and 1989, respectively. Ozone dose, expressed as AOT40, was calculated from 1 April to the beginning of November, using data from a nearby (500 m) EMEP station for the non-seasonal period. However, the AOT40 values were not corrected for missing hours during the exposure periods, and since the number of missing hours for each season varied between 604 and 1461 h, some caution is recommended.

The suggested critical ozone dose (AOT40), 10 $\mu l\ l^{-1}$ h, was exceeded each year in the NF+ treatment, and at the end of the 5-year experiment the trees had been exposed to an AOT40 of 99.3 $\mu l\ l^{-1}$ h. In the NF treatment, the corresponding sum for the 5 years was 44.3 $\mu l\ l^{-1}$ h. In this treatment, the suggested critical dose level was only exceeded during the season of 1985 (15.8 $\mu l\ l^{-1}$ h), although it was almost reached during the second season (9.9 $\mu l\ l^{-1}$ h). In the CF treatment, AOT40 was 15.3, 8.0, 1.7, 1.2 and 3.1 $\mu l\ l^{-1}$ h in 1985, 1986, 1987, 1988 and 1989, respectively. The AOT40 in 1985 was much higher than in the following years, when charcoal filtering was started earlier. After 5 years the trees in the CF treatment had been exposed to an AOT40 of 29.3 $\mu l\ l^{-1}$ h.

The Rörvik open-top chamber experiment was finished in 1989, and in 1991 the station was moved inland to Östad, 50 km northeast of Göteborg (N 57° 54′, E 12° 24’). A new experiment with Norway spruce was started in 1991, part of which was finished in November 1995. While the Rörvik experiment was set up to identify effects of ozone on the structure and function of shoots of Norway spruce, the overall aim of the new experiment is to quantify the effects of ozone, alone or in combination with drought or phosphorous deficiency, on whole tree productivity and physiology. Chambers similar to those of Heagle et al. (1973) are used. In order to facilitate nutrient control, as well as root studies, each sapling was planted in a 120-l barrel filled with sand. All pots are connected to a computer-controlled system for irrigation and fertilization. Altogether 11 treatments are used, with six replicates of each. The treatments are three levels of ozone, two levels of ozone in combination with two levels of phosphorus supply, two levels of ozone in combination with two levels of water supply and ambient air plots, with and without pots. The treatments started in 1992, except for the treatment with phosphorus deficiency which was started in 1993.

The 24-h seasonal mean ozone concentrations in 1992, 1993 and 1994 (1 April – 31 October) were 4.6, 6.4 and 8.4 nl l^{-1} in the CF treatment, 29.8, 34.2 and 38.2 nl l^{-1} in the NF+ treatment and 22.8, 24.3 and 28.6 nl l^{-1} in the ambient air (AA), respectively. The corresponding AOT40s were 0, 0.3 and 1.1 μl l^{-1}h in CF, 8.6, 13.4 and 38.6 μl l^{-1}h in NF+ and 1.3. 4.0 and 9.7 μl l^{-1}h in AA (Karlsson et al. 1995). The AOT40 was corrected for a few missing hours. The AOT40 in 1992 was lower than in the 2 following years; however, only values from 1 July are included. The suggested critical dose for ozone, 10 μl l^{-1}h, was exceeded in 1993 and 1994 in the NF+ treatments and very nearly exceeded in the ambient air in 1994.

In a rural area 40 km west of Copenhagen, Norway spruce was exposed to elevated cocentrations of O_3, SO_2 and NO_2 for 2 years in eight open-top chambers in order to study effects on growth and antioxidants (Nast et al. 1993). In January 1986, buds from a 48-year-old tree were grafted onto 4-year-old trees. In October 1987 the trees were planted in 6-l pots containing a mixture of peat moss and sand (2:1, v/v). In October 1988 the trees were repotted in 8-l pots. The treatments were charcoal-filtered air, non-filtered air, non-filtered air and ozone, and non-filtered air and ozone, SO_2 and NO_2. A constant amount of ozone (30 nl l^{-1}) was added daily from mid-April until the end of September, while 20 nl l^{-1} of SO_2 and NO_2 was added only during winter time. During April-September, 8-h (9.00–17.00) mean ozone cocentrations in the four treatments were 29, 40 73 and 70 nl l^{-1} in 1988 and 31, 38, 59 and 58 nl l^{-1} in 1989, respectively.The corresponding daily mean ozone concentrations were 21, 30, 42, 40, and 23, 30, 37, 36 nl l^{-1}. Daily mean SO_2 and NO concentrations were never above 3 nl l^{-1} and that of NO_2 was never above 9 nl l^{-1} during the vegetation period.

In Norway, open-top chambers were used to expose Norway spruce to ozone for 1 year (Ogner 1993a). Two sites were used; six chambers were located at Ås, outside Oslo, and six at Bergen, on the west coast of Norway. Thus, the climate of the two sites was considerably different, Ås, representing an inland climate, Bergen a coastal climate. The walls of the open-top chambers were made of polyethylene which had a light transmission of 85–90% between 380 and 800 nm. Six clones of Norway spruce were used, two clones originating from each of two Norwegian locations (66° N, 60° N), and one from Ukraine (49° N). Scions from mature trees were grafted on 4-year-old rooted cuttings in 1983. The trees were grown in peat containing 25% perlite and repotted the last time in 1988 in 50-litre pots. Exposure to charcoal filtered air, and to 100 and 200 nl l^{-1} ozone was started in May 1990 when the trees were about 2.5 m tall. Ozone was added daily between 8.00 and 16.00 from 7 May to 10 September 1990. Except for during precipitation periods the soil was drip-watered during the daytime. Nutrients were added manually once a week. As usual in Fennoscandia, ambient concentrations of SO_2 and NO_2 were low, especially at the Bergen site.

The only open-release exposure system currrently in use is located in central Finland outside Kuopio. In 1990, one area was operating (Wulff et al. 1992) but since 1991 two replicate systems have been available. Potted trees are exposed to twice ambient concentrations of ozone. Ozone, which is produced from pure oxygen, is released from perforated tubes upwind of the plants. Fumigation is carried out day and night to maintain the diurnal fluctuations of ozone concentration.

8.5 Ozone-Induced Changes in Structure

8.5.1 Controlled Exposure Experiments - Conifers

Visual examination of whole trees and needles as well as estimations of needle loss are the most frequently used methods to assess forest health. However, these methods are not suitable for establishing causal relationships. Needle loss is a non-specific response of trees, as several different stress factors can cause needle loss. Furthermore, needle loss represents the final stage in a sequence of reactions to stress. By using light and electron microscopy it is possible to make an early diagnosis of an injury that eventually may lead to needle loss. More detailed information regarding changes of the ultrastructure of needles will also give indications of where to look for disturbed functions, as well as making it easier to attribute the response to a specific cause.

Early experiments established the negative effects of ozone on plant structure using high concentrations of ozone for short periods of time (Thomson et al. 1966, 1974; Evans and Miller 1972; Pell and Weissberger 1976). A number of different species, as well as different cultivars, was exposed in laboratory exposure systems. Of the different cell organelles, chloroplasts were the most sensitive and exposure to high concentrations of ozone for short times caused chloroplast destruction (Thomson et al. 1974; Pell and Weissberger 1976). However, when plants were exposed to concentrations of ozone, more close to ambient levels, a decrease in chloroplast size was observed both in bean plants (*Phaseolus vulgaris*; Vogels and Masuch 1984) and in Norway spruce (Sutinen 1987a).

Ozone-induced structural changes in needles of Norway spruce were studied in the open-top chamber experiment at Rörvik, Sweden. The structure of green needles from the flush of 1985 was followed from November 1985, 1 month after the end of the first exposure period, until November 1987 (Sutinen et al. 1990).

Regardless of tree age, normal chloroplasts of spruce and pine needles are lensshaped, the stroma is light and the amount of plastoglobuli usually small (Walles et al. 1973; Soikkeli 1978; Sutinen 1987a; Meyberg et al. 1988; Holopainen and Nygren 1989). Throughout the period the chloroplasts in the stomata, mesophyll and endodermis in needles from trees exposed to filtered air were normal (Fig 8.2a). The structures of other cell organelles, such as microbodies and mitochondria, as well as that of the cytoplasm, were also intact. From spring (April) to late autumn (October) the chloroplasts were located along the cell walls. During the growing season chloroplasts had well-developed thylakoid systems and starch grains. During the winter the chloroplasts were clumped together and no starch was present; otherwise the structure was similar to that in the summer.

Ozone-induced changes were observed first in the chloroplasts of the mesophyll cells located in the upper part of the needles (adaxial side); the size of chloroplasts was reduced, the thylakoids and envelopes were difficult to resolve and the electron-dense stroma was heavily granulated (Fig. 8.2b). The amount of plastoglobuli was increased, especially in cells located below the epidermis. At a later stage the microbodies also became affected. Changes in ultrastructure similar to those occurring in NF+-

Fig. 8.2. Transmission electron and light micrographs of spruce needles from trees exposed to filtered air (**a** and **c**) and to non-filtered air with extra ozone (**b** and **d**) in open-top chambers at Rörvik, Sweden. Ozone exposure was started in July 1985. The needles were collected in May 1987. **a** Transmission electron micrography showing the fine structure of a healthy cell where chloroplasts are lens-shaped, the starch grains are large, the stroma is light and the amount of plastoglobuli small. **c** Light micrography of the same structure as in **a**. Typical of the beginning of the growing season, the chloroplasts are near the cell walls and have large starch grains seen as white granules in the chloroplasts. **b** Transmission electron micrograph showing ozone-induced symptoms characterized by a reduction in chloroplast size and electron-dense stroma rich in granular particles. In all other respects the cell organelles are intact. **d** The same structure as in **b** at the light microscopy level. The chloroplasts are located at the cell walls but few have starch grains. *Bars* 1 μm (**a** and **b**); 12 μm (**c** and **d**) treated needles appeared in NF-treated needles in July 1986, 9 months after the end of the first exposure period.

Over time these changes typically progressed within the mesophyll tissue in both NF- and NF+-treated needles. Changes always appeared first in the adaxial mesophyll cell layers and then began to appear in the underlying tissue. In Fig. 8.2c, a light micrograph of a typical needle from

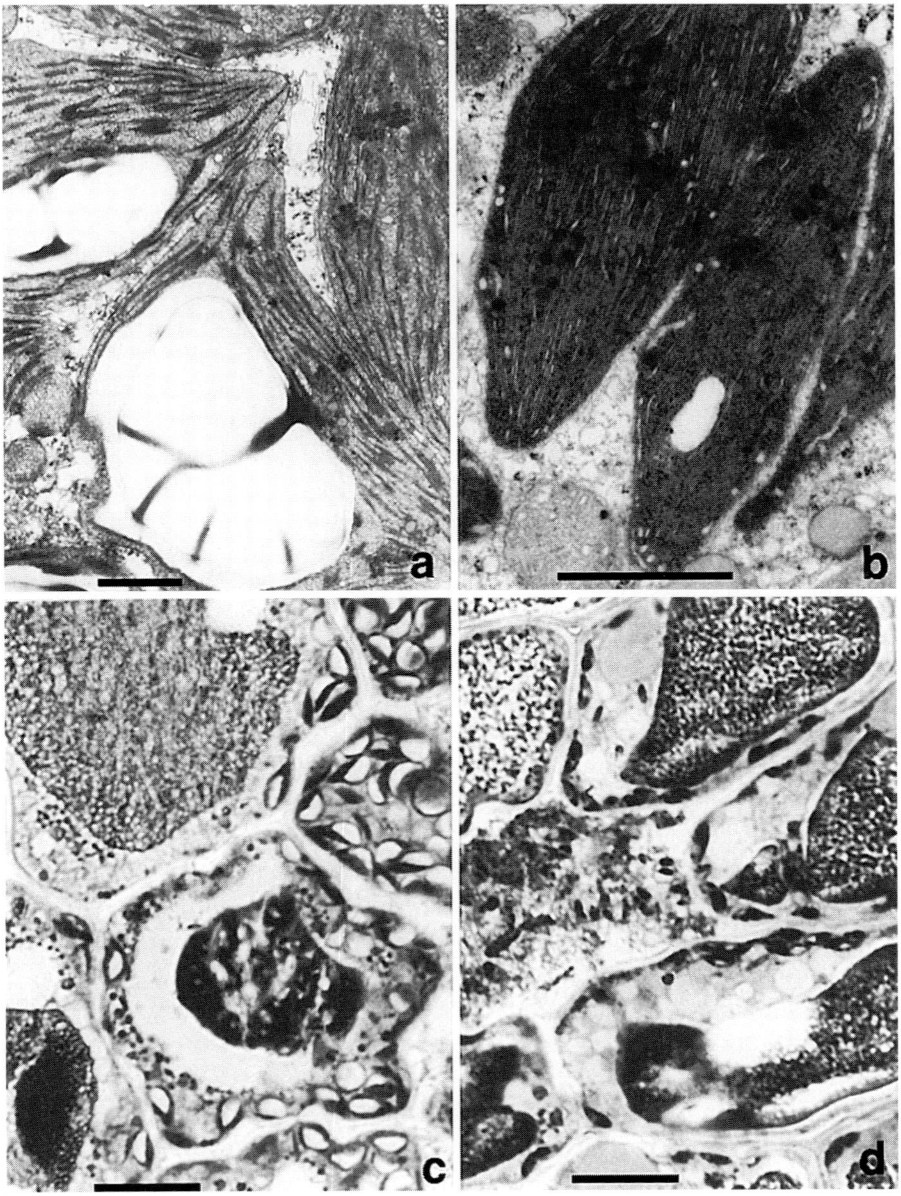

the CF treatment is presented. The length of the chloroplasts in the mesophyll tissue close to the upper epidermis is similar to that of underlying mesophyll tissue. The amount of starch is also similar. In the NF+ treatment, smaller chloroplasts and starch grains were observed in the upper mesophyll tissue (Fig. 8.2d), while the largest starch grains were located close to the endodermis of the lower part of the needle. The gradual advance of structural changes within the mesophyll tissue was quantified using light microscopy by repeated measurements of chloroplast lengths. When comparing needles from the CF and NF+ treatments, a significant difference in chloroplast length was observed in June 1986. Chloroplasts in needles from the NF treatment also decreased in length. When these were compared with chloroplasts from the CF treatment, a significant difference was observed in June 1987.

When the changes of the mesophyll tissue had reached the endodermis, the endodermal chloroplasts became affected in the same way as the chloroplasts of the mesophyll tissue. Changes in ultrastructure of the albuminous cells from NF- and NF+-treated needles could be seen approximately at the same time as changes were seen in the endodermis. The changes occurring in albuminous cells differed from those of the mesophyll and the endodermis. In the albuminous cells, a general breakdown of all cell organelles together with a disintegration of the cytoplasm was observed. However, totally degenerated albuminous cells were rare and could be seen only in the needles where the mesophyll tissue was severely damaged. A collapse of cell walls in the phloem was not observed at any time.

Ultrastructural changes in needles of Scots pine and Norway spruce, similar to those observed at Rörvik, were also found when seedlings were exposed to ozone in the open-release exposure system in Finland(Holopainen et al. 1993). In addition to exposure to ozone, the seedlings were exposed to nutrient stress (N, K and P). After the second growth period in low nutrients, the stroma of chloroplasts in current year shoots from ozone-exposed seedlings was increased in density and was granular. In the older needle generations, the size of the chloroplasts and starch grains had declined. No such changes were observed in seedlings grown at an optimum nutrient level. Instead ozone exposure resulted in a slight increase in the size of starch grains. These results clearly indicated that a low availability of N, P and K made the seedlings more sensitive to ozone.

On the other hand, when Scots pine and Norway spruce seedlings, deficient in Ca and/or Mg, were exposed to ozone for two consecutive growing seasons, no clear ozone-induced structural changes were observed in the current and 1-year-old needles. However, although the nutrient stress was not considered to have been severe, an accumulation of starch was observed in chloroplasts from ozone-fumigated seedlings grown with sufficient nutrients, but not in chloroplasts from nutrient deficient trees (Rantanen et al. 1994).

8.5.2 Controlled Exposure Experiments - Deciduous Trees

Using the open-release exposure system, saplings of three clones (A, B, C) of European white birch (*B. pendula)* were exposed in 1991 and 1992 to ozone concentrations 1.2 and 1.5 times the ambient, concentration respectively (Pääkkönen et al. 1993). In 1992, five clones (A, B, C, D, E) were exposed (Pääkönen et al. 1995). One clone (D) was sensitive and one (C) insensitive to ozone, while the other three (A, B, E) were of intermediate sensitivity. Ozone exposure decreased the thickness of the leaf in clone D. The decrease was due to a reduced thickness of the palisade, as well as the spongy parenchyma. Leaf thickness in clone A was not affected by ozone. However, while total leaf thickness was not affected in clones B, C and E, the palisade parenchyma was thinner and the spongy parenchyma thicker in clone C, while the opposite was the case for clones B and E. Furthermore, ozone decreased stomatal conductance but increased stomatal density in the emerging leaves. Similar changes in leaf differentiation were observed when birch (*B. pendula*) was exposed to moderately elevated concentrations of ozone (Günthardt-Goerg et al. 1993). Leaf area and leaf width were reduced while the density of stomata, scales and hairs increased. Since these changes, apart from the reduced leaf width, are similar to those of mild drought, it was suggested that these changes could be due to an ozone-induced drought stress during elongation phase

Electron microscopy revealed that ozone exposure increased the number of irregular, spherical and electron dense chloroplasts in the palisade and spongy parenchyma of all clones (Pääkkönen et al. 1995). The swelling and curling of thylakoids increased, as well as the granulation of the stroma. In the sensitive clone D, ozone exposure decreased the length of the chloroplasts and the amount of starch. These changes were not observed in the tolerant clone C.

The supply of nitrogen is of importance for the response of birch to ozone (Pääkönen and Holopainen 1995). Birch clones (*B. pendula*) were exposed to 1.5-1.7 times the ambient ozone concentration in the open-release system in 1992 and 1993. The results showed that nitrogen-induced changes in mean leaf size, autumn yellowing and leaf anatomy were different to those induced by ozone. Ozone accelerated senescence when the nitrogen supply was low(37 kg N ha^{-1}). When the supply of nitrogen was good(150 Kg N ha^{-1}), the senescence of ozone-exposed trees was retarded. The conclusion of these experiments was that a sufficient supply of nitrogen can increase plant resistance to ozone.

8.5.3 Influence of Growth Strategy on Ozone-Induced Structural Changes

Plant habitats vary with the frequency of stress and disturbances, and the success of a plant in a habitat depends on how well it is adapted to cope with environmental changes. In the C-S-R system of plant strategies (Grime 1979) plants are classified in relation to resource capture, resource allocation and responses to stress and disturbances. Stress is defined as factors that limit photosynthetic production, ie. shortage of light, water and mineral nutrition, while disturbances are the total or partial destruction of biomass by for example grazing, floods, drought or wind.

Competitors (C) are adapted to productive habitats with a low frequency of stress and disturbances while stress-tolerators (S) are found in habitats with a comparatively high frequency of stress but with a low frequency of disturbances. Competitors are characterized by high growth rates and high morphogentic plasticity. They are able to respond rapidly to changes in the distribution of resources within the habitat. Stress-tolerators are characterized by long life spans, slow growth rates, low morphogenetic plasticity and sequestration and slow turnover of resources. Ruderals (R) are adapted to habitats with low stress and a high frequency of disturbances. They have short life spans and usually respond to stress by premature senescence. Thus the overall stress response of competitors is to maximize the capture of resources, that of stress-tolerators is to conserve captured resources while the typical stress response of ruderals is to ensure seed set (Grime 1979). All kinds of intermediate strategies between these three main types exist, for example stress-tolerant competitors.

Climax species, such as Norway spruce, Scots pine, beech and oak, can be classified as stress-tolerant competitors, while tree species appearing early in the succession, e.g. birch and aspen (*Populus*), can be classified as competitors (Grime 1979). It has recently been suggested that the overall response of plants to ozone is dependent on plant strategy (Laurernce et al. 1994; Selldén and Pleijel 1995). Furthermore, it is probable that ozone can be regarded both as a stress factor and as a disturbance (Selldén and Pleijel 1995). It is well established that moderate ozone concentrations reduce net photosynthesis of many plant species (Reich 1987; Darrall 1989). When reducing net photosynthesis, ozone acts as a stress factor. However, high concentrations often cause visible injury to leaves, i.e.

partial biomass destruction. In this case ozone can be regarded as a restricted disturbance. The ability of a species to cope with disturbances would then determine the sensitivity to high ozone concentrations or sudden ozone episodes.

Deciduous trees such as birch (*B.pendula, B.pubescens and B. verrucosa*) and aspen (*Populus tremuloides*) respond to ozone exposure with chlorophyll loss and premature leaf fall which can be interpreted as premature senescence (Mortensen and Skree 1990; Pääkönen et al. 1993; Greitner et al. 1994). Structural investigation of ozone-exposed birch clones revealed an increase in cytoplasmic lipids, translucency of the mitochondrial matrix and spherical chloroplasts with protrusions, changes which are typical of senescence (Pääkönen et al. 1995) However, ozone-induced changes do not always mimic senescence. When one clone of birch (*B.pendula)* was exposed to ozone in field fumigation chambers for one season, cabohydrate metabolism was affected. In senescing leaves starch was entirely retranslocated while in leaves from ozone-exposed trees starch accumulated along the small leaf veins and in the guard cells, the amount increasing with increasing ozone concentration (Günthardt-Goerg et al. 1993). However, in both investigations leaf morphology was affected. In the Finnish investigation leaf morphology changed in four out of five clones. In the ozone-tolerant clone C, leaf area decreased and leaf number increased, which was suggested to be an acclimation to ozone (Pääkönen et al. 1993).Since rapid reallocation of captured resources to sustain new growth is a typical stress-response of competitors, premature senescence of older leaves and morphological changes should not necessarily be regarded as damage but may reflect attempts to compensate for the stress.

Ozone exposure of Norway spruce decreased the chlorophyll content of the needles after about one exposure season (Skärby et al. 1995). However, in conifers, the experimentally induction of structural changes by ozone, apart from the accumulation of plastoglobuli, were not typical of senescence (Sutinen 1987b; Reutze and Schmitt 1988). Furthermore, the structural changes (Meyberg et al. 1988) observed in needles from declining spruce trees were not similar to those of senescence. When analyzing hexane-and dichloromethane-soluble compounds in needles of Norway spruce exposed to ozone in open-top chambers in Norway, it was found that four compounds from the hexane extract changed with ozone concentrations (Ogner 1993b). One compound was later identified as free phytol (Ekeberg et al.1995). Although senescence is accompanied by a degradation of chlorophyll, free phytol does not normally accumulate until the very last stage of senescence. Other stress factors such as high and low temperatures, drought and frost are known to affect phytol containing compounds, however the phytol remains esterified to fatty acids. The mechanisms behind the increase in free phytol are unclear. It may be a result of the chlorophyll degradation usually observed after ozone exposure.

In addition, Guillemaut et al. (1992) found that the chloroplast protein pattern in needles of declining trees of Norway spruce and fir (*Abies alba* Mill.) was different from that in senescent needles. Ozone is considered to be a main stress factor in this area of France (Vosges). Finally, when Norway spruce was exposed to ozone in solar domes during the summer, a decrease in chlorophyll content was observed in 1- and 2-year-old needles (Robinson and Wellburn 1991). However, the response occurred every 2 years. A decrease in pigment one year was followed by an increase the next year, although the ozone exposure did not differ. This indicates that the chlorophyll loss was not due to premature senescence.

However, Norway spruce, Scots pine and fir have strong stress-tolerating characteristics and since long leaf life spans are typical of stress-tolerators, it is not surprising that the ozone-induced changes do not resemble those of normal senescence. In this context it should be noted that in wheat (*Triticum aestivum*) a ruderal plant adapted to respond to stress by premature senescence to ensure rapid seed set, all ozone-induced changes in ultrastructure (Ojanperä et al. 1992), membrane lipids (Sandelius et al. 1995) and amount and activity of Rubisco (Lehnherr et al. 1987) mimicked that of normal senescence.

8.5.4 Field Investigations – Conifers

Results from the open-release and open-top chamber experiments have shown that ambient and moderately elevated concentrations of ozone in the long term may cause structural changes in the foliage of seedlings and saplings. The changes, first seen in the chloroplasts, were similar in spruce and pine needles. Furthermore, these structural changes were also observed when young spruce seedlings were exposed to ozone in a climate chamber (Sutinen 1987a). Some of the changes, such as an increased density of the chloroplast stroma and reduced chloroplast size, also occurred in birch foliage when the trees were exposed to ozone in the field (Pääkönen et al. 1995)

To test if similar changes could be observed in adult field-grown trees, needles were collected from 135 plots in southern, central and northern Finland between 1989 and 1992. The structure of spruce and pine needles was studied using light and electron microscopy (Sutinen et al. 1995). Healthy chloroplasts of Scots pine from Finnish Lapland are shown in Fig. 8.3a,c. The structure is typical of the winter condition, the lens-

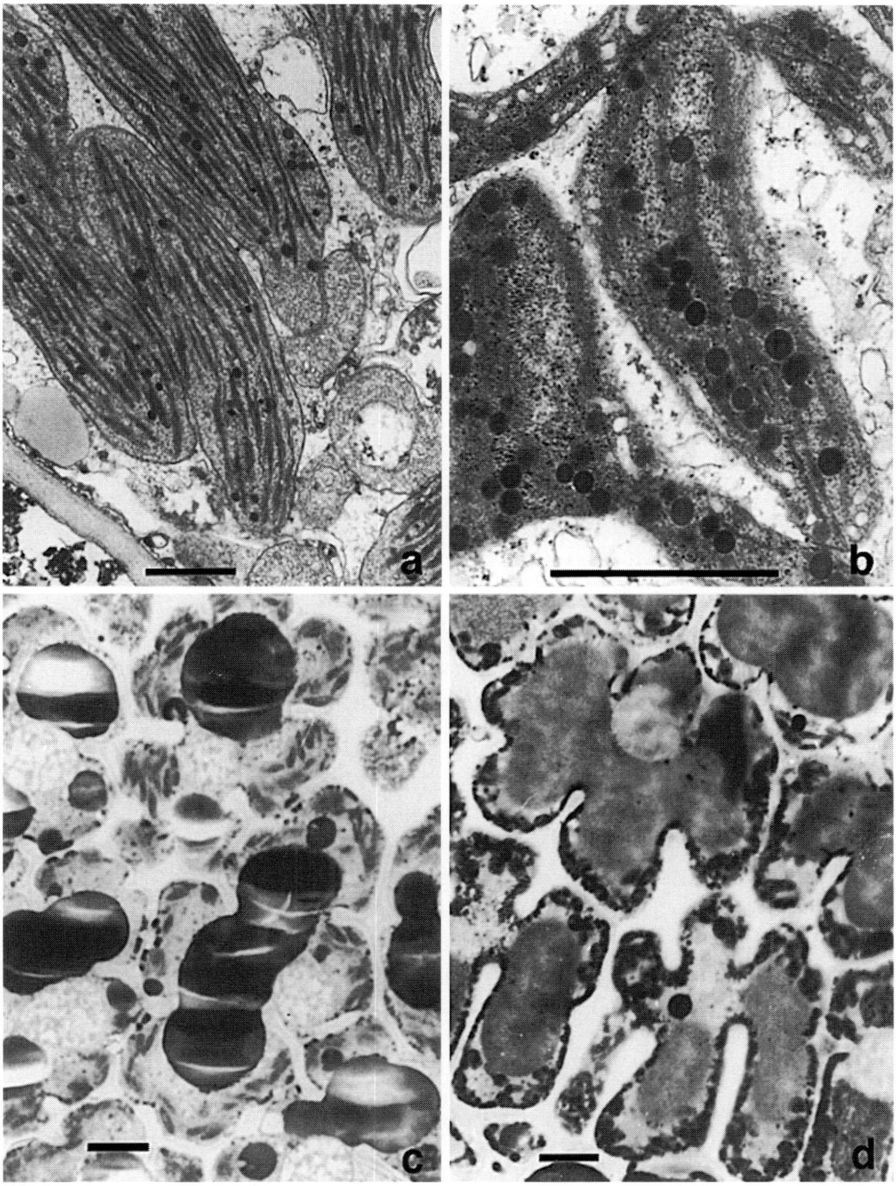

Fig. 8.3. Transmission electron and light micrographs of needles of Scots pine trees from Finnish Lapland in winter, 1988. **a** Fine structure of a healthy cell with lens-shaped chloroplasts and electron-opaque stroma. The chloroplasts have no starch which is typical during winter. **c** The same structure as in **a** at the light microscopy level. The lens shaped chloroplasts are clumped together as is typical in winter. **b** Changes, typical of ozone effects. The chloroplasts are small and the electron-dense stroma is rich in granules. **d** The same structure as in **b** at the light microscopy level. *Bars* 1 μm (**a** and **b**); 12 μm (**c** and **d**)

shaped chloroplasts are clumped together and no starch is present. In addition to the normal structure of needles, structural changes, similar to those induced experimentally by ozone, were observed in the needles. The chloroplasts are smaller and the stroma electron-dense and granular (Fig. 8.3b). Light microscopy shows that the chloroplasts are less clumped together (Fig. 8.3d). The trees with needles showing such symptoms were found all over the country.

Other symptoms, such as those experimentally induced by simulated acid rain deposition, were observed also. The symptoms of acid rain deposition are characterized by a vacuolization of the plasmalemma (Holopainen and Nygren 1989) and in a later phase also by a vacuolization of the cytoplasm. These changes were observed frequently in Finnish Lapland, north of Lake Inari. Typical symptoms of sulphur dioxide are the precipitation of the tannin and simultaneous darkening of the cytoplasm (Soikkeli 1981). Trees with needles showing these symptoms were found mainly in plots located near the emitting sources or cities rather than in more remote areas.

In earlier field studies, structural changes typical of ozone have been observed in spruce needles from the Taunus mountains, West Germany (Sutinen 1987c) as well as in conifer needles collected in southern Sweden and in southern and northern Finland (Sutinen 1989, 1990). At all locations the ozone concentrations during the growing seasons were similar to or higher than the concentrations used in the controlled exposure experiments.

Visible symptoms were observed occasionally in needles sampled from the field. Yellowing of the needles was typical of the spruces studied in Germany (Sutinen 1987c) and a dirty greyish-green colour of the needles as well as necrotic and waterlogged spots were typical of spruces studied in southern Finland (Sutinen 1990). However, at present it is not possible to attribute these symptoms to a particular stress such as ozone.

8.5.5 Structural Changes as a Diagnostic Indicator

The usefulness of ultrastructural changes as a diagnostic indicator has been much debated. Forest trees are exposed to natural stress factors, such as drought, extreme temperatures and excess light, in addition to stress factors of anthropogenic origin. Gaseous air pollutants enter the needles through the stomata and structural changes are first seen in the subepidermal mesophyll (Sutinen 1987c; Fink 1989; Schmitt et al. 1989; Sutinen 1989; Sutinen et al. 1990). Over time, underlying tissue becomes affected. However, while the cells of the vascular bundle remain unaffected by ozone until very late stages of mesophyll damage, mineral deficiency (Mg, K and Ca) first induces a collapse of the phloem (Fink 1989). Thus, Fink (1989) suggested that the dynamics of the damage could assist in distinguishing between direct effects of pollutants and direct effects of nutrient stress.

Other air pollutants, such as SO_2 and NO_2 (Rantanen et al. 1994), acid rain deposition (Holopainen and Nygren 1989) and fluoride (Soikkeli and Tuovinen 1979) induce structural changes differing from each other, as well as from those induced by ozone. Furthermore, the structural changes caused by drought (Palomäki et al. 1994) and frost stress as well as by phosphorus deficiency (Palomäki and Holopainen 1994) also differ from those induced by ozone. Thus, structural changes in the foliage of the trees seem to be a valuable indicator for estimating the impacts of ozone in field-grown conifers.

8.6 Ozone-Induced Changes in Gas Exchange

It has been well established that ozone reduces net photosynthesis (Pye 1988; Darrall 1989). However, the effects of ozone on the gas exchange of plants are complex and the results from the Swedish open-top chamber experiment at Rörvik can be used to illustrate this. Net photosynthesis at saturating PPFD (photosynthetic photon flux density) in shoots of different ages was studied in 1987, 1988 and 1989. In the 2- and 3- year-old shoots, it was generally observed that photosynthesis declined with increasing ozone concentration and age (Fig. 8.4). In 1987, a significant reduction in the photosynthesis of the 2-year-old shoots exposed to NF compared with those exposed to CF was observed. Furthermore, in the current-year shoots, net photosynthesis was generally lower with the CF treatment than with the NF and the NF+ treatments in 1989 (Wallin et al. 1990). Similar observations of transient ozone-induced stimulatory effects on net photosynthesis have been made in other studies of Norway spruce (Eamus et al. 1990) as well as in studies of other species, e.g. Ponderosa pine (Beyers et al. 1992) and aspen (Greitner et al. 1994). At present there is no explanation for the ozone-induced stimulation in net photosynthesis. It may be a manifestation of an ozone-induced acceleration of the rate of development (Greitner et al. 1994; Skärby et al. 1995) or it may be an attempt by the trees to compensate for the ozone-induced reduction in net photosynthesis of older leaves and needles (Beyers et al. 1992; Selldén and Pleijel 1995) or a combination of both.

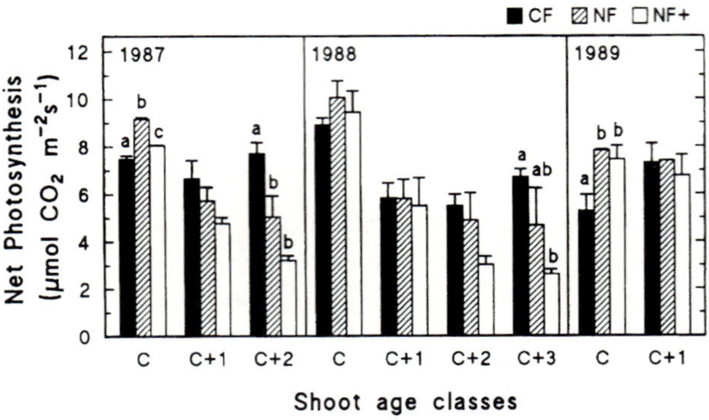

Fig. 8.4. Net photosynthetic capacity of shoots of Norway spruce from trees exposed to different concentrations of ozone in open-top chambers at Rörvik, Sweden. The measurements were made at saturating PPFD and at approximately 330 µmol mol^{-1} CO_2 at the end of the growing seasons of 1987, 1988 and 1989. *Bars* marked with different letters are significantly different at $p<0.05$. *Error bars* show standard deviation (n-1). *CF* charcoal filtered air; *NF* non-filtered air; *NF+* non-filtered air and 25 nl l^{-1} ozone for 7 h a day; *C* current year shoots; *C+1, C+2, C+3* 1-, 2- and 3 year-old shoots, respectively. (Wallin 1990)

The stimulation of net photosynthesis in the current-year shoots could at least partly be explained by an observed increase in gas phase conductance with respect to CO_2. However, it could not explain the inhibition of net photosynthesis in the older shoots. Instead, the increase in C_i concealed some of the effects on the photosynthetic apparatus. The increase in C_i is also a result of a higher gas phase conductance relative to the photosynthetic rate, suggesting that ozone will negatively influence the water economy of these needles. This was also shown in a separate study on the effects of soil moisture deficit on trees exposed to different ozone concentrations (Wallin and Skärby 1992).

Both the apparent quantum yield and the carboxylation efficiency in the older shoots were negatively affected in NF and NF+, with the largest decrease being in the carboxylation efficiency, which is determined mainly by the Rubisco activity (Wallin et al. 1992a,b). The ozone-induced reduction in the carboxylation efficiency affects the photosynthetic rate negatively under most conditions, while the reduced apparent quantum yield affects photosynthesis mainly under conditions of low PPFD (Wallin 1990).

It was suggested by Wallin (1990) that the effects of ozone on canopy photosynthesis of Norway spruce will vary with tree age since older trees have a larger proportion of older needles than young trees. Using a model approach it was found that in young trees the overall effect of ozone on canopy photosynthesis was small since the stimulation of net photosynthesis in current-year shoots compensated for the decrease in older shoots. Furthermore, since the negative effects of ozone on net photosynthesis were more prominent at low light intensities, the trees were less able to compensate when exposed to high concentrations of ozone and low light intensities (Skärby et al. 1995).

Grafted cuttings of Norway spruce were exposed to charcoal-filtered air and non-filtered air plus 30 nl 1^{-1} of ozone from May to October, 1988–1990. Net photosynthesis and chlorophyll fluorescence (F_v/F_m) were measured on selected days at different times of the day (Mikkelsen and Ro-Poulsen 1994). Before the fumigation started (9.00) and after the end of fumigation (17.00), net photosynthesis of NF+-treated trees was not significantly different from that CF- treated trees. During the day, when both light and ozone concentration increased, significant decreases in net photosynthesis were observed in NF+-treated. In view of the interactions between light and ozone concentration, the authors suggested that ozone affects photoinhibition. This suggestion was further supported by the results on chlorophyll fluorescence. In current-year shoots a significant increase in the measurement was observed at 11.30 in trees from the NF+ treatment, compared to trees from the CF treatment. The increase was followed by a depression lasting until 14.00. Measurements of net photosynthesis were also made in October, 5 days after the end of the fumigation period. Although not significantly different, net photosynthesis of NF+-treated current year shoots was higher than that of CF-treated shoots, suggesting that the trees attempted to compensate for the ozone stress by physiological change, as is typical of stress-tolerators.

A similar interaction between light and ozone concentration was observed in beech, another species with strong stress-tolerating chracteristics. Beech seedlings were exposed to ozone in open-top chambers from April to October in 1988 and 1989 and net photosynthesis was measured on 2 consecutive days in June 1989 (Mikkelsen 1990). Before the fumigation started at 9.00 a.m. no difference in net photosynthesis was observed between the NF+- and CF-treated trees. However, later during the day net photosynthesis decreased in NF+-treated trees. Since no difference was observed in net photosynthesis in the mornings before fumigation started, the trees were obviously able to repair any damage to their photosynthetic apparatus.

8.7 Ozone-Induced Changes in Growth

Photosynthesis is necessary for growth; nevertheless, the relationship between net photosynthesis and growth is complex and a decrease in net photosynthesis does not necessarily mean that growth is reduced. Ozone-induced decreases in net photosynthesis may reduce growth. However, as stored resources can be used to compensate for the stress, at least for short periods of time, species with strong stress-tolerating characteristics should be less sensitive to ozone than competitors. On the other hand, when ozone acts as a disturbance, causing leaf or needle loss, the growth of stress tolerators will probably be more reduced than that of competitors (Selldén and Pleijel 1995). Although both competitors and stress-tolerators are adapted to habitats with comparatively low frequencies of disturbances, the low growth rates of stress-tolerators will not allow rapid replacement of needles or leaves. Thus, we should expect that the response of trees to ozone will vary with species and plant strategy. In this context it should be remembered that success in a habitat is not equal to high timber production, but rather to survival and successful reproduction.

Although competitors such as birch, aspen and poplar tend to compensate for ozone stress, ozone-induced reductions in growth are usually observed already after one season of exposure to moderately elevated concentrations of ozone (Günthardt-Georg et al. 1993; Pääkönen et al. 1993). The growth reductions are usually accompanied by increases in the shoot/root ratio. Even short periods of exposure can result in growth reductions. When young seedlings of birch (*B. pubescens*) were exposed to 62 nl l^{-1} ozone for 34 days in exposure chambers placed in a greenhouse, total plant biomass decreased significantly. Although the shoot/root ratio was unaffected, the branch/main shoot ratio increased, indicating that the seedlings reacted to ozone by changes in carbon allocation (Mortensen 1995). Growth reductions were also observed when birch (*B. pubescens* and *B. verrucosa*) and alder (*Alnus incana*) were exposed to ozone (<82 nl l^{-1}) for up to 50 days (Mortensen and Skree 1990). However, although enhanced leaf senescence was observed after exposure to 35 nl l^{-1} and visible injury after exposure to 53 nl l^{-1}, shoot/root and leaf/stem ratios were not affected.

The growth of conifers is less sensitive to ozone, and growth reductions are rarely observed after only one exposure season. For example, ozone exposure for 4 months reduced net photosynthesis but not growth in white pine (*Pinus strobus*; Reich et al. 1987). In red spruce (*Picea rubens*) growth was not affected until after the end of the second exposure season (Amundson et al. 1991) and when Norway spruce was exposed for two seasons in open-top chambers, growth was not reduced (Nast et al. 1993). The results from the Swedish ongoing study confirm that several seasons of ozone exposure are needed before the growth of Norway spruce is affected. A significant ozone-induced reduction of total plant biomass was not observed until September 1994, about 3 years after the start of the exposure (Karlsson et al. 1995). Furthermore, the decrease in biomass was mainly due to a decrease in stem biomass, showing that carbon allocation was affected.

However, conifers differ in ozone sensitivity. The influence of ozone on growth and dry matter partitioning of young pot-grown conifer seedlings has been studied by Mortensen (1994). Young seedlings were exposed to moderately elevated concentrations of ozone (70 nl l^{-1}) for 81-116 days in exposure chambers placed in a greenhouse. Ozone-induced decreases in plant dry weight was observed in Sitka spruce (*Picea sitchensis*), Douglas-fir (*Pseudotsuga menziesii*), and in one of three provenances of Scots pine. Plant dry weight of Japanese larch (*Larix leptolepis*), Siberian larch (*Larix sibirica*), lodgepole pine (*Pinus contorta*), mountain pine (*Pinus mugo*), white pine (*Pinus strobus*) and the three provenances of Norway spruce were unaffected by ozone. Although total plant biomass was mostly unaffected by ozone, dry matter partitioning was affected. The shoot/root ratio was increased in the three provenances of Scots pine, in one of the provenances of Norway spruce and in lodgepole pine, while the needle/stem ratio was increased in seven of the nine species, maybe as attempts to compensate for the ozone stress.

No ozone-induced reductions in plant dry matter were observed in Norway spruce when the seedlings were exposed to 70 nl l^{-1}. However, growth reductions were observed in two out of four clones of Norway spruce when the seedlings were exposed to 80–100 nl l^{-1} ozone for 100 days (Mortensen 1990). In another experiment two clones of Norway spruce were exposed to 27, 57 and 88 nl l^{-1} for 55 days, respectively. Growth reductions, as well as increased needle/stem ratios, were observed in the highest ozone concentration (Mortensen 1990). Whether these differences depend on differences in ozone concentrations, in ozone uptake or in sensitivity to ozone is not possible to decide.

8.8 Conclusions

Ozone concentrations in Fennoscandia are lower than in central Europe, and the number of ozone episodes is also lower, particularly in the north. Nevertheless, results from Fennoscandia, from controlled exposure experiments, as well as from the field, indicate that ozone concentrations are sufficiently high to reduce growth of field-grown conifers and deciduous trees, such as birch.

The overall response of plants to ozone seems to depend on plant strategy. Competitive plants, such as birch, adapted to maximize resource capture, respond rapidly to ozone by changes in morphology. Experiments using open-top chambers or laboratory cuvettes show that conifers respond less rapidly to ozone. Since stress-tolerators, which are adapted to exploit temporal variation in resources rather than spatial variation, respond to stress by rapid physiological changes and storage of captured resources, one would expect stress-tolerating competitors, such as Scots pine and Norway spruce, to respond less rapidly than typical competitors. However, in the long-term, we can expect negative effects of ozone on the growth of conifers, particularly if the environmental conditions are such that the availability of stored resources for repair become limiting. Furthermore, almost all investigations found that ozone affected carbon allocation. An ozone-induced decrease in carbon allocation to roots and stem may have serious consequences for the long-term survival of the trees. An increase in the shoot/root ratio will make the trees more susceptible to drought stress. Smaller roots and larger canopies may also increase the sensitivity to wind, a serious matter if climate change leads to an increased number of stormy days.

Acknowledgements. We would like to acknowledge gratefully the support of Nils and Dorthi Troëdssons Research Foundation, the Ministries of Environment, Agriculture and Forestry, Finland, the Kvarken Council of the Gulf of Bothnia, The Swedish Council for Forestry and Agricultural Research, Swedish National Environmental Protection Agency, IVL, the Joint Committee of Power and Heating Producers on Environmental Issues, The Swedish Pulp and Paper Association, Volvo and Saab.

References

Amundson RG, Alscher RG, Fellows S, Rubin G, Fincher J, van Leuken P, Weinstein LH (1991) Seasonal changes in the pigments, carbohydrates and growth of red spruce as affected by exposure to ozone for two growing seasons. New Phytol 118:127–137

Barklund P, Ericsson A, Gemmel P, Johansson U, Olsson M, Walheim M, Åhman G (1995) Bark och vedskador hos gran med kådflöde "Kådflödesjukan hos gran". SLU Info/Skog Rapport 15 (in Swedish)

Beyers JL, Riechers GH, Temple PJ (1992) Effects of long-term ozone exposure and drought on the photosynthetic capacity of ponderosa pine (*Pinus ponderosa* Laws.) New Phytol 122:81–90

Darrall NM (1989) The effects of air pollutants on physiological processes in plants. Plant Cell Environ 12:1–30

Eamus D, Barnes JD, Mortensen L, Ro-Poulsen H, Davison AW (1990) Persistent stimulation of CO_2 assimilation and stomatal conductance by summer ozone fumigation in Norway spruce. Environ Pollut 63:365–379

Ekeberg D, Jablonska AM, Ogner G (1995) Phytol as a possible indicator of ozone stress by *Picea abies*. Environ Pollut 89:55–58

Evans LS, Miller PR (1972) Ozone damage to ponderosa pine: a histological and histochemical appraisal. Am J Bot 59:297–304

Fink S (1989) Pathological anatomy of conifer needles subjected to gaseous air pollutants or mineral deficiencies. Aquilo Ser Bot 27:1–6

Greitner CS, Pell EJ, Winner (1994) Analysis of aspen foliage exposed to multiple stresses: ozone, nitrogen deficiency and drought. New Phytol 127:579–589

Grime JP (1979) Plant strategies and vegetation processes. John Wiley, Chichester

Guillemaut P, Weber-Lotfi F, Blache D, Prost M, Rether B, Dietrich A (1992) Conifer decline in north-east of France: characteristic changes in chloroplast protein pattern and absence of anti-oxidative defence capability point to an involvement of ozone. Physiol Plant 85:215–222

Günthardt-Goerg MS, Matyssek R, Scheidegger C, Keller T (1993) Differentiation and structural decline in the leaves and bark of birch (*Betula pendula*) under low ozone concentrations. Trees 7:104–114

Heagle A, Body D, Heck W (1973) An open-top field chamber to assess the impact of air pollution on plants. J Environ Qual 2:365–368

Holopainen T, Nygren P (1989) Effects of potassium deficiency and simulated acid rain alone and in combination on the ultrastructure of Scots pine needles. Can J For Res 19:1402–1411

Holopainen T, Palomäki V, Rantanen L (1993) Responses of conifer seedlings to slightly elevated ozone level in Kuopio open-field fumigation experiment. In: Tuomisto J, Ruuskanen J (eds) Proc 1st Finnish Conf on Environmental Sciences, Kuopio, 8-9 Oct 1993. Environmental Research in Finland Today. Finnish Society for Environmental Sciences, Kuopio Univ Publ. Nat Environ Sci 14:117–120

Jukola-Sulonen EL, Mikkola K, Nevalainen S, Yli-Kojola H (1987) Havupuiden elinvoimaisuus Suomessa vuosina 1985-1986. Metsäntutkimuslaitoksen tiedontoja 256 (in Finnish)

Karlsson PE, Medin E-L, Wickstöm H, Selldén G, Wallin G, Ottosson S, Skärby L (1995) Ozone and drought stress – interactive effects on the growth and physiology of Norway spruce (*Picea abies* (L.) Karst.) Water Air Soil Pollut 85:1325–1330

Laurence JA, Amundson RG, Friend AL, Pell EJ, Temple PJ (1994) Allocation of carbon in plants under stress: an analysis of the ROPIS experiments. J Environ Qual 23:412–417

Laurila T, Lättilä H (1994) Surface ozone exposures measured in Finland. Atmos Environ 28:103–114

Lehnherr B, Grandjean A, Mälcher A, Fuhrer J (1987) The effect of ozone in ambient air on ribulosphosphate carboxylase/oxygenase activity decreases photosynthesis and grain yield in wheat. J Plant Physiol 130:189–200

Lindgren M, Salemaa M (1994) Metsiemme elinvoimaisuus 1986-1993. In: Mälkönen E, Sivula H (eds) Suomen metsien kunto metsien terveydentilan tutkimusohjelman väliraportii, Helsinki. Metsäntutkimuslaitoksen tiedonantoja 527:11–24 (in Finnish).

Meyberg M, Lockhausen J, Kristen U (1988) Ultrastructural changes in mesophyll cells of spruce needles from a declining forest in northern Germany. Eur J For Pathol 18:169–175

Mikkelsen TN (1990) Physiological effects of ozone on beech (*Fagus sylvatica* L.) under semi-field conditions. Aquilo Ser Bot 29:45–49

Mikkelsen TN, Ro-Poulsen H (1994) Exposure of Norway spruce to ozone increases the sensitivity of current year needles to photoinhibition and desiccation. New Phytol 128:153–163

Miljø- og Energiministeriet (1995) De danske skoves sundhetstilstand. Resultater af overvågningen i 1994 (English summary). Skov- og Naturstyrelsen, Informationssektionen, Haraldsgade 53, 2100 Copenhagen Ø ISBN 87-601-4773-4

Mortensen L (1990) Effects of ozone on growth and dry matter partitioning in different provenances of Norway spruce (*Picea abies* (L.) Karst). Norw J Agric Sci 4:61–66

Mortensen L (1994) The influence of carbon dioxide or ozone concentration on growth and assimilate partitioning in seedlings of nine conifers. Acta Agric Scand Sect B Soil Plant Sci 44:157–163

Mortensen L (1995) Effect of carbon dioxide concentration on biomass production and partitioning in *Betula pubescens* Ehrh. Seedlings at different temperature regimes. Environ Pollut 87:337–343

Mortensen L, Skree O (1990) Effects of low ozone concentrations on growth of *Betula pubescens* Ehrh., *Betula verrucosa* Ehrh., and *Alnus incana* (L.) Moench. New Phytol 115:165–170

Myhre A, Aarnes H, Forberg E, Nilsen S (1988) Distribution of phyto-toxic ozone in Norway summer 1986 assayed by damage on Tobacco Bel W3. Dept of Biology, The Phytotron, University of Oslo, Blindern, 0316 Oslo 3

Nast W, Mortensen L, Fischer K, Fitting I (1993) Effects of air pollutants on the growth and antioxidative system of Norway spruce exposed in open-top chambers. Environ Pollut 80:85–90

NIJOS (1995) Landsomfattande representative registreringer av skogens vitalitet 1994. Rep 1/95, ISBN 82-7464-077-2 (English summary)

Ogner G (1993a) No general effect of ozone on foliar nutrient concentrations in mature scions of grafted *Picea abies* trees. Environ Pollut 82:197–200

Ogner G (1993b) No general ozone-specific indicator among hexane- and dichloromethane-soluble components of *Picea abies* needles exposed to ozone in open-top chambers. Environ Pollut 82:223–229

Ojanperä K, Sutinen S, Pleijel H, Selldén G (1992) Exposure of spring wheat *Triticum aestivum* L cv Drabant to different concentrations of ozone in open-top chambers: effects on the ultrastructure of flag leaf cells. New Phytol 120:39–48

Pääkkönen E, Holopainen T (1995) Influence of nitrogen supply on the response of clones of birch (*Betula pendula* Roth.) to ozone. New Phytol 129:595–603

Pääkkönen E, Paasisalo S, Holopainen T, Kärenlampi L (1993) Growth and stomatal responses of birch (*Betula pendula* Roth.) clones to ozone in open-air and chamber fumigations. New Phytol 125:615–623

Pääkkönen E, Holopainen T, Kärenlampi L (1995) Ageing-related anatomical and ultrastructural changes in leaves of birch (*Betula pendula* Roth.) clones as affected by low ozone exposure. Ann Bot 75:285–294

Palomäki V, Holopainen T (1994) Effects of phosphorus deficiency and recovery fertilisation on growth, mineral concentration, and ultrastructure of Scots pine needles. Can J For Res 24:2459–2468

Palomäki V, Holopainen JK, Holopainen T (1994) Effects of drought and waterlogging on ultrastructure of Scots pine and Norway spruce needles. Trees 9:98–105

Parviainen J, Seppänen P (1994) Metsien ekologinen kestävyys ja metsänkasvatusvaihtoehdot. Metsäntutkimuslaitoksen tiedontoja 511:110

Pell EJ, Weissberger WC (1976) Histopathological characterization of ozone injury to soybean foliage. Phytopathology 66:856–861

Pleijel H, Wallin G, Karlsson PE, Skärby L, Selldén G (1995) Gradients of ozone at a forest site and over a crop field – consequences for the AOT40 concept of critical level. Water Air Soil Pollut 85:2033–38

Pye JM (1988) Impact of ozone on the growth and yield of trees: a review. J Environ Qual 7:347–360

Rantanen L, Palomäki V, Holopainen T (1994) Interactions between exposure to O_3 and nutrient status of trees: effects on nutrient content and uptake, growth, mycorrhiza and needle ultrastructure. New Phytol 128:679–687

Reich PB (1987) Quantifying plant response to ozone: a unifying theory. Tree Physiol 3:63–91

Reich PB, Schoettle AW, Stroo HF, Troiano J, Amundson RG (1987) Effects of ozone and acid rain on white pine (*Pinus strobus*) seedlings grown in five soils. I. Net photosynthesis and growth. Can J Bot 65:977–987

Reutze M, Schmitt U (1988) Histologie der Alterung von Fichtennadeln (*Picea abies* (L) Karst). Angew Bot 62:9–20

Robinson DC, Wellburn AR (1991) Seasonal changes in pigments of Norway spruce, *Picea abies* (L.) Karst, and the influence of summer ozone exposures. New Phytol 119:251–259

Ro-Poulsen H, Andersen B, Mortensen L, Moseholm L (1981) Elevated ozone levels in ambient air in and around Copenhagen indicated by means of indicator plants. Oikos 36:171–176

Sandelius AS, Näslund K, Carlsson A, Pleijel H, Selldén G (1995) Exposure of spring wheat (*Triticum aestivum* L., cv. Drabant) to ozone in open-top chambers. Effects on acyl lipids and chlorophyll content of flag leaves. New Phytol 131:231–239

Selldén G, Pleijel H (1995) Photochemical oxidant effects on vegetation – response in relation to plant strategy. Water Air Soil Pollut 85:111–122

Skärby L, Jönsson B (1985) Skador av ozon på potatis, spenat, ärter och bönor. Växtskyddsnotiser 49:9–16 (in Swedish)

Skärby L, Grennfeldt P, Lunneryd S-G (1979) Elevated ozone levels at the Swedish west coast and in southern Sweden (Skåne), using tobacco as an indicator plant. IVL Report B461, Göteborg

Skärby L, Wallin G, Selldén S, Karlsson PE, Ottosson S, Sutinen S, Grennfelt P (1995) Tropospheric ozone – a stress factor for Norway spruce. In: Staaf H, Tyler G (eds) Effects of acid deposition and tropospheric ozone on forest ecosystems in Sweden. Ecol Bull (Copenhagen) 44:133–146

Soikkeli S (1978) Seasonal changes in mesophyll ultrastructure of needles of Norway spruce (*Picea abies*). Can J Bot 56:1932–1940

Soikkeli S (1981) Comparison of cytological injuries in conifer needles from several polluted industrial environments in Finland. Ann Bot Fenn 18:47–61

Soikkeli S, Tuovinen T (1979) Damage in mesophyll ultrastructure of needles of Norway spruce in two industrial environments in central Finland. Ann Bot Fenn 16:50–64

Sutinen S (1987a) Ultrastructure of mesophyll cells of spruce needles exposed to O_3 alone and together with SO_2. Eur J For Pathol 17:362–368

Sutinen S (1987b) Cytology of Norway spruce needles. I. Changes during ageing. Eur J For Pathol 17:65–73

Sutinen S (1987c) Cytology of Norway spruce needles. II. Changes in yellowing spruces from the Taunus mountains, West Germany. Eur J For Pathol 17:74–85

Sutinen S (1989) Ultrastructure of conifer needles exposed to O_3 and $O_3 + SO_2$ compared to the needle structure in the field in Finland, Sweden and in west Germany. Aquilo Ser Bot 27:29–34

Sutinen S (1990) Structural changes in needle tissue of spruce trees from a damaged stand in southern Finland. Scand J For Res 5:403–412

Sutinen S, Skärby L, Wallin G, Selldén G (1990) Long-term exposure of Norway spruce, *Picea abies* (L.) Karst., to ozone in open-top chambers. II. Effects of the ultrastructure of needles. New Phytol 115:345–355

Sutinen S, Mäenpää M, Koivisto L (1995) Structure of mesophyll cells of pine and spruce needles from several areas in Finland. In: Acid reign '95? Abstract book. Kluwer, Dordrecht, DIR 28, p 98

Thomson WW, Dugger WM, Palmer Rl (1966) Effects of ozone on the fine structure of the palisade parenchyma cells of bean leaves. Can J Bot 44:1677–1682

Thomson WW, Nagahashi J, Platt K (1974) Further observations on the effects of ozone on the ultrastructure of leaf tissue. In: Dugger M (ed) Air pollution effects on plant growth. American Chemical Society, Washington DC

Vogels K, Masuch G (1984) Metabolic and substructural changes of bean leaves caused by separate and simultaneous ozone and sulphur dioxide exposures. In: Proceedings of an international workshop on the evaluation and assessment of the effects of photochemical oxidants on human health, agricultural crops, forestry, materials and visibility. Feb 29 - March 2, Göteborg

Walles B, Nyman B, Aldén T (1973) On the ultrastructure of needles of *Pinus silvestris* L. Stud For Suec 106:4–26

Wallin G (1990) On the impact of tropospheric ozone on photosynthesis and stomatal conductance of Norway spruce, *Picea abies* (L.) Karst. Ph D Thesis, University of Göteborg, Göteborg, ISBN 91-86022-60-1

Wallin G, Skärby L (1992) The influence of ozone on the stomatal and non-stomatal limitation of photosynthesis in Norway spruce, *Picea abies* (L.) Karst., exposed to soil moisture deficit. Trees 6:128–136

Wallin G, Skärby L, Selldén G (1990) Long-term exposure of Norway spruce, *Picea abies* (l.) Karst., to ozone in open-top chambers. I. Effects on the capacity of net photosynthesis, dark respiration and leaf conductance of shoots of different ages. New Phytol 115:335–344

Wallin G, Skärby L, Selldén G (1992a) Long-term exposure of Norway spruce, *Picea abies* (L.) Karst., to ozone in open-top chambers. III. Effects on the light response of net photosynthesis in shoots of different ages. New Phytol 121:387–394

Wallin G, Ottosson S, Selldén G (1992b) Long-term exposure of Norway spruce, *Picea abies* (L.) Karst., to ozone in open-top chambers. IV. Effects on the stomatal and non-stomatal limitation of photosynthesis and on the carboxylation efficiency. New Phytol 121:395-401

Wijk S, Berghäll S, Wulff S (1994) Skogsskador i Sverige 1993 (English summary). Skogsstyrelsen VI 94. Skogsstyrelsens förlag, S-551 83 Jönköping, Sweden ISBN 110-0295

Wulff A, Hänninen O, Tuomainen A, Kärenlampi L (1992) A method for open-air exposure of plants to ozone. Ann Bot Fenn 29:253–262

9 Methods for Controlled and Field Ozone Exposures of Forest Tree Species in North America

R.C. Musselman and B.A. Hale

9.1 The Forest and Ozone

9.1.1 Progress in Air Pollution Research

Research on the effects of pollutants on vegetation has concentrated primarily on agricultural crops. Pollutant effects on crops have been examined since the 1950s, after initial observations of oxidant stipple injury to field crops which were attributed to O_3 (Richards et al. 1958; Heggestad and Middleton 1959). Research began with field observation and description of leaf injury, and duplication of injury symptoms with laboratory O_3 fumigations. Early research examined cultivar differences in response to O_3, and effects of O_3-free (charcoal-filtered) air and antioxidant chemicals on the prevention of oxidant injury. Research on O_3 effects on plants matured to studies of mechanisms of injury, and dose-response studies of pollutant effects on productivity. These studies often used chambers to control fumigation. Specific methodologies have been developed to study air pollution effects in laboratory, greenhouse, or field experimental settings. Experiments have been conducted on numerous crop species. Recent research has examined the effects of specific components of O_3 exposure (peak concentration, duration, respite times [intervals when plants are not exposed to O_3]) on plants during different stages of development. Research is continuing to determine the best way to characterize the O_3 exposure parameter. Results of O_3 fumigation–plant response studies, as well as the methods used to obtain them, have been recently reviewed (Lefohn 1992b).

Much of the research on the effects of O_3 on plants in North America has been conducted with crops, and it has been documented that O_3 can cause large economic losses to crop growers (Adams and Crocker 1989). Most of the data which led to this conclusion came from the US Environmental Protection Agency National Crop Loss Assessment Network (NCLAN) experiments conducted exclusively on agricultural crops in the

1980s (Heck et al. 1990). However, early observations of effects of O_3 on forest trees in natural ecosystems were reported in eastern US for white pine (Berry and Riperton 1963) and in southwestern US for ponderosa pine (Miller et al. 1963).

Research into the effects of O_3 on trees presents challenges not associated with agricultural crops. Trees are exposed to pollutant stress for longer periods of time than are annual crops (Reich 1987) so that the effects might be cumulative. The exposure might reduce accumulation of reserves, and require utilization of reserves for detoxification or repair of O_3-injured tissue. Trees often have a large biomass which can act as a buffer to the stress from pollutants. Large root systems may temporarily provide storage nutrients to overcome the loss of productivity from pollutant stress. Since ecosystems are large and diverse, the impact of pollutants on an individual tree may have a minimal effect on ecosystem structure of function at the community or landscape level as other individuals replace the sensitive ones. The large spatial area occupied by forest communities and the nonuniformity of the terrain and soil on which they occur also makes identification of pollutant effects difficult.

Investigations of forest responses to O_3 are being conducted in both field and controlled environment settings, using both young and mature trees. Many of the methods for study are adapted from research on the effects of O_3 on crops, which in some cases are modified to address the differences between agro- and forest ecosystems. The present description of exposure methods includes a discussion of their limitations and strengths.

9.1.2 Limitations to the Application of Crop Research Methods to Forest Ecosystems

Forest response to pollutants cannot be inferred from crop pollutant response research. There are many complex differences between crops and forests, and extrapolation of findings from crop research to forests is problematic. Certain basic biochemical or physiological mechanisms of plant response to pollutants may be the same; however, the magnitude of tree response may be different than that of crops, or may differ more between field and controlled environment studies. Thus, forest and tree O_3 response research should not be considered as an extension of agricultural crop research, as the structure and function of forests are far more complex than those of crops. The larger genetic and environmental variability within the community, the extended period of exposure to the pollutant, and the greater potential for interaction with biotic and abiotic stresses are characteristics of forests. The size of individual plants in forest ecosystems also complicates many crop research protocols.

Variability. Forest ecosystems are highly variable, both genetically and environmentally. A wide environmental variability can be present at several levels of scale, including the microsite, plant community, and the landscape level in the forest ecosystem. Trees often grow in complex terrain, which increases the environmental variability from site to site. Changes in bedrock, soil, slope, aspect, temperature, moisture, light intensity, and other environmental variables can change abruptly and frequently in forest ecosystems, with corresponding changes in plant growth and in species diversity and abundance. As a result, plant community structures often change abruptly over short distances rather than gradually as is evident in more uniform environments. This is very important in the assessment of forest response to O_3, as species susceptibility to O_3 can be modified by environmental factors such as temperature, soil moisture, and relative humidity. Thus, it may be difficult to generalize community sensitivity from data gathered from individual plants or from tightly controlled experimental conditions.

Genetic variability of forest trees is often larger than that for crop species which have been specifically bred for uniform growth and productivity under defined environmental conditions. Within forest tree species, genetic variability can be very high. Even within clonally propagated species, which are expected to have less genetic variability, variation between clones in susceptibility to pollutants can be high (Karnosky et al. 1992a,b; Berrang et al. 1991). The wide range of environmental conditions under which forest trees must survive requires large genetic diversity for population adaptation to such changing conditions. Genetic variability enables forest ecosystems to be resilient in reaction to air pollution and other stresses, since O_3 tolerant genotypes existing in the population may replace sensitive genotypes. Oleksyn et al. (1994) have demonstrated that *Pinus sylvestris* provenances with higher genetic diversity were more tolerant to air pollutants than those with low genetic diversity. An increase in frequency of tolerant individuals in a population in response to selection pressure in the form of air pollutants (population plasticity) can only occur if the population is genetically variable for tolerance to pollutant stress (Taylor and Pitelka 1992). Population plasticity in response to air pollutants could be due to a greater frequency of individuals with mechanisms of pollutant exclusion (e.g., reduced stomatal conductance) or cellular detoxification. However, loss of O_3 sensitive genotypes from a population may be accompanied by a loss of other more desirable attributes if these factors are linked. A large variability in plant response to pollutants increases the difficulty of detecting subtle pollutant responses relative to other environmental effects on tree growth or productivity.

Continuous Multiyear Exposure. Trees often live for many years and this longevity results in a long and complex exposure to pollutants. The growing season exposure is often longer for trees than for annual crops, since trees either retain their leaves all year round (conifers) or leaf out quickly in the spring from stored reserves (deciduous trees). In addition, individual trees are exposed every year throughout their lifetimes to varying O_3 exposure. Because of its high reactivity, O_3 breaks down on contact and is not accumulated within plant tissue, in contrast to fluoride or oxides of sulfur and nitrogen. However, the negative stress-induced response of plants to O_3 may accumulate. Trees can be exposed to O_3 for tens or hundreds of years, depending on the longevity of the species. Mature trees may be more (Samuelson and Edwards 1993) or less (Grulke and Miller 1994) sensitive to O_3 than juvenile trees. The position within the canopy can also influence plant response to O_3 (Fredericksen et al. 1995). This may be due to differences in the physiological activity of plants at different ages.

Stress Interactions. Plant responses to other environmental stresses may influence their response to O_3, or vice versa. Because of the large variability and diversity in natural ecosystems, it is often difficult to detect the response of ecosystem components to pollutant stress. Combined with the response of trees to other, perhaps more damaging stresses such as low moisture, extreme temperatures, and nutrient deficiencies, the identification of the response to O_3-induced stress is very difficult (Colls and Unsworth 1992). Complex natural ecosystems would typically have a number of stresses, in addition to air pollutants, impacting the ecosystem on varying temporal and spatial scales. How plants react to these other stresses influences how plants react to pollutant stress. A precise detection and quantification of the response to pollutants, in the presence of plant reaction to other stresses, is extremely difficult if not impossible (Woodwell et al. 1989).

9.2 Basis for Comparison of Field vs Chamber Research

The methodologies used in vegetation research have become more sophisticated over the years as new technologies have been developed. New exposure systems have been devised with computer-controlled pollutant dispensing systems that make it possible to better simulate the exposures which plants receive in the field. Studies of plant response to O_3 exposure require the exposure of vegetation to several different, definable levels of the pollutant that can be specifically described and the measurement of plant response to that exposure. Standard protocols for measuring and reporting environmental conditions should be followed (Krizek and McFarlane 1983). The variation in O_3 regimes may be achieved by controlled fumigation, chemical or mechanical exclusion, or by natural gradients of O_3. Controlled O_3 fumigation systems are designed to maintain a modified gaseous atmosphere around a plant for a defined period of time, during which observations of plant response can be conducted. All fumigation systems share some common features: general plant growth conditions (light, temperature, humidity, carbon dioxide, soil moisture) must be met with and held consistent between treatments, and differential concentrations of O_3 must be supplied to the vegetation and maintained during the exposure period. Exposure systems have been established in controlled environments, greenhouses, and the field. More recent reviews of wet and dry deposition exposure systems have refined our knowledge of the strengths and limitations of experimental approaches for studying the effects of O_3 on crops and trees (CEC 1987; US EPA 1987a,b; Strain et al. 1991; Manning and Krupa 1992; Weigel et al. 1992).

Controlled fumigation systems range from cuvettes which enclose leaves or branches (Ennis et al. 1990) to field plots surrounded by a series of tubes with calibrated orifices spatially distributed to emit gaseous pollutants to a plant canopy (Lee and Lewis 1978). Systems which exclude O_3 by mechanical or chemical means have been used, as have natural gradients of O_3, to evaluate vegetation response to ambient O_3 (Oshima et al. 1976; Kostka-Rick and Manning 1992). Physical characteristics and environmental conditions of air pollution fumigation systems have been reviewed (Heagle and Philbeck 1979; Heck et al. 1979; Kilz 1987; US EPA 1987a,b; Winner and Greitner 1989; Payer et al. 1990; Manning and Krupa 1992). A comparison of the usefulness of the various O_3 fumigation or exclusion systems for plant response studies is presented in Table 9.1. Generally, a greater control of environmental conditions is associated with a greater possible range of treatment levels, the replication of experimental units, and a lower likelihood of extraneous environmental factors affecting treatment response, but also a lower potential to simulate field losses. We present a brief description of the various fumigation systems.

Table 9.1. Comparison of fumigation systems for studies of plant response to ozone exposure

Ozone fumigation or exclusion system	Simulation of field losses	Replication of experimental unit	Range of treatment levels	Likelihood of extraneous factors affecting response
Controlled environment chambers	Low	High	High	Medium
Greenhouse chambers	Low	High	High	Medium
Closed top field chambers	Medium	Medium	High	Medium
Open-top field chambers	Medium to high	Medium	High	Medium
Mechanical field exclusion	High	Low to medium	Low	Medium
Open field fumigation	High	Low	Low	High
Natural gradients	High	Low	Low	High
Chemical protectants	Medium to high	High	Low	Medium to high

9.2.1 Fumigation Systems – Controlled Environment

Whole Plant Chambers. Controlled environment exposure systems are those in which light sources and control of temperature and relative humidity are artificial. Light quality and quantity are likely to be lower than in ambient environments, usually resulting in lower photosynthetically active radiation (PAR) and spectral patterns that are different from ambient. Temperature and relative humidity are likely to be more stable and reproducible in a controlled environment than in ambient air. Controlled environment exposure systems are typified by the widely used continuous stirred tank reactor (CSTR; Fig. 9.1), originally designed for mass balance studies of O_3 flux to vegetation (Heagle and Philbeck 1979). They often are placed inside controlled environment growth chambers, in growth rooms with controlled lighting, temperature, and humidity, or in greenhouses. CSTR chambers have distinct advantages for gas exchange studies because fluxes can be readily calculated when controlling for environmental and pollutant conditions. The rapid air mixing minimizes horizontal and vertical gradients within chambers as well as leaf boundary layer resistance. An advantage of controlled environment facilities is the more precise identification of the environmental conditions under which the plants are treated, such as the tem-

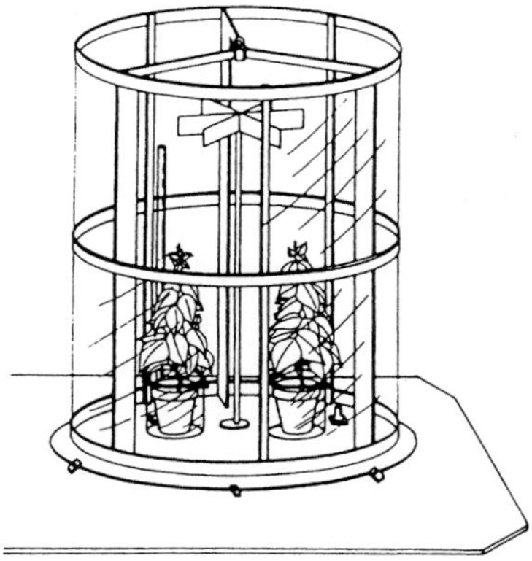

Fig. 9.1. A continuous stirred tank reactor (CSTR) chamber. Air containing a specific concentration of ozone enters the chamber near the top and exits at the bottom of the chamber (not shown). Supplemental lighting is often provided above the chamber. (Heagle and Philbeck 1979)

perature, light, and humidity. These conditions can have a large effect on how plants respond to pollutants; thus, precise identification of the levels of these covariants is important in experimentation and data analysis.

Disadvantages of controlled environment chambers include (1) the artificial pollution and growing conditions (especially temperature, humidity, and soil) do not represent natural ecosystem conditons; (2) lighting systems rarely provide the spectrum and intensity of PAR present in ambient environments; (3) the rapid air movement may cause wind injury to sensitive plants and reduce leaf boundary layer resistance to O_3 uptake, thus inflating plant response to external dose; (4) the size of chambers, as constructed to date, limits research to the study of small potted plants or seedling of larger plants. While CSTR chambers in controlled environments are useful for evaluating O_3 effects on air pollution response mechanisms and physiological processes, it is difficult to extrapolate the data to field situations.

CSTRs have been used in greenhouse environments to examine the response of tree seedlings to O_3 (Shafer et al. 1993). When CSTR chambers are placed in greenhouse environments, light, temperature, and relative humidity conditions fluctuate with those occurring in the greenhouse. Generally, light intensity is lower and temperature is higher in greenhouses than outside the greenhouse. Environmental conditions in greenhouses are more closely related to field studies than to controlled environments; but plant culture and environmental conditions are still quite different from those of field exposure chambers, making direct extrapolation difficult. These studies are, however, directly applicable to the phytotoxicity of O_3 for greenhouse-grown ornamental and floriculture crops.

Fumigation systems for controlled environment chambers are now often controlled using O_3 supplied through mass flow controller valves operated from a feedback system linked to a computer which is programmed for specific levels of O_3, or programmed to follow ambient or above ambient O_3 levels. Hence, O_3 levels can be changed and the change monitored and recorded within seconds. Fumigations using these systems can more realistically simulate and monitor ambient O_3 concentrations than the old system which used in the past square wave experiments where O_3 was manually turned on, held as near as possible to a specific concentration for a specific time period, and then turned off. These new systems also allow computer data logging of environmental conditions such as temperature, humidity, and light intensity, and analysis of actual O_3 concentration for a more accurate calculation of the exposure parameter.

A whole tree cuvette for conifer seedlings has recently been developed. The cuvette system controls light, temperature, vapor pressure, and carbon dioxide concentration. A computer is used for system control and for data acquisition. The system has been used to determine the response of western red cedar and white spruce seedlings to environmental stress (Livingston et al. 1994).

9.2.2 Fumigation Systems – Field

Open-Top Chambers. Open-top field chambers are used in most field studies of plant response to gaseous pollutants (Fig. 9.2). They were first designed for studies on annual herbaceous crop plants (Heagle et al. 1973; Mandl et al. 1973). These chambers have diameters and heights which currently limit their use for larger plants. However, both standard and enlarged versions have been used successfully in tree seedling and sapling studies (Tingey et al. 1976; Shafer et al. 1987; Hogsett et al. 1989; Laurence et al. 1989; Sasek and Richardson 1989; Shafer and Heagle 1989; Adams et al. 1990; Billen et al. 1990; Chappelka et al. 1990; Andersen et al. 1991; Edwards 1991; Eissenstat et al. 1991; Tjoelker and Luxmoore 1991; Byres et al. 1992; Dean and Johnson 1992; Edwards et al. 1992; Friend and Tomlinson 1992; Kress et al. 1992; Mudano et al. 1992; Qui et al. 1992; Simini et al. 1992; Stow et al. 1992; Thornton et al. 1992; Karnosky et al. 1995; Nast et al. 1993; Olszyk et al. 1992; Temple et al. 1993; Amundson et al. 1995; Clark et al. 1995). Because the results from tree species studies are often extrapolated to predict O_3 effects on forests, these studies require good exposure control in order to replicate ambient O_3 profiles characteristic of many low-elevation, rural areas of eastern North America. This condition could not have been met with using an open field exposure system. Open-top chambers which are large enough for mature

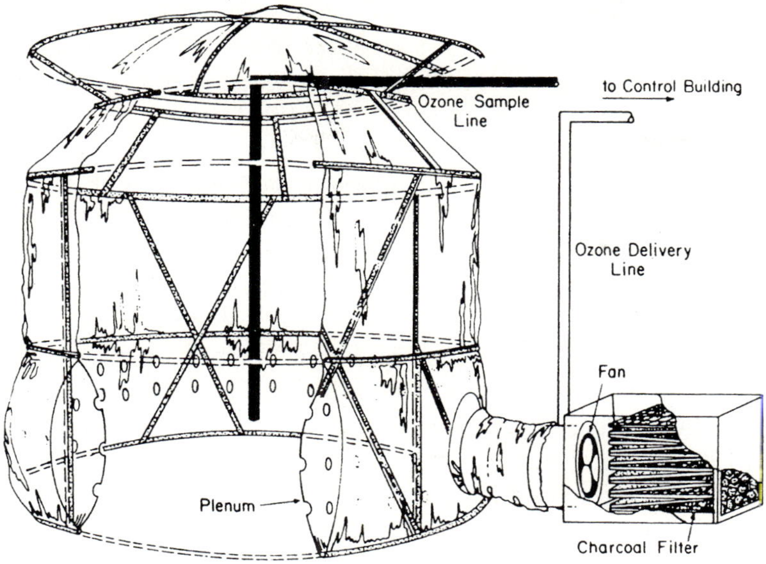

Fig. 9.2. An open top chamber (OTC) commonly used in field fumigation systems. (Hogsett et al. 1985)

trees have been developed but are expensive (Kats et al. 1985; Albaugh et al. 1992; Mandl et al. 1989; Olszyk et al. 1992). Sufficient air flow is difficult to obtain in these large chambers, and environmental conditions, particularly light, temperature, and humidity, are modified from that outside the chambers.

A seasonal study of tree response to O_3 used open-top chambers with roll-up sides (Wiltshire et al. 1992). Except for exposure days, the plants were maintained in ambient climatic conditions. The exposure episodes numbered between 27 and 30 throughout the growing season, and occurred on days with ambient meteorology associated with naturally occurring O_3 episodes, i.e., high incident radiation and temperature with little air movement (Wiltshire et al. 1992). The maintenance of near-ambient meteorological conditions during both growth and exposure periods ensured that this study represents field-grown plant responses to O_3 better than traditional open-top chamber studies.

The main advantage of open-top field chambers is the ability to provide an enclosed environmental area for an increased range of pollutant treatments at near-ambient environmental conditions while excluding ambient pollutants. Most current open-top field chamber designs have been widely and successfully used for studying the impact of O_3 on crops over a growing season, such as the US Environmental Protection Agency NCLAN program (Preston and Tingey 1988; Heck et al. 1990). While the open-top chambers provide for the least amount of environmental modification of any outdoor chamber, small chamber effects should be cumulative over long-growing seasons or multiple-season exposure to pollutants necessary for perennial plants. It is well documented that fumigation chambers can have a significant effect on plant response. Chamber responses have been summarized by Manning and Krupa (1992). Microclimatic changes within chambers can have small, but cumulative, effects on plant growth, although it has been documented that the relative plant response to O_3 is not influenced by the chamber (Heagle et al. 1988).

Open-top chamber effects on microclimate include reductions in light intensity, wind velocity, rainfall, dew formation, and persistent increases in air temperature and possibly relative humidity (US EPA 1987a,b; Heagle et al. 1988; McLeod and Baker 1988; Heck et al. 1994; Fuhrer 1994). Open-top chambers can have a large rain-shadow effect, with reductions in rainfall in chambers of nearly 50% (Fuhrer 1994). Decreased vapor pressure deficit in chambers results in higher evapotranspiration which can cause further reductions in soil moisture within chambers. For plants taller than 120 cm, there is more air movement near the bottom of the plant canopy than near the top during calm periods (Heagle et al. 1979; Weinstock et al. 1982). Recent research has shown that open-top chambers may affect deposition velocity for O_3 (Fuhrer 1994). However, measurements of chamber effects on gas exchange rates suggest that O_3 uptake by the foliage may be similar between open-top chambered plants and non-chambered plants (US EPA 1987b; Unsworth et al. 1984). These findings do not reflect the apparent differences in environment and O_3 deposition.

Extensive comparisons have been made among plants growing in carbon-filtered chambers (CF), non-filtered chambers (NF), and similarly sized and located non-chambered, ambient air plots (AA). Much attention has been given to the fact that AA plants are often different in productivity than NF plants despite their exposure to air of similar quality, although this is hardly surprising given the modification of microclimate by the open-top chamber (Manning and Krupa 1992). In NCLAN studies, plants in NF chambers are commonly taller than those in AA (Heagle et al. 1979; Olszyk et al. 1980; Kress and Miller 1985). Chamber effect on height is the only consistently observed response (Heagle et al. 1988). Observed plant yield in AA can be equal to, greater than, or less than that in NF chambers (Heagle et al. 1988). Evaluations of differences is plant yield between plants growing in NF chambers and those in AA, from NCLAN studies, suggest that chamber effects on plant response to O_3 are relatively small (Heck et al. 1994; Heagle et al. 1988). Further, microclimate differences within an open-top chamber can cause a significant effect on yield; but there was rarely a significant interaction between the effects of position and O_3 on plant yield (Heagle et al. 1988).

Some modification of microclimate can occur without changing plant response to O_3. A comparison of plant growth and plant response to O_3 exposure in open-top chambers, closed-top chambers and air exclusion systems was conducted by Olszyk et al. (1986c). The authors discovered that there was interaction between plant response to O_3 and the type of exposure system for less than 10% of the growth parameters measured, suggesting that plant response to O_3 (i.e, the relationship among O_3 treatments) was the same regardless of exposure system. Plants from exclusion systems were shorter than those grown in open-top chambers and generally weighed more. Of all the three groups of plants, those in the control plots of the exclusion system (i.e., receiving ambient O_3 exposure) were most similar in size to plants grown in field plots. While this and a study on 'Valencia' orange trees (Olszyk et al. 1992) indicate that environmental modification caused by chambers will affect plant growth, there is no evidence that there is a large effect of chambers on plant response to O_3. It is assumed that, because of the decreasing relative effects on plant environment caused by controlled environment chambers, greenhouses, closed-top field chambers, open-top field chambers, open air systems, and ambient gradients in that order, the system effects on plant response to O_3 will decrease in the same order.

Concern has been raised about plant response to trace pollutants in open-top chambers, specifically N_2O_5, HNO_3, and NO in chambers receiving O_3 generated from dry air, and NO_2 in chambers receiving ambient air. These trace pollutants may have a direct effect (positive or negative) on plant processes, or may change how plants respond to O_3, and without careful evaluation, these effects may go undistinguished from those of O_3. A comparison of alfalfa response to the same O_3 exposure, generated either electrostatically from air or through non-filtration of ambient air, indicated that the generated O_3 treatment was more phytotoxic than the ambient O_3 treatment, likely due to the cogeneration of N_2O_5 and NO along with O_3 from dry air (Olszyk et al. 1990). Brown and Roberts (1988) reported N_2O_5 and HNO_3 production in chambers with O_3 generated from electric discharge of ambient air. Bytnerowicz et al. (1995) reported HNO_3 in chambers with O_3 generated from compressed air. They also reported increased concentrations of HNO_2 in charcoal-filtered chambers.

Open-top chamber studies which use filtered vs non-filtered ambient air have been proposed to avoid the problem of the cogeneration of nitrogenous compounds when generating O_3. The drawback of this or any two-treatment approach is that plant responses to low ambient levels of O_3 as might occur during many years are quite subtle. Furthermore, determination of the response of plants to O_3 requires multiple levels of the pollutant and levels high enough to cause more than the subtle response often seen with ambient air. These conditions occur naturally only in areas with high levels of ambient O_3, such as southern California and perhaps some areas of eastern US and Canada.

Branch and Leaf Chambers. Most of the recent developments in controlled environment exposure systems have been modifications of existing systems. The tremendous interest in evaluation of mature tree response to O_3 has prompted the development of large branch chambers (Fig. 9.3) for

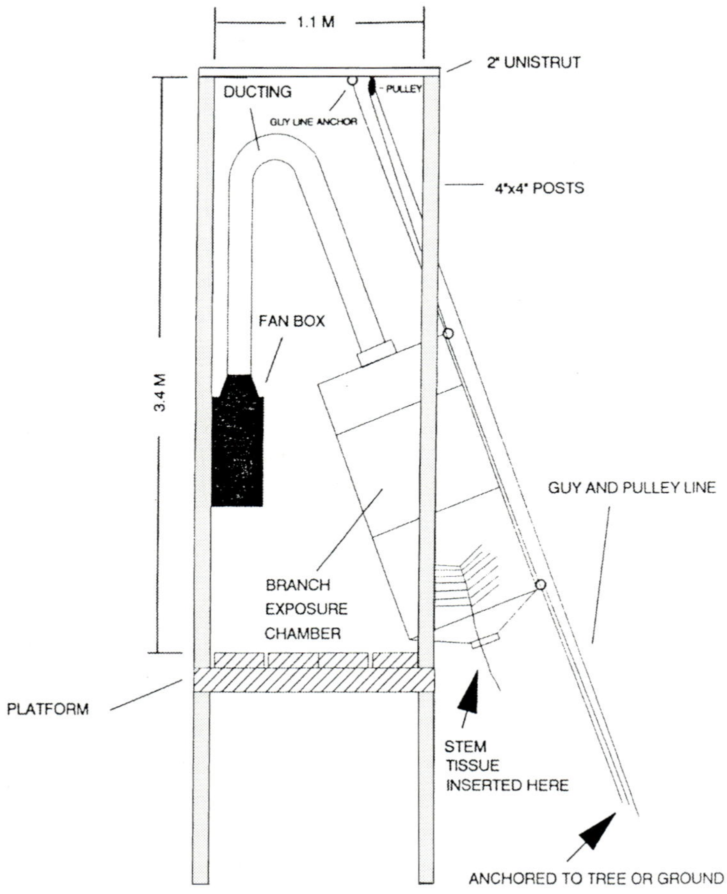

Fig. 9.3. A branch exposure chamber (BEC) with support system. The support system is an elevated platform adjacent to the tree from which the BEC is anchored. Standard metal scaffolding could be used in place of the wooden platform shown. (Houpis et al. 1991)

estimating O_3 flux to trees. These branch chambers share many of the design characteristics of a CSTR. An early design utilized a wind-activated cylinder with various amounts of charcoal filtration (Hernandez et al. 1986). Recent improvements to this design have chamber walls consisting of a transparent film spread over a supporting frame, a fan to reduce boundary layer resistance across the foliar surface, and an air inlet and outlet so that differential O_3, carbon dioxide (photosynthesis), and water vapor (leaf diffusive resistance) measurements can be taken (Evans and Dougherty 1986; Ennis et al. 1990; Houpis et al. 1991; Teskey et al. 1991). The advantages of the branch system include: (1) the ease with which the Teflon bag can be replaced; (2) uniform light transmission can be maintained; (3) the branch chamber can be moved from plant to plant; (4) it can be used in situ; and (5) it can be modified for different-sized branches. One of the disadvantages of the branch chamber, and indeed of any such cuvette which places one part of the plant under different environmental conditions to the rest of the plant, is that it is uncertain whether the isolation leads to a response which is different from that which would have been observed if the rest of the plant were under the same environmental conditions as the branch. In addition, total tree growth cannot be estimated using branch chambers, as only part of the plant is treated with O_3. It must also be recognized that the sample branch may respond differently than other branches of the tree; the entire tree response is the integrative response of all the individual branches of the tree.

Closed-Top Field Chambers. Several designs of field exposure chambers have been developed to overcome some of the disadvantages of the open-top chambers, especially the incursion of ambient air. Closed-top chambers were first developed in the 1950s; generally, their use diminished in favor of open-top chambers. The problems with these earlier closed-top systems, which were generally greenhouse-type structures, were higher temperatures, lower light intensity, and lack of adequate air flow in the chambers. Although there was a considerable chamber effect on plants grown in these facilities, they were useful in determining that ambient O_3 could reduce growth and yield of plants (Thompson and Taylor 1969). Precise quantification of the O_3 effects were complicated by the drastic chamber effects on light and temperature and subsequent growth. Closed-top chambers have been more recently constructed in California to assess crop loss as a result of O_3. Closed-top chambers were chosen because the authors wished to characterize the pollutant concentration very precisely. Pollutant gradients within these chambers (Fig. 9.4) are minimal (Musselman et al. 1986c). The chambers were octagonal in shape and covered

Fig. 9.4. A closed-top field chamber. Filtered and non-filtered air is delivered to the mixing box and into the chamber. The chamber is described in detail in Musselman et al. (1986c)

with Teflon film; the soil was completely replaced with standard greenhouse mix. Temperatures in the chamber were higher (by 2–4 °C at midday and 1–2 °C at night) than in the ambient air, and PAR light levels were reduced by 11%. While the chambers were not suitable for studies requiring close approximation of field conditions, they were very useful when tight control over soil moisture and pollutant concentration was needed.

Closed-top chambers were constructed and installed in the UK to study responses of shrubs and large herbaceous species to long-term, low (chronic) concentrations of sulphur dioxide, nitrogen dioxide and O_3 (Rafarel and Ashenden 1991). They were a smaller version of an earlier design, as the larger chambers required pure gas sources of nitrogen dioxide and sulphur dioxide to be diluted into the ventilating air stream which resulted in exposure concentrations that varied during fumigation. The flow rate of the smaller chambers meant that premixed gases were sufficient to maintain steady control of treatment concentrations. Because the gases were discharged from the source at constant concentrations, different treatments were achieved by placing one or more pollutant supply tubes in the exposure chambers. Good air circulation (and moderate ambient temperatures) maintained the domes at near ambient conditions. However, whether or not results can be extrapolated to predict plant response to O_3 under ambient culture conditions has not been determined.

Closed-top field chambers have been developed for carbon dioxide enrichment studies which could be adapted for O_3 fumigations. These chamber have been used for crops (Jones et al. 1984) and in natural ecosystems with small plants (Oechel et al. 1992). Temperature, humidity, and carbon dioxide concentration are controlled and monitored in these systems.

Self-Ventilating Field Chambers. A field air exclusion system was developed for use in remote areas without electric power (Miller and Yoshiyama 1973). The chambers are self-ventilating, and are used to exclude ambient air pollutants. Significant increases in temperature and decreases in light intensity within the chambers limit their use for estimating growth responses of plants to O_3. This chamber design could perhaps be modified to maximize light interception at the plant canopy.

Linear Gradient Systems. Unchambered fumigation can be accomplished using O_3 dispensing tubes lying on the ground between plant rows (linear gradient systems; Fig. 9.5; Laurence et al. 1982; US EPA 1987b). Since this method is based on relatively uncontrolled dilution (dependent on ambient wind direction, wind speed, and turbulence), and the amount of pollutant decreases along the tube from the fumigation source at the blower end of the tube, precise monitoring of pollutant concentration at various distances from the source tube is required. An open air O_3 exclusion system was used by Olszyk et al. (1986a,b,c) to exclude O_3 from field plots in a high O_3 environment. The system was used to provide a gradient of O_3 exposures to row crops without chambers. Although the linear gradient systems were designed for short height agricultural row crop situations, they perhaps could be adapted for use in natural ecosystems.

Open Air Field Fumigation. Open air field exposure systems have the potential to most closely estimate field losses due to O_3, as the plants are grown and exposed under ambient field environmental conditions. However, of all the exposure systems, this is the least controllable and repeatable. It has been used in the past to expose plants to "static" concentrations (desired concentration is the same throughout the ex-

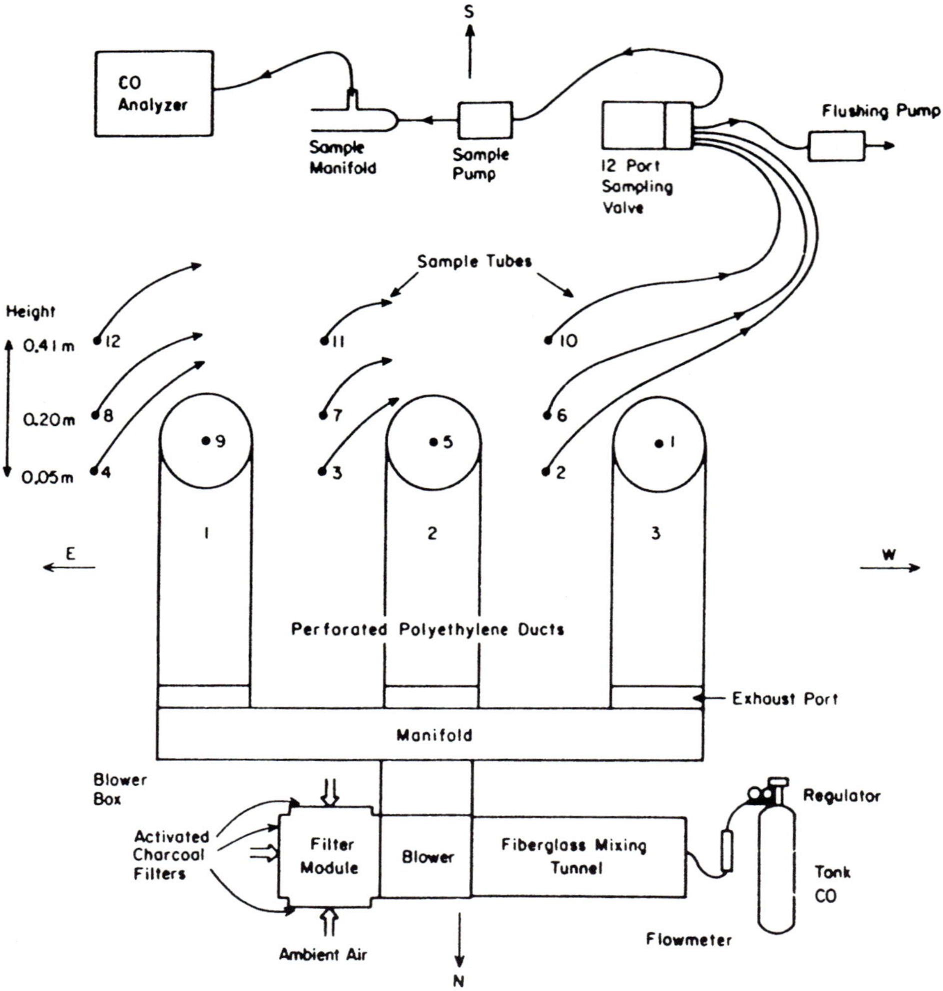

Fig. 9.5. A linear gradient field fumigation system. Three fumigation ducts are shown. Bulleted numbers indicate ozone gas sampling points (Olszyk et al. 1986a)

posure period) of pollutants such as sulphur dioxide or hydrogen fluoride (US EPA 1987b). The Zonal Air Pollution System (ZAPS) has been vastly modified and improved to enable the exposure of plants to the diurnally varying pattern of O_3 concentration typical of ambient O_3 exposure (Runeckles et al. 1990). The system represents a significant advancement over earlier open-air field exposure systems, in that twelve discrete seasonal treatments which simulate ambient patterns are achieved rather than the usual two or three. Ozone was supplied to plots, which were laid out in groups of four, through a manifold suspended over the plant canopy. The wind speed and direction determined the actual seasonal O_3

exposures, although the O_3 was released in concentrations proportional to that observed at the time in the ambient environment. While the 12 treatments are not repeatable over time, a regression relationship between pollutant exposure and plant response can be established for each growing season. The regression relationships, in particular the slopes describing plant response to O_3, can then be used to describe the average (± error) response plus the maximum and minimum responses within the years of study. Open air fumigation systems require precise spatial and temporal characterization of pollutant concentrations within the plots, computer controlled operation of the fumigation systems based on rapid feedback from monitored data within the plots, and computer data logging and summarization.

The Liphook study in England of long-term responses of *Picea sitchensis* (Bong.) Carr, *Picea abies* (L.) Karst. and *Pinus sylvestris* L. to sulphur dioxide and O_3 in combination consisted of seven growth plots, 50 m in diameter, five of which were surrounded by 64 vertical pipes from which pollutant gasses were emitted (McLeod et al. 1992). The system is described in detail by Wellburn et al. in Chapter 3 (this Vol.). The pattern of gas dispersion resulted in a uniform horizontal distribution of hourly mean gas concentration across each central 25-m diameter experimental plot. Measurement of consecutive 2-min mean values at five locations across the plots demonstrated high uniformity (McLeod et al. 1985). This exposure system, like all open-air exposure systems, simulates field plant growth conditions far better than open- or closed-top chambers. The usefulness of the data is limited, however, by the low number of treatments and lack of replication (within the season) of those treatments. With five "enclosures", and two non-enclosed ambient plots, this is by far the largest of the very few of these systems which have been in operation. Similar systems have been in operation in the US for pollutant fumigation (Hendrey et al. 1992) and carbon dioxide enrichment experiments (Lewin et al. 1994). These systems are now called free-air (chamberless) carbon dioxide enrichment, or FACE systems (Hendrey and Kimball 1994; Lewin et al. 1994).

An open air branch chamber system has been developed for use on forest canopy branches (Tjoelker et al. 1994). The system exposed branches to O_3 through a manifold of teflon tubes fitted beneath each branch. The system was able to deliver O_3-enriched air to the branch with little change in micrometeorological conditions.

9.2.3 Ambient Gradients

Naturally occurring gradients of O_3 concentration can be used to evaluate plant response to O_3 under ambient conditions of temperature, humidity, soils, and soil moisture. By this method, plants are grown along a transect of known differential pollutant concentrations, usually downwind of a major point source or urban area. The concentration of pollutant is diluted as distance from the source increases. This technique was used by Oshima et al. (1976, 1977), who reported yield losses for alfalfa and tomato along an O_3 gradient in southern California. A study of four cultivars of red clover (*Trifolium repens* L) and spring barley (*Hordeum vulgare* L) was conducted along a transect with a gradient of sulphur dioxide, nitrogen dioxide, and O_3 concentrations in the UK (Ashmore et al. 1988). O_3 concentration was inferred from injury to Bel W3 and Bel B cultivars of tobacco, but was found to have very little relationship to cultivar performance. The authors cautioned that these results must be interpreted with an understanding that differences among sites in other environmental parameters could contribute to the detection of (or failure to detect) O_3 effects on the crops. Tonneijck and Bugter (1991) also found that foliar oxidant stipple injury of Bel W3 does not correlate well with ambient O_3 concentrations, and that the foliar response is highly related to environmental conditions. Although plants can often be used as bioindicators to verify phytotoxic response to ambient O_3, much work needs to be done before plants can be effectively used as biomonitors to indicate the presence of specific concentrations of O_3.

Winner et al. (1989) examined foliar oxidant stipple injury on three species in the Shenandoah National Park, Virginia. They reported that both foliar injury and ambient O_3 concentration increased with elevation. Chappelka et al. (1992) also reported an increase in O_3 with elevation in southeastern US and a corresponding increase in the amount of oxidant stipple injury of O_3-sensitive tree species. These studies suggest that ambient gradients may be a good approach to identifying tree sensitivity to O_3, although the relationship between leaf stipple and ecosystem function is not known.

In order to interpret ambient gradient studies, good characterization of site parameters (rainfall, temperature, radiation, soil type) is needed. A basic understanding of how plants respond to these factors in the absence of O_3 is necessary for an understanding of how they will respond in the presence of O_3. Response to O_3 is highly dependent on how plants respond to other environmental parameters. Although Manning and Krupa (1992) assert that natural gradients are the "ideal way to conduct O_3/plant response studies in ambient air in field plots", they concede that few gradients which meet statistical requirements for intermediate O_3 concentrations exist outside southern California (Krupa 1984). It is possible, however, that more gradients will be identified as rural air monitoring increases. They also recognize that while using artificial soils removes a significant source of variation in plant response to O_3, pot-grown plants do not closely simulate the rooting environment found in the field (Manning and Krupa 1992). Finally, their statement with respect to gradients that "many replicates are possible" is difficult to understand, as the gradients can surely not be well replicated in space, and, given the variation in O_3 concentration (frequencies) patterns from year to year, also not well replicated in time. Ambient gradients used repeatedly in time and space could yield valuable information about the magnitude of natural ecosystem response to O_3 relative to other sources of variation.

Many investigators consider that ambient gradients have great potential for quantifying plant response to ambient O_3. However, there may be very few ambient gradients without large differences in microclimatic and edaphic conditions which affect plant growth and may well confound plant response to O_3. Because of these problems, little research is being conducted using ambient gradients.

9.2.4 Cultivar Comparisons

The comparison of isogenic lines of a particular species which differ only in their tolerance to O_3 is "the ideal way to determine the effect of ambient O_3 on plants in the field" (Manning and Krupa 1992). While there are a number of species with cultivars of differing O_3 sensitivities, there are no isogenic lines for O_3 sensitivity available at this time. A sensitive cultivar of tobacco, Bel-W3 has frequently been used as a bioindicator or biomonitor. The use of O_3-sensitive cultivar Bel-W3 in field studies of O_3 response has been reviewed (Heggestad 1991). Recent reports have indicated that Bel-W3, and perhaps other O_3-sensitive bioindicators, may not be useful as biomonitors of O_3 (Tonnejck and Bugter 1991). Ozone-sensitive and O_3-tolerant clones of poplar (Karnosky et al. 1992a), white pine (Dochinger and Seliskar 1970; Houston 1974), loblolly pine (Chappelka et al. 1990), and clover (Heagle et al. 1992) have also been identified and used as bioindicators. The use of O_3-sensitive cultivars in determining the magnitude of O_3 effects on plants may be difficult, since sensitive cultivars are likely to differ in their responses to other ambient pollutants and/or environmental stresses. Separating the response of these cultivars to O_3 stress from response to non-O_3 stresses may not be possible under normal field experimental conditions.

9.2.5 Protective Chemicals

Chemicals which protect plants from O_3 have been in use since the 1970s to evaluate plant response to O_3. Ethylene diurea (EDU) has been used in studies to modify the O_3 sensitivity of several species (Manning 1988; Manning and Krupa 1992). EDU is a systemic antioxidant chemical which protects plant tissue from oxidant stipple injury and delays O_3-induced early senescence. Although it contain urea, it does not act as a plant nutrient, nor does it show growth regulator or pesticide activity (Manning 1992). EDU is not commercially available and has been used only on an experimental basis. EDU and perhaps other as yet undiscovered chemicals have potential as tools to evaluate field crop losses due to O_3 in the absence of chambers without the modification of microclimate. As a low cost and simple technique, EDU can be applied to larger plot sizes than are currently possible with open-top chambers, thus reducing some of the uncertainty of extrapolating experimental results to a large scale. Currently, field protocols for the use of EDU have not been well established. The frequency and rate of application which protect plants varies with species and edaphic and atmospheric conditions; EDU can have toxic effects (Manning and Krupa 1992). Heagle (1989) and Manning and Krupa (1992) point out that EDU is probably more useful in conjunction with open-top chambers or plume fumigations so that a factorial range of O_3 can be administered to the plants. It is not clear whether EDU protection can be sufficiently finely tuned into a range of discrete levels suitable for regression analysis, although it may be well suited for studies which compare ambient and O_3-containing filtered air. The mechanism by which EDU protects plants from O_3 is largely unknown. The commercial unavailability of EDU prohibits the routine use of the chemical as an antioxidant.

Other chemicals have been utilized as antioxidants, including fungicides such as benomyl, triadimefon, and diphenylamine (Gilbert et al. 1975; Musselman 1985; Musselman and Tashenberg 1985). There appears to be no current research being conducted on the usefulness of fungicides or other chemicals as antioxidants for O_3 injury to vegetation. Further comparisons of plant response using open-top chambers and chemical exclusion should be made.

9.3 Limitations to Comparisons

The various systems for exposing plants to pollutants have been selected by researchers for use in determining plant response to O_3. The particular system selected may depend on nature of the plant species being tested, or the availability of a system for experimental use. However, the interpretation of data from experiments on plant response to O_3 is limited by non-system factors such as scaling from plant to population, temporal and spatial differences, and types of data analysis.

9.3.1 Scaling Problems

A problem with determining the effects of air pollutants on mature forest trees concerns scaling from leaf or branch to tree and stand, and scaling from juvenile to mature. These differences are important in plant response to pollutants. Most controlled environmental experiments use exposure chambers which, from a functional and cost effective standpoint, are relatively small in size. Thus, the experimental unit is generally small – a seedling, sapling, or tree branch. The relationship between juvenile seedling response to pollutants and mature tree response to pollutants is difficult to study and is not well understood. Research has shown that the response of mature trees can be different from that of seedlings or saplings (Pye 1988; Samuelson and Edwards 1993; Grulke and Miller 1994; Hanson et al. 1994; Fredericksen et al. 1995). Physiological and biochemical reactions in seedlings and mature trees differ, and these reactions can greatly affect response to pollutants.

Extrapolation of results with a leaf or branch to a whole plant, even in similarly aged individuals, can also be difficult. Sampling protocols are important in the selection of leaves or branches which are representative of canopy response. Models have been developed to extrapolate leaf response to a canopy for effects of elevated CO_2 on plants (Reynolds et al. 1992). The model requires detailed spatial and temporal environmental data for the leaves within the canopy. Similar models could be developed for canopy response to O_3.

9.3.2 Temporal/Spatial Problems

Both temporal and spatial differences need to be considered when comparing results from controlled experiments to field situations. Controlled fumigation experiments are conducted in an extremely restricted spatial setting on very small microsites within forest ecosystems, with environmental conditions which are likely modified by the fumigation facilities. Extrapolation of data from those sites where experiments are conducted to other sites can be troubling. First, the small size of the experimental site makes selection of a representative population difficult. Second, the experimental site is often modified, sometimes drastically, by fumigation chambers or simply by traffic from researchers. Even in areas with relatively uniform environments, extrapolation of data from a small experimental unit to forests can be difficult, particularly if the variability in the population is large, as is common with tree species growing under natural conditions.

It is difficult to demonstrate that the experimental unit is representative of the larger population in its response to pollutants. In natural ecosystems with high genetic and environmental variability, the likelihood that the sample population may not be representative of the whole population increases. Thus, proper experimental design and sample unit selection becomes crucial in natural ecosystems. Another problem exists in extrapolation if the error term in the experimental response is large. A large error term in the experimental population will result in a larger uncertainty for the population it represents.

Temporal differences are also important. Most fumigation experiments are short-term, with treatment carried out for part of a season, a full season, or at best for several years. These time periods are very short in relation to the lifetime of a forest tree; trees rather act as an integrative, and to some extent cumulative, response to the pollutant exposure. Environmental conditions during the fumigation period may not be representative of the environmental conditions to which the trees are most frequently exposed. Nor do they represent the normal length of exposure under ambient conditions. Fumigation experiments are often conducted on seedlings, saplings, or small trees, and response of these trees may not be representative of their response when they are mature. The response of a tree during its lifetime cannot be captured by fumigation experiments during one short period of time early in the life of the tree.

9.3.3 Data Analysis

The statistical design of an experiment is crucial in defining the types of analyses to which the data can be subjected, and consequently the kinds of information obtained about plant response to O_3. Therefore, the goals of a study must be clearly defined in order to choose an appropriate experimental design. Studies on plant response to O_3 exposure have most frequently used experimental designs which generate data suitable for determination of regression-type cause–effect relationships or analysis of variance tests to which means separation tests can be applied. These exposure–response relationships generalize the mathematical relationship between the plant parameter of interest and O_3 exposure. Plant response to concentrations other than those used in the experiment can be interpolated from regression relationships and thresholds of plant response can be determined (Ormrod et al. 1988). Extrapolation beyond the range of experimental data is also possible, but with caution regarding variability and linearity of the relationship.

To detect statistically significant differences between plants grown in filtered and nonfiltered air in chambers requires a high number of replications (Rawlings et al. 1988). Heagle (1989) illustrated the inefficiency of the two-treatment approach by suggesting that 40 replicates of a two-treatment design (80 EDU plots or open-top chambers) would be needed to detect a 10% O_3 effect at $P = 0.05$, assuming a coefficient of variation of 12%. Coefficient of variation values for plant response to O_3 are often higher than 12% for crops, and would be expected to be even higher for forest species. Since seasonal O_3 concentrations vary greatly with geographic location and year, the collection of data for yield loss estimates using only two O_3 treatments will be much slower than if O_3-enriched chamber exposure systems are used, and will not contribute to the development of predictive models. The establishment of dose–response relationship using regression analyses requires multiple treatments, including treatment levels above ambient. To develop a statistically robust surface requires at least three or better still five or more treatment levels, with some replication.

Because of the need to contain each O_3 treatment by a chamber, there is always a problem in obtaining good replication for statistical analysis in field chamber experiments. Nonstandard statistical techniques are available to alleviate this deficiency. Incomplete factorial designs can be used for multifactor studies (Allen et al. 1987). Response surface designs require fewer replications, but care must be taken when assuming similar variability among treatments and in data interpretation. Nonparametric statistical techniques can be used in experimental design and data analysis. Air pollution response research has used the K-S test (Musselman et al. 1986b), time series analysis (Younglove et al. 1988), contingency tables (McCool et al. 1986), and response surfaces (Allen et al. 1987) for determination of plant response to pollutants. Beyond these examples, the use of these methods in controlled environment or field air pollution research experimental design and analysis has been rare.

Experimental designs for exposure–response relationships can easily be expanded so that plant response to O_3 and another pollutant at multiple levels can be determined. Incomplete factorial designs are a more precise and efficient approach to multifactor studies, leading to exposure–response surfaces for estimation of the relative yield loss at ambient concentrations of pollutants (Allen et al. 1987). Choosing the appropriate incomplete factorial design for a response surface study requires forethought on whether all areas of the surface are of equal interest, and necessitates the use of treatment levels from near zero to well above ambient. However, response surface designs have rarely been used in pollutant mixture studies, nor have they been used to study the interaction between pollutants and environmental parameters such as light, temperature, or moisture.

Because most O_3 air pollution research has placed greater emphasis on damage (i.e., effects that reduce the intended human use of the plant), than on presence or absence of injury, studies have more frequently used experimental designs which generate data suitable for regression analysis, or ANOV and subsequent treatment mean separation analyses, to model and test the impact of O_3 on plant response. While the impact of current ambient levels of O_3 is of primary interest and can be studied using only two O_3 levels generated by chambers with charcoal-filtered air to remove O_3 and non filtered air (albeit with some limitations; Heagle 1989), the development of exposure–response models necessitates the use of additional treatments at above ambient concentrations. The optimal number of treatments, and the treatment levels, depends on the particular exposure–response model. Most polynomial models obtain greater precision when the lowest treatment is near zero O_3 and the highest treatment level is well above ambient. Use of only two treatments, charcoal-filtered and nonfiltered air, is inappropriate for determination of the dynamics of plant response to O_3 exposure.

For experiments in which treatments are quantitative, the traditional analysis is regression which relates O_3 exposure to plant response. Regression has been used to fit a common model to data combined from replicate studies of the same species when it is reasonable to assume that the primary cause of biological response is pollutant exposure and that differences in environmental, edaphic, and/or agronomic conditions among sites do not significantly change the shape of the regression relationships. When pooling data across sites and years, additional terms for the site and year effects are often included in the model as either fixed or random components, depending on the population of interest. Inferences over random environments implies that the environments sampled by the experiments are representative of the population of regions of interest under a variety of environmental conditions. In this case, site and year effects are incorporated as random components in the model when fitting a common model. The appropriate analysis is a mixed model, fitting an exposure response model with variance components. This analysis has been recently used to combine data from replicate studies of varying durations to test the importance of length of exposure in influencing plant response (Olszyk et al. 1993). Accounting for site and year effects may be especially important in field studies of forest response to O_3.

The precise identification of O_3 injury response is difficult since O_3 injury symptoms are not necessarily unique and other stresses can cause injury that resembles O_3 injury symptoms. It takes much training to specifically identify oxidant stipple injury from other similar symptoms; and O_3-induced chlorosis may not be distinguishable from other environmentally induced chlorosis. Additionally, the absence of oxidant stipple injury often may not indicate a lack of effect of O_3 on plants, and the quantity of oxidant stipple injury may not relate directly to effects of O_3 on plant productivity.

Peterson and Mickler (1994) presented an evaluation of statistical limitations of tree seedling fumigation studies. They conclude that few studies discuss statistical power, experimenter bias, variability in treatment exposure regimes, variability in measurement of experimental variables, or inference of mature tree and forest responses from experimental seedling populations. This information is important in interpretation of data from O_3 fumigation experiments.

9.4 What Is Needed for Future Studies?

Several items need to be addressed when designing systems to examine plant response to O_3. A close examination of the biological and physiological response to O_3 is necessary to understand plant response. The functional O_3 exposure parameters which are used in plant response determinations are important. Finally, perhaps some new methods of determining plant response to O_3 which are not limited to the current fumigation systems need to be examined.

9.4.1 Biological and Physiological Response

Although O_3 can injure cuticles in some plants (Barnes et al. 1988; Kerstiens and Lendzian 1989a,b), most O_3 injury to plant tissue is internal to the leaf. The palisade cells just below the upper surface of the leaf are often the first to show injury. The primary point of entry of O_3 into the leaf is through the stomata. Thus, factors which influence stomatal aperture are important for O_3 tolerance. Ozone itself may induce stomatal closure and subsequent protection in some species (Gabelman 1970). The actual flux of O_3 into leaves is more closely related to plant response than external O_3 exposure (Grünhage et al. 1993, Musselman et al. 1994a).

Although uptake of O_3 is necessary for tissue injury and subsequent plant stress which can lead to reductions in growth and productivity, there is a wide range in inherent biochemical tolerance to O_3 injury of internal plant tissue. Saxe (1991) reviewed the literature on stomatal response of plants to O_3, and concluded that stomatal closure is not consistently related to O_3 tolerance and that biochemical detoxification and repair mechanisms can be more important in protecting plant tissue from O_3. Taylor (1994) discusses O_3 avoidance mechanisms which prevent uptake, transport, or metabolism of O_3 and tolerance mechanisms which repair or compensate for O_3 injury. He concludes that although pollutant uptake is critical, it is often linked to sensitivity to other stress mechanisms such as drought, while tolerance mechanisms are specifically linked to O_3 toxicity. Laurence et al. (1994) indicate that plants with low stomatal conductance can take up substantial amounts of O_3 over time, which may result in increased O_3–induced senescence in species with older foliage such as conifers.

Forest tree species which have large ranges in genetic variability are likely to have wide ranges in their stomatal and biochemical response to pollutants. Although we know that for many crop species there is a wide range in variability among cultivars of responses to pollutants, little research has been conducted on such variability in populations of native trees. Different genotypes of a few tree species have been tested and a wide range in O_3 tolerance was found among genotypes of loblolly pine (Taylor 1994), poplar (Karnosky et al. 1992 a,b), and white pine (Bennett et al. 1994). Although some plant species are listed as more susceptible to O_3 than others, there is a wide range of variability in susceptibility within species, particularly within species having sensitive genotypes.

9.4.2 Functional Ozone Parameters for Exposure

Optimum spacing of treatment levels is important in order to relate exposure to response. The value of the treatment level depends on how the O_3 exposure is measured. Various parameters of O_3 exposure have been used in defining the O_3 treatment levels. The methods for determining the

O_3 exposure parameters have been discussed elsewhere (Lefohn 1992a). Ozone exposure parameters should logically order increasing levels of the O_3 value with increasing response. The daytime mean concentration index (i.e., either M7, 7-h daily mean or M12, 12-h daily mean) has been used by NCLAN to determine the effects of O_3 on plant response. The seasonal mean of the 7-h and 12-h daytime average O_3 concentration has been used by the NCLAN program to relate plant response to O_3 exposure. It is generally recognized that seasonal O_3 averages are not useful parameters to describe the O_3 exposure response of plants (Lefohn 1992a; Musselman et al. 1994a).

Recent reports have suggested that moderate levels of O_3 may be more important than high levels and that high levels are relatively unimportant in plant response to O_3 (Krupa et al. 1993, 1994, 1995). These reports have been critically evaluated elsewhere (Lefohn et al. 1994; Musselman et al. 1994a). When plant yield is considered, plant response is affected by the concentration of O_3 and other exposure factors such as duration, frequency, threshold, respite time, as well as by physiological, biochemical, and environmental factors which may influence treatment effect over the growing period. Research aimed at understanding the importance of exposure factors utilizes experimental designs that apply two or more different patterns of exposure that are equal on some scaling of total exposure (Musselman et al. 1983, 1986a, 1994b, Hogsett et al. 1985). Experiments designed specifically to address the importance of the components of exposure may require the use of exposure regimes and concentration levels that are not typical of the ambient environment.

It is apparent that O_3 flux into leaves is important in plant response (Grünhage et al. 1993). However, establishing an O_3 exposure index for air quality standards requires the use of ambient air quality data. Research designed to examine O_3 exposure dynamics has shown that the components of O_3 in ambient air which are most important for plant response include (1) weighting of peak concentrations; (2) cumulative O_3 data; (3) phenological stage of development of the exposed vegetation; and (4) flux of the O_3 into the plant (Musselman et al. 1994a). It is also apparent that a single parameter O_3 exposure index may be inadequate to relate to plant response.

9.4.3 Future Field Research

Pollutant treatment of plants at the ecosystem level may be necessary to determine the response at the community or landscape level. Fumigation treatment of large trees or small plots containing several trees may be more useful in determining ecosystem response to pollutants than the commonly used experimental method of fumigating small trees or seedlings in chambers. Whole system fumigation chambers would allow experimental manipulation of large trees or large field plots with O_3, with better control of other environmental variables. Ecocosms, plastic covered frames with a surface area of 200 m^2 and a height of 10 m, have been suggested for the experimental manipulation of tundra and boreal ecosystems (Strain et al. 1991). However, these systems would require large expenditures for equipment, maintenance, and labor. They would also cause changes in environmental conditions which would need to be considered in the experimental analysis. Open air fumigation systems are likely to be best for whole ecosystem exposures; but these systems have no control of environmental variables. Open air fumigation of whole ecosystems have been proposed for global change experiments (Walker et al. 1991).

Pollutant concentrations vary both temporally and spatially within the experimental area; hence intensive monitoring of environmental and treatment variables is required. A tradeoff will need to be made between (1) the use of whole system chambers with precise control of O_3 fumigation and precise identification of other environmental variables and (2) the use of open air fumigation systems with low or no chamber influence on the system but difficulty in the ability to precisely identify treatment concentrations and environmental variables affecting plant response.

New methods have been proposed for estimating ecosystem response to pollutants which are not limited to single or multiple factor experimentation at specific field sites during specific time periods. These techniques utilize process-level modeling which simulate forest and ecosystem structural and functional response to the interaction of a multiple of factors (Taylor et al. 1994). These models must include large databases of information concerning plant biochemical and physiological processes, and atmospheric processes affecting these plant processes. The goal is to simulate ecosystem response to various environmental stresses, including pollutants, by looking at individual plant responses to these stresses, and scaling these to community and landscape levels. The models assume the ability to scale tissue and plant level physiological responses to stand and landscape level productivity responses. The large genetic and environmental variability in forest ecosystems and the errors involved in scaling with this approach may limit the usefulness of the method. Although this line of approach shows promise and should continue, much work needs to be accomplished before these models can be used to estimate O_3 effects on forest ecosystems. Information from single or multiple factor experimentation at specific sites can provide valuable input data for inclusion in these models. Whole system manipulations (Mooney et al. 1991; Strain et al. 1991; Walker et al. 1991; Woodward 1992) may be useful for obtaining data for ecosystem response modelling.

9.5 Summary and Conclusions

Although research on forest response to O_3 has used standard crop fumigation systems, or systems modified from those used for crops, air pollution research on natural ecosystems presents unique experimental problems. A large variability in the genetics of plant populations examined makes detection of O_3 response difficult, and a large variability in environmental conditions under which natural ecosystems exist limits experimental detection of O_3 effects. Many forest species are long-lived perennials, which have a longer growing season for O_3 exposure, and are exposed to O_3 year after year for decades.

Each type of fumigation system is particularly well suited to certain types of studies of plant response to O_3; no one system is appropriate for all types of studies of plant responses to O_3. Each has advantages and limitations which must be evaluated in terms of the research objectives it was designed to meet (Table 9.1). Controlled environment chambers are well suited for mechanistic type studies at the molecular or cellular level, where good control and definition of environmental conditions are needed. However, growth responses to O_3 determined from controlled environment chamber studies cannot be extrapolated to predict field losses due to O_3 as the cultural conditions in the two systems are just too dissimilar. This is particularly critical in natural forest situations where environmental conditions are highly variable.

Open-top chambers represent the best method for developing exposure–response functions and for determining plant yield responses to O_3; concentration and duration of the pollutant are well controlled and the plants are grown under near field culture conditions. There are limitations and uncertainties associated with the data: the plot size is small relative to a field, microclimate differences may influence plant sensitivity to O_3, and air quality after passage through the charcoal filter may be altered. Exclusion methods, particularly those using chemicals such as EDU, are the least disruptive of ambient cultural conditions in the field, and so these approaches most closely estimate "real" losses due to O_3. Their application cannot be used for developing exposure–response relationships because it is impossible to quantify the degree of protection actually offered by the chemical in the field. Chemical protectants have not to our knowledge been used for forest species or in natural ecosystems.

Other nonchamber methods can be used to determine plant response to O_3. Open field fumigation systems are less disruptive to natural ecosystems and cause fewer environmental changes. The differential response of plants growing under conditions of ambient O_3 gradients have been used in the determination of plant response to pollutants. In general, open field exposure systems or natural gradients are not replicable nor can a range of treatments be imposed to enable construction of response functions necessary for interpolation of O_3 concentrations which cause plant response.

Scaling from seedlings in fumigation plots to whole trees or to a forest stand under natural growing conditions is difficult. Fumigation plots are limited both spatially and temporally compared to the natural environment of forest trees. These factors may limit the determination of forest response to pollutants using small fumigation plots. The choice of the appropriate experimental design and statistical analysis is critical in identifying the uncertainties regarding extrapolation and in determining the response. There is concern that these uncertainties may be forgotten in the scaling up of the plant response data to the forest stand and landscape level and in the integration of fumigation plot data into larger cost-benefit models. There is an urgent need to estimate these uncertainties so that the small plot data can be fully used with little doubt as to how well it represents "real" ecosystem changes.

Individual plant response to O_3 is reflected in changes in ecosystem structure or function. Future forest response research must focus on exposure response mechanisms of plants. Effects of O_3 on physiological and biochemical events in plant cells should be examined since they relate to plant growth, productivity, and survival. The proper descriptor of the O_3 exposure parameter should be the flux of O_3 into the leaf rather than the ambient O_3. However, the determination of O_3 flux is difficult and thus may be impractical to routinely use in exposure response analysis.

The usefulness of whole ecosystem O_3 fumigation or exclusion systems and computer modelling of forest response to stresses should be examined in future determinations of forest response to O_3. Larger scale exclusion studies could contribute to quantifying the uncertainties involved in extrapolating small plot response to forest response. A better characterization of atmospheric chemistry within open-top chambers is also necessary. Once these uncertainties are fully characterized and quantified, existing models of loss can be constructed more precisely and then incorporated with greater confidence into national O_3 response models.

References

Adams RM, Crocker TD (1989) The agricultural economics of environmental change: some lessons from air pollution. J Environ Manage 28:295–307

Adams MB, Edwards NT, Taylor GE Jr, Skaggs BL (1990) Whole-plant ^{14}C-photosynthate allocation in *Pinus taeda*: seasonal patterns at ambient and elevated ozone levels. Can J For Res 20:152–158

Albaugh TJ, Mowry FL, Kress LW (1992) A field chamber for testing air pollution effects on mature trees. J Environ Qual 21:476–485

Allen OB, Marie BA, Ormrod DP (1987) Relative efficiency of factorial designs for estimating response surfaces with references to gaseous pollutant mixtures. J Environ Qual 16:316–320

Amundson RG, Kohut RJ, Laurence JA (1995) Influence of foliar N on foliar soluble sugars and starch of red spruce saplings exposed to ambient and elevated ozone. Tree Physiol 15:167–174

Andersen CP, Hogsett WE, Wessling R, Plocher M (1991) Ozone decreases spring root growth and root carbohydrate content in ponderosa pine the year following exposure. Can J For Res 21:1288–1291

Ashmore MR, Bell JNB, Mimmack A (1988) Crop growth along a gradient of ambient air pollution. Environ Pollut 53:99–121

Barnes JD, Davison AW, Booth TA (1988) Ozone accelerates structural degradation of epicuticular wax on Norway spruce needles. New Phytol 110:309–318

Bennett JP, Anderson RL, Mielke ML, Ebersole JJ (1994) Foliar injury air pollution surveys of eastern white pine (*Pinus strobus* L.): a review. Environ Monit Assess 30:247–274

Berrang P, Karnosky DF, Bennett JP (1991) Natural selection for ozone tolerance in *Populus tremuloides*: an evaluation of nationwide trends. Can J For Res 21:1091–1097

Berry CR, Riperton LA (1963) Ozone: a possible cause of white pine emergence tipburn. Phytopathology 53:552–557, 239–251

Billen N, Schätzle H, Seufert G, Arndt U (1990) Performance of some growth variables. Environ Pollut 68:419–434

Brown KA, Roberts TM (1988) Effects of ozone on foliar leaching in Norway spruce (*Picea abies* L. Karst): Confounding factors due to NO_x production during ozone generation. Environ Pollut 55:55–73

Byres DP, Johnson JD, Dean TJ (1992) Seasonal response of slash pine (*Pinus elliottii* var *elliottii* Engelm.) photosynthesis to long-term exposure to ozone and acidic precipitation. New Phytol 122:91–96

Bytnerowicz A, Tran M, Anderson P (1995) Effects of charcoal air filtration and ozone generation on concentrations of some N and S compounds in open-top field chambers. Atmos Environ 29:1355–1358

Chappelka AH, Kush JS, Meldahl RS, Lockaby BG (1990) An ozone-low temperature interaction in loblolly pine (*Pinus taeda* L.). New Phytol 114:721–726

Chappelka AH, Hildebrand E, Skelly JM, Mangis D, Renfro JR (1992) Effects of ambient ozone concentrations on mature eastern hardwood trees growing in Great Smoky Mountains National Park and Shenandoah National Park. Paper 92–150.04 presented at 85th Annu Meet June 21–26, 1992, Kansas City, Missouri, Air and Waste Management Association, 12 pp

Clark CS, Weber JA, Lee EH, Hogsett WE (1995) Accentuation of gas exchange gradients in flushes of ponderosa pine exposed to ozone. Tree Physiol 15:181–189

Colls JJ, Unsworth MH (1992) Air pollution interactions with natural stressors. In: Barker JR, Tingey DT (eds) Air pollution effects on biodiversity. Van Nostrand Reinhold, New York, pp 93–108
Commission of the European Communities (CEC) (1987) Microclimate and plant growth in open-top chambers. Air Pollution Research Report 5, EUR 11257. Proc Worksh organized by the Commission of the European Communities and the Kernforschungszentrum Karlsuhe, Freiburg, 17–19 Sept 1986
Dean TJ, Johnson JD (1992) Growth response of young slash pine trees to stimulated acid rain and ozone stress. Can J For Res 22:839–848
Dochinger LS, Seliskar CE (1970) Air pollution and the chlorotic dwarf disease of Eastern White Pine. For Sci 16:46–55
Edwards GS, Friend AL, O'Neill EG, Tomlinson PT (1992) Seasonal patterns of biomass accumulation and carbon allocation in *Pinus taeda* seedlings exposed to ozone, acidic precipitation, and reduced soil Mg. Can J For Res 22:640–646
Edwards NT (1991) Root and soil respiration responses to ozone in *Pinus teada* L. seedlings New Phytol 118:315–321
Eissenstat DM, Syvertsen JP, Dean TJ, Yelenosky G, Johnson JD (1991) Sensitivity of frost resistance and growth in citrus and avocado to chronic ozone exposure. New Phytol 118:139–146
Ennis CA, Lazrus AL, Kok GL, Zimmerman PR (1990) A branch chamber system and techniques for simultaneous pollutant exposure experiments and gaseous flux determinations. Tellus 42B:170–182
Evans LS, Dougherty P (1986) Exposure systems and physiological measurements. Forest response program quality assurance methods manual. EPA/600/86/193a Raleigh, NC
Fredericksen TS, Joyce BJ, Skelly JM, Steiner KC, Kolb TE, Kouterick KB, Savage JE, Snyder KR (1995) Physiology, morphology, and ozone uptake of leaves of black cherry seedlings, saplings, and canopy trees. Environ Pollut 89:273–283
Friend AL, Tomlinson PT (1992) Mild ozone exposure alters ^{14}C dynamics in foliage of *Pinus taeda* L. Tree Physiol 11:215–227
Fuhrer J (1994) Effects of ozone on managed pasture; I. Effects of open-top chambers on microclimate, ozone flux, and plant growth. Environ Pollut 86:297–305
Gabelman WH (1970) Alleviating the effects of pollution by modifying the plant. HortScience 5:250–252
Gilbert MD, Maylin GA, Elfving DC, Edgerton LJ, Gutenmann WH, Lisk DJ (1975) The use of diphenylamine to protect plants against ozone injury. HortScience 10:228–231
Grulke NE, Miller PR (1994) Changes in gas exchange characteristics during the life span of giant sequoia: implications for response to current and future concentrations of atmospheric ozone. Tree Physiol 14:659–668
Grünhage L, Dämmgen U, Haenel H-D, Jäger H-J, Holl A, Schmitt J, Hanewald K (1993) A new potential air-quality criterion derived from vertical flux densities of ozone and from plant response. Angew Bot 67:9–13
Hanson PJ, Samuelson LJ, Wullschleger SD, Tabberer TA, Edwards GS (1994) Seasonal patterns of light-saturated photosynthesis and leaf conductance for mature and seedling *Quercus rubra* L. foliage: differential sensitivity to ozone exposure. Tree Physiol 14:1351–1366
Heagle AS (1989) Ozone and crop yield. Annu Rev Phytopathol 27:397–423
Heagle AS, Philbeck RB (1979) Exposure techniques. In: Heck WW, Krupa SV, Linzon SN (eds) Methodology for the assessment of air pollution effects on vegetation, Chapt 6, pp 6–1 to 6–19. Air Pollution Control Association, Pittsburg
Heagle AS, Philbeck RB, Heck WW (1973) An open-top chamber to assess the impact of air pollution on plants. J Environ Qual 2:365–368

Heagle AS, Philbeck RB, Rogers HH, Letchworth MB (1979) Dispensing and monitoring ozone in open-top field chambers for plant effects studies. Phytopathology 69:15–20

Heagle AS, Kress LW, Temple PJ, Kohut RJ, Miller JE, Heggestad HE (1988) Factors influencing ozone dose–yield response relationships in open-top field chamber studies. In: Heck WW, Taylor OC, Tingey DT (eds) Assessment of crop loss from air pollutants. Elsevier London, pp 141–180

Heagle AS, McLaughlin MR, Miller JE, Joyner RL (1992) Response of two white clover clones to peanut stunt virus and ozone. Phytopathology 82:254–258

Heck WW, Krupa SV, Linzon SN (1979) Handbook of methodology for the assessment of air pollution effects on vegetation. Air Pollution Control Association, Pittsburgh.

Heck WW, Heagle AS, Miller JE, Rawlings JO (1990) A national program (NCLAN) to assess the impact of ozone on agricultural resources. In: Berglund RL, Lawson DR, McKee DJ (eds) Tropospheric ozone and the environment: papers from an international conference, March 1990, Los Angeles, pp 224–254

Heck WW, Heagle AS, Miller JE (1994) Factors affecting plant response to ozone – an overview. In: Kuttler W, Jochimsen M (eds) Immissionsökologische Forschung im Wandel der Zeit. Festschrift für Robert Guderian. Westarp Wissenschaften, Uhlichstrasse pp 113–130

Heggestad HE (1991) Origin of Bel-W3, Bel-C and Bel-B tobacco varieties and their use as indicators of ozone. Environ Pollut 74:264–291

Heggestad HE, Middleton JT (1959) Ozone in high concentrations as cause of tobacco leaf injury. Science 129:208–210

Hendrey GR, Kimball BA (1994) The FACE Program. Agricult For Meteorol 70:3–14

Hendrey GR, Lewin KF, Kolber Z, Evans LS (1992) Controlled enrichment system for experimental fumigation of plants in the field with sulfur dioxide. J Air Waste Manage Assoc 42:1324–1327

Hernandez TT, deBaurer LI, Delgado MLO (1986) Identification y determinacion de los principales pigmentos fotosinteticos de hojas de *Pinus hartwegii* affectadas por gases oxidantes. Agrosciencia 66:71–82

Hogsett WE, Tingey DT, Holman SR (1985) A programmable exposure control system for determination of the effects of pollutant exposure regimes on plant growth. Atmos Environ 19:1135–1145

Hogsett WE, Tingey DT, Hendricks C, Rossi D (1989) Sensitivity of western conifers to SO_2 and seasonal interaction of acid fog and ozone. In: Olson RK, Lefohn AS (eds) Effects of air pollution on western forests. Air and Waste Management Association. Pittsburg, pp 469–491

Houpis JL, Costella MP, Cowles S (1991) A branch exposure chamber for fumigating ponderosa pine to atmospheric pollution. J Environ Qual 20:467–474

Houston DB (1974) Response of selected *Pinus strobus* L. clones to fumigations with sulfur dioxide and ozone. Can J For Res 4:65–68

Jones P, Jones JW, Allen LH Jr, Mishoe JW (1984) Dynamic computer control of closed environmental plant growth chambers. Design and verification. Trans ASAE 27(3): 879–888

Karnosky DF, Gagnon ZE, Reed DD, Witter JA (1992a) Growth and biomass allocation of symptomatic and asymptomatic *Populus tremuloides* clones in response to seasonal ozone exposures. Can J For Res 22:1785–1788

Karnosky DF, Gagnon ZE, Reed DD, Witter JA (1992b) Effects of genotype on the response of *Populus tremuloides* Michx. to ozone and nitrogen deposition. Water Air Soil Pollut 62:189–199

Karnosky DF, Gagnon ZE, Dickson RE, Coleman MD, Lee EH, Isebrands JG (1995) Changes in growth, leaf abscission, and biomass associated with seasonal tropospheric ozone exposures of *populus tremuloides* clones and seedlings. Can J For Res 26:23-37

Kerstiens G, Lendzian KJ (1989a) Interactions between ozone and plant cuticles. I. Ozone deposition and permeability. New Phytol 112:13–19

Kerstiens G, Lendzian KJ (1989b) Interactions between ozone and plant cuticles. II. Water permeability. New Phytol 112:21–27

Kats G, Olszyk DM, Thompson CR (1985) Open-top experimental chambers for trees. J Air Pollut Control Assoc 26:1298–1301

Kilz E (1987) Microclimate and plant growth in open-top chambers. Air Pollution Reports 5. Commission of the European Communities, Brussels

Kostka-Rick R, Manning WJ (1992) Partitioning of biomass and carbohydrates in field-grown radish under ambient concentrations of ozone, and treated with the anti-ozonant ethylene-diurea. New Phytol 121:187–200

Kress LW, Miller JE (1985) Impact of ozone on field corn yield. Can J Bot 63:2408–2415

Kress LW, Allen HL, Mudano JE, Stow TK (1992) Impact of ozone on loblolly pine seedling foliage production and retention. Environ Toxicol Chem 11:1115–1128

Krizek DT, McFarlane JC (1983) Controlled-environment guidelines. HortScience 18:662–664

Krupa SV (1984) Field exposure methodology for assessing the effects of photochemical oxidants on crops. In: Proc 77th Annu Meet Air Pollution Control Association. Preprint no 88–1042. Air Pollution Control Association. Pittsburgh

Krupa SV, Manning WJ, Nosal M (1993) Use of tobacco cultivars as biological indicators of ambient ozone pollution: an analysis of exposure–response relationships. Environ Pollut 81:137–146

Krupa SV, Nosal M, Legge AH (1994) Ambient ozone and crop loss: establishing a cause–effect relationship. Environ Pollut 83:269–276

Krupa SV, Grünhage L, Jäger H-J, Nosal M, Manning WJ, Legge AH, Hanewald K (1995) Ambient ozone (O_3) and adverse crop response: a unified view of cause and effect. Environ Pollut 87:119–126

Laurence JA, MacLean DC, Mandl RH, Schneider RE, Hansen KW (1982) Field tests of a linear gradient system for exposure of row crops to SO_2 and HF. Water Air Soil Pollut 17:399–407

Laurence JA, Kohut RJ, Amundson RG (1989) Response of red spruce seedlings exposed to ozone and simulated acidic precipitation in the field. Arch Environ Contam Toxicol 18:285–290

Laurence JA, Amundson RG, Friend AL, Pell EJ, Temple PJ (1994) Allocation of carbon in plants under stress: an analysis of the ROPIS experiments. J Environ Qual 23:412–417

Lee JJ, Lewis RA (1978) Zonal air pollution system: design and performance. US EPA Publ No 600/3-78-021, Corvallis, Oregon

Lefohn AS (1992a) The characterization of ambient ozone exposures. In: Lefohn AS (ed) Surface level ozone exposures and their effects on vegetation. Lewis Publ Chelsea, Michigan, pp 93–156

Lefohn AS (1992b) Surface level ozone exposures and their effects on vegetation. Lewis Publ, Chelsea, Michigan, 366 pp

Lefohn AS, Edwards PJ, Adams MB (1994) The characterization of ozone exposures in rural West Virginia and Virginia. J Air Waste Manage Assoc 44:1276–1283

Lewin KF, Hendrey GR, Nagy J, LaMorte RL (1994) Design and application of a free-air carbon dioxide enrichment facility. Agric For Meteorol 70:15–29

Livingston NJ, Davies GJ, Eby BM, Filek G, Fuchs EE, Pepin S, Percy RE (1994) A whole-plant cuvette system to measure short-term responses to conifer seedlings to environmental change. Tree Physiol 14:759–768

Mandl RH, Weinstein LH, McCune DC, Keveny M (1973) A cylindrical, open-top chamber for the exposure of plants to air pollutants in the field. J Environ Qual 2:371–376

Mandl RH, Laurence JA, Kohut RJ (1989) Development and testing of open-top chambers for exposing large, perennial plants to air pollutants. J Environ Qual 18:534–540

Manning WJ (1988) EDU: a research tool for assessment of the effects of ozone on vegetation. Paper 88/69.2 presented at the 81st Air Pollution Control Assoc Annu Meet, Dallas, 8pp

Manning WJ (1992) Assessing the effects of ozone on plants: use and misuse of ethylenediurea (EDU), Paper presented at the 85th Annu Meet and Exhibit June 21–26, 1992, Kansas City, Air and Waste Management Association, Pittsburg, Pap No 92–150.02. 11 pp

Manning WJ, Krupa SV (1992) Experimental methodology for studying the effects of ozone on crops and trees. In: Lefohn AS (ed) Surface level ozone exposures and their effects on vegetation. Lewis Publ Chelsea, Michigam, pp 93–156

McCool PM, Younglove T, Musselman RC, Teso RR (1986) Plant injury analysis: contingency tables as an alternative to analysis of variance. Plant Dis 70:357–360

McLeod AR, Baker CK (1988) The use of open field systems to assess yield response to gaseous pollutants. In: Heck WW, Taylor OC, Tingey DT (eds) Assessment of crop loss from air pollutants. Elsevier, London, pp 181–210

McLeod AR, Fackrell JE, Alexander K (1985) Open-air fumigation of field crops: criteria and design for a new experimental system. Atmos Environ 19:1639–1649

McLeod AR, Shaw PJA, Holland MR (1992) The Liphook forest fumigation project: studies of sulphur dioxide and ozone effects on coniferous trees. For Ecol Manage 51:121–127

Miller PR, Yoshiyama RM (1973) Self-ventilated chambers for identification of oxidant damage to vegetation at remote sites. Environ Sci Technol 7:66–68

Miller PR, Parmeter JR Jr, Taylor OC, Cardiff EA (1963) Ozone injury to the foliage of ponderosa pine. Phytophathology 53:1072–1076

Mooney HA, Drake BG, Luxmoore RJ, Oechel WC, Pitelka LF (1991) Predicting ecosystem responses to elevated CO_2 concentrations. BioScience 41: 96–104

Mudano JE, Allen HL, Kress LW (1992) Ozone impacts on seasonal foliage dynamics of young loblolly pine. For Sci 38:324–335

Musselman RC (1985) Protecting grapevines from ozone injury with ethylenediurea and benomyl. Am J Enol Vitic 36 (1):38–42

Musselman RC, Tashenberg EF (1985) Usefulness of vineyard fungicides as antioxidants for grapevines. Pant Dis 69:406–408

Musselman RC, Oshima RJ, Gallavan RE (1983) Significance of pollutant concentration distribution in the response of ‘red kidney’ beans to ozone. J Am Soc Hortic Sci 108: 347–351

Musselman RC, Huerta AJ, McCool PM, Oshima RJ (1986a) Response of beans to simulated ambient and uniform ozone distributions with equal peak concentrations. J Am Soc Hortic Sci 111:470–473

Musselman RC, McCool PM, Younglove T, Teso RR, Oshima RJ (1986b) Statistical analysis of differences in response of beans to ozone. Sci Hortic 30:165–176

Musselman RC, McCool PM, Oshima RJ, Teso RR (1986c) Field chambers for assessing crop loss from air pollutants. J Environ Qual 15:152–157

Musselman RC, McCool PM, Lefohn AS (1994a) Ozone descriptors for an air quality standard to protect vegetation. J Air Waste Manage Assoc 44:1383–1390

Musselman RC, Younglove T, McCool PM (1994b) Response of *Phaseolus vulgaris* L. to differing ozone regimes having identical total exposure. Atmos Environ 28(16):2727–2731

Nast W, Mortensen L, Fischer K, Fitting I (1993) Effects of air pollutants on the growth and antioxidative system of norway spruce exposed in open-top chambers. Environ Pollut 80:85–90

Oechel WC, Riechers G, Lawrence WT, Prudhomme TI, Grulke N, Hastings SJ (1992) 'CO_2LT' an automated, null-balance system for studing the effects of elevated CO_2 and global climate change on unmanaged ecosystems. Funct Ecol 6:86–100

Oleksyn J, Prus-Giowacki W, Giertych M, Reich PB (1994) Relation between generic diversity and pollution impact in a 1912 experiment with East European *Pinus syivestris* provenances. Can J For Res 24:2390–2394

Olszyk DM, Tibbitts TW, Hertzberg WM (1980) Environment in open-top field chambers utilized for air pollution studies. J Environ Qual 9:610–615

Olszyk DM, Kats G, Dawson PJ, Bytnerowicz A, Wolf J, Thompson CR (1986a) Characteristics of air exclusion systems vs. chambers for field air pollution studies. J Environ Qual 15:326–334

Olszyk DM, Bytnerowicz A, Kats G, Dawson PJ, Wolf J, Thompson CR (1986b) Effects of sulfur dioxide and ambient ozone on winter wheat and lettuce. J Environ Qual 15:363–369

Olszyk DM, Bytnerowicz A, Kats G, Dawson PJ, Wolf J, Thompson CR (1986c) Crop effects from air pollutants in air exclusion systems vs. field chambers. J Environ Qual 15:417–422

Olszyk DM, Dawson PJ, Morrison CL, Takemoto BK (1990) Plant response to nonfiltered air vs. added ozone generated from dry air or oxygen. J Air Waste Manage Assoc 40:77–81

Olszyk DM, Takemoto BK, Kats G, Dawson PJ, Morrison CL, Preston JW, Thompson CR (1992) Effects of open-top chambers on 'Valencia' orange trees. J Environ Qual 21:128–134

Olszyk D, Bytnerowicz A, Kats G, Reagan C, Halce S, Kerby T, Millhouse D, Roberts B, Anderson C, Lee GH (1993) Cotton yield losses and ambient ozone concentrations in California's San Joaquin Valley. J Environ Qual 22:602–611

Ormrod DP, Marie BA, Allen OB (1988) Research approaches to pollutant crop loss functions. In: Heck WW, Taylor OC, Tingey DT (eds) Assessment of crop loss from air pollutants. Elsevier, London, pp 27–44

Oshima RJ, Poe MP, Braegelmann PK, Baldwin DW, Van Way V (1976) Ozone dosage–crop loss function for alfalfa: a standardized method for assessing crop losses from air pollutants. J Air Pollut Control Assoc 26:861–865

Oshima RJ, Braegelmann PK, Baldwin DW, Van Way V, Taylor OC (1977) Reduction of tomato fruit size and yield by ozone. J Am Soc Hortic Sci 102:289–293

Payer HD, Pfirrmann T, Mathy P (1990) Environmental research with plants in closed chambers. Air Pollution Res Rep 26, Commission of the European Communities, Brussels

Peterson CE Jr, Mickler RA (1994) Considerations for evaluating controlled exposure studies of tree seedlings. J Environ Qual 23:257–267

Preston EM, Tingey DT (1988) The NCLAN program for crop loss assessment. In: Heck WW, Taylor OC, Tingey DT (eds) Assessment of crop loss from air pollutants. Elsevier London, Chapt 3, pp 45–62

Pye J (1988) Impact of ozone on the growth and yield of trees: a review. J Environ Qual 17:347–360

Qui Z, Chappelka AH, Somers GL, Lockaby BG, Meldahl RS (1992) Effects of ozone and simulated acidic precipitation on above- and below-ground growth of loblolly pine (*Pinus taeda*). Can J For Res 22:582–587

Rafarel CR, Ashenden TW (1991) A facility for large-scale exposure of plants to gaseous atmospheric pollutants. New Phytol 117:345–349

Rawlings JO, Lesser VM, Dassel KA (1988) Statistical approaches to assessing crop losses. In: Heck WW, Taylor OC, Tingey DT (eds) Assessment of crop loss from air pollutants. Elsevier, New York, pp 389–416

Reich PB (1987) Quantifying plant response to ozone: a unifying theory. Tree Physiol 3:63–91

Reynolds JF, Chen J, Harley PC, Hilbert DW, Dougherty RL, Tenhunen JD (1992) Modeling the effects of elevated CO_2 on plants: extrapolating leaf response to a canopy. Agric For Meteorol 61:69–94

Richards BL, Middleton JT, Hewitt WB (1958) Air pollution with relation to agronomic crops: V. Oxidant stipple of grape. Agron J 50:559–561

Runeckles VC, Wright EF, White D (1990) A chamberless field exposure system for determining the effect of gaseous air pollutants on crop growth and yield. Environ Pollut 63:61–77

Samuelson LJ, Edwards GS (1993) A comparison of sensitivity to ozone in seedlings and trees of *Quercus rubra* L. New Phytol 125:373–379

Sasek TW, Richardson CJ (1989) Effects of chronic doses of ozone on loblolly pine: photosynthetic characteristics in the third growing season. For Sci 35:745–755

Saxe H (1991) Photosynthesis and stomatal responses to polluted air, and the use of physiological and biochemical responses for early detection and diagnostic tools. Adv Bot Res 18:1–128

Shafer SR, Heagle AS (1989) Growth responses of field-grown loblolly pine to chronic doses of ozone during multiple growing seasons. Can J For Res 19:821–831

Shafer SR, Heagle AS, Camberato DM (1987) Effects of chronic doses of ozone on field-grown loblolly pine: seedling responses in the first year. J Air Pollut Control Assoc 37:1179–1184

Shafer SR, Reinert RA, Eason G, Spruill SE (1993) Analysis of ozone concentration–biomass response relationships among open-pollinated families of loblolly pine. Can J For Res 23:706–715

Simini M, Skelly JM, Davis DD, Savage JE (1992) Sensitivity of four hardwood species to ambient ozone in north central Pennsylvania. Can J For Res 22:1789–1799

Stow TK, Allen HL, Kress LW (1992) Ozone impacts on seasonal foliage dynamics of young loblolly pine. For Sci 38:102–119

Strain BR, Allen LH Jr, Baldocchi D, Bazzaz F, Burke J, Dahlman R, Denmead T, Hendrey G, McLeod A, Melillo J, Oechel W, Risser P, Rogers H, Rozema J, Wright R (1991) Available technologies for field experimentation with elevated CO_2 in global change research. In: Mooney HA, Medina E, Schindler DW, Schulze E-D, Walker BH (eds) Ecosystem experiments, Chapt 15, SCOPE 45. John Wiley, New York

Taylor GE Jr (1994) Role of genotype in the response of loblolly pine to tropospheric ozone: effects at the whole-tree, stand, and regional level. J Environ Qual 23:63–82

Taylor GE, Pitelka LF (1992) Genetic diversity of plant populations and the role of air pollutants. In: Barker JR, Tingey DT (eds) Air pollution effects on biodiversity. Van Nostrand Reinhold New York, pp 111–130

Taylor GE Jr, Johnson DW, Andersen CP (1994) Air pollution and forest ecosystems: a regional to global perspective. Ecol Applic 4:662–689

Temple PJ, Riechers GH, Miller PR (1993) Growth responses of ponderosa pine to long-term exposure to ozone, wet and dry acidic deposition, and drought. Can J For Res 23:59–66

Teskey RO, Dougherty PM, Wiselogel AE (1991) Design and performance of branch chambers suitable for long-term ozone fumigation of foliage in large trees. J Environ Qual 20:591–595

Thompson CR, Taylor OC (1969) Effects of air pollutants on growth, leaf drop, fruit drop, and yield of citrus trees. Environ Sci Technol 3:934–940

Thornton FC, Pier PA, McDuffie C Jr (1992) Red sprue response to ozone and cloudwater after three years exposure. J Environ Qual 21:196–202

Tingey DT, Wilhour RG, Standley C (1976) The effect of chronic ozone exposures on the metabolite content of ponderosa pine seedlings. For Sci 22:234–241

Tjoelker MG, Luxmoore RJ (1991) Soil nitrogen and chronic ozone stress influence physiology, growth, and nutrient status of *pinus taeda* L. and *Liriodendron tulipifera* L. seedlings. New Phytol 119:69–81

Tjoelker MG, Volin JC, Oleksyn J, Reich PB (1994) An open-air system for exposing forest-canopy branches to ozone pollution. Plant Cell Environ 17:211–218

Tonneijck AEG, Bugter RJF (1991) Biological monitoring of ozone effects on indicator plants in the Netherlands: initial research on exposure–response functions. VDI Berichte 901:613–624

US EPA (1987a) Air pollution exposure systems and experimental protocols, vol 1. A review and evaluation of performance. EPA/600/3–87/037a, Corvallis, Oregon

US EPA (1987b) Air pollution exposure systems and experimental protocols. vol 2: Description of facilities. EPA/600/3–87/037b, Corvallis, Oregon

Unsworth MH, Heagle AS, Heck WW (1984) Gas exchange in open-top field chambers. I. Measurement and analysis of atmospheric resistances to gas exchange. Atmos Environ 18:373–380

Walker BH, Attiwill P, Graetz D, MacMahon J, Medina E, H, Schulze E-D, Tamm CO, Trillmich F, Wright R (1991) Whole-terrestrial ecosystem experiments in the context of global change: working group report. In: Mooney HA, Medina E, Schindler DW, Schulze E-D, Walker BH (eds) Ecosystem Experiments, Chapt 14. SCOPE 45. John Wiley, New York

Weigel HJ, Colls JJ, Geisser PA (1992) Microclimate in open-top chambers. In: Jager HJ, Unsworth M, De Temmerman L, Mathy P (eds) Effects of air pollution on agricultural crops in Europe: results of the European open-top chambers project, Air Pollution Research Report 46, Commission of the European Communities, Brussels, Belgium, pp 618

Weinstock L, Kender WJ, Musselman RC (1982) Microclimate within open-top air pollution chambers and its relation to grapevine physiology. J Am Soc Hortic Sci 107:923–926

Wiltshire JJ, Wright CJ, Unsworth MH (1992) A new method for exposing mature trees to ozone episodes. For Ecol Manage 51:115–120

Winner WE, Greitner C (1989) Field methods used for air pollution research with plants. In: Pearcy W, Ehleringer J, Mooney HA, Rundel PW (eds) Plant physiological ecology: field methods and instrumentation. Chapman and Hall, New York, pp 399-425

Winner WE, Lefohn AS, Cotter IS, Greitner CS, Nellessen J, McEvoy LR Jr, Olson RL, Atkinson CJ, Moore LD (1989) Plant responses to elevational gradients of O_3 exposures in Virginia. Proc Natl Acad Sci USA 86:8828–8832

Woodward FI (1992) Predicting plant responses to global environmental change. New Phytol 122:239–251

Woodwell GM, Cook DR, Cowling EB, Johnson AH, Kimmerer TW, Matson PA, McLaughlin SS, Raynal DJ, Swank WT, Waring RH, Winner WE, Woodman JN (1989) Biological markers of air pollution stress and damage in forests. Part I. The committee's report. National Research Council, National Academy Press Washington, DC

Younglove T, McCool PM, Musselman RC, Smith E (1988) The use of transfer models to describe the relationship between pollutant exposure and leaf stomatal opening. Environ Exp Bot 28:307–313

10 Physiological Processes and Plant Responses to Ozone Exposure

R.L. HEATH and G.E. TAYLOR Jr.

10.1 Introduction

To maintain plant growth and productivity in any ecosystem, plants must respond to all stresses in a carefully controlled manner. However, oxidant air pollutants constitute a stress that seems to lower plant productivity and to make the plant more sensitive to other diverse stresses, thus seemingly overwhelming the plant's resistance (Heck et al. 1988; US EPA 1988; also see other chapters in this volume). On the other hand, nitrogen-containing air pollutants at low concentrations, e.g., NO_2, may add to the plant's productivity when nitrogen is limiting (US EPA 1992). Current problems are (1) at what level do these pollutants individually cause the plant's productivity and resistance to be altered; (2) what are the mechanisms by which the pollutants induce plant alterations; and ultimately (3) how does the plant respond to multiple, combined stresses?

Troposphere air pollution, especially ozone, continues to be a problem throughout the world (Taylor et al. 1994). Relatively high ozone levels are often linked to a decline in crop productivity and changes within ecosystems far removed from urban settings. Within the developed nations, the government has made a clear commitment to reduce the levels of polluted air, primarily by enforcement of emission standards based upon scientific research (US EPA 1988). Indeed, the mandated standards of major pollutants are, and should be, based on an understanding of effects on biological organisms. While the primary standard is based upon human health considerations, the secondary standard in the US is based upon injury to plants and ecosystems (Tingey et al. 1990). Yet we do not have a clear understanding of the basic events that cause injuries for even a single plant. It seems difficult, if not impossible, to understand events within an ecosystem if we do not know how a single pollutant stress interacts with an individual plant.

Plants respond to air pollutants in a similar manner to other stresses; the mechanisms include exclusion, tolerance, compensation, and repair

Ecological Studies, Vol. 127
Sandermann et al. (eds) Forest Decline and Ozone

(Levitt 1980). The response depends upon the concentration, duration and exposure of the pollutant, environmental conditions, and the developmental and metabolic state of the plants (Guzy and Heath 1993). Any response is detrimental in some sense to plant productivity because it costs the plant metabolic resources. In some cases the stomata close under the pollutant exposure excluding the pollutant from the interior of the plant, and thus preventing damage. However, if this happens frequently when CO_2 fixation is rate limiting, plant productivity will suffer.

Although plant exposure to air pollution may inhibit overall anabolic processes (Dugger and Ting 1970a,b), recent evidence indicates that specific air pollutants, like many other abiotic stresses (see Sachs and Ho 1986 and references cited therein), stimulate the synthesis of some proteins (Castillo et al. 1987; Matters and Scandalios 1987; Beckett et al. 1989, 1990; Pietila et al. 1990; Ranieri et al. 1990; Schmitt and Sandermann 1990). For example, tolerance may occur if endogenous antioxidants or reductive compounds intercept the oxidative pollutants and lower their levels near the cell's reactive surface. Yet the regeneration of those antioxidants constitutes an energy demand and a drain upon the organism. The belief is that the elucidation of stress compensation mechanisms will permit the development of genetic and horticultural strategies to control specific ozone injury in ecological settings and to understand general stress strategies of plants as well as allow the elaboration of a more solid basis for the setting of air quality standards.

This paper focuses on ozone and its mechanisms of injury at the physiological level of the individual plant. In order to understand how the ecosystem can respond to ozone, we feel that an understanding of the plant's response is critical. At the very least, the ecosystem is the sum of its parts, and plants are a large portion of its parts. Without an understanding of the development of mechanisms of injury at any level, we are reduced to observing and tabulating the responses. Unable to predict what any future change in the environment will do to plant growth or ecosystem development, we will be made helpless to understand what to do currently.

10.1.1 Air Pollution Injury

Injury induced by air pollutants is difficult to define (Woodwell et al. 1989). Early work used the formation of visible changes of leaves to determine injury. The observation of this injury is generally used in ecosystems to determine detrimental air quality. Yet over the years other forms of injury have been described in addition to the formation of visible patterns on the leaf surface, such as (1) loss of plant productivity as measured by food, fibre, or wood production; (2) shifts in metabolic pathways; (3) increased inability to withstand other stresses presented either concurrently or sequentially; and (4) acceleration of normal plant senescence as measured by a loss of older leaves or needles.

Often visible foliar injury is the only method available to ascertain damage. But the manifestation of foliar injury through chlorosis and necrosis are commonly observed among many different types of stresses, and different environmental conditions may alter the development of those injury patterns. Other foliar injury patterns (see Jacobson and Hill 1970) are "water-logging", lower surface "bronzing", upper surface "silvering", and formation of anthocyanin (Koukol and Dugger 1967). While these visible markers can be used to assess injury, the identification of the stress causing the visible injury still often relies upon an educated opinion.

Plant productivity is often the method of choice for assessing air pollution injury as it directly measures the loss of plant material. In agricultural terms this loss is directly translatable into an economic loss, which is also the case for forestry. For general ecosystems, the loss of productivity is not easily linked to a deleterious problem. While the decline in litter on the forest floor leads to a long-term loss in availability of nutrients, that decline may merely slow the growth of the forest. In addition, loss of productivity is associated with a decline in the ability of plants to withstand pathogen attack (e.g., the death of trees in the San Bernadino forest is generally not caused by air pollutants, but rather by the attack of bark beetles upon weakened trees; Miller et al., this Vol.).

Of the above indicators of air pollutant the most problematical is that of altered metabolism. Too often those alterations cannot be easily traced to air pollutant exposure directly, but rather would appear to be due to another less obvious stress (e.g., drought or excessive heat). While the loss of photosynthetic capacity is generally linked to a loss of productivity, it is not always the case and more details of the general processes must be sought (Elmore 1980). Yet physiological processes which can be measured quantitatively, such as photosynthesis and transpiration, have been used to monitor injury (Patterson 1980; Darrell and Jäger 1984; Hogsett et al. 1988; Taylor et al. 1988), but only on single leaves.

More importantly, the altered biochemical states within the tissue lead to the inability of the plant to respond properly to existing environmental conditions and to other stresses (Davison et al. 1988; Manning and Keane 1988). At the lowest, perhaps most natural, level of ozone exposure, an acceleration of natural aging is believed to occur, resulting in earlier leaf abscission and other developmental changes (Rodriquez et al. 1990).

10.1.1.1 Types of Oxidant Stress

Heath (1988) has argued that there are three forms of oxidant induced injury – acute, chronic, and accelerated senescence; however, it is perhaps better to reclassify these injury forms into two types based upon a clear line of demarcation from observations. A review of the literature will clearly show two forms of injury – that which produces a visible injury and that which does not, but which exhibits some biochemical or physiological change.

At higher concentrations of atmospheric pollutants for short periods of time, distinct visible injury generally occurs due to some cellular death within tissues, leading to a decline in the total area of metabolically active tissue. It is believed that death is due to a loss of membrane integrity and the formation of oxidative products leading to a loss of metabolites and ions into the extracellular tissue space. While acute stress generally gives rise to visible injury, that is not always the case. In more recent experiments, one finds conditions where a single pulse of ozone at a high level can generate changes in metabolism without giving rise to visible changes on the leaf (see Heath 1994b, 1996a). Further, lower concentrations of pollutants for longer periods of time may produce no visible injury, but lead to a lowered rate of photosynthesis, a decline in transpiration, and release of ethylene. Under these circumstances, metabolism is altered and the pool sizes of many metabolites are changed.

Thus, acute and chronic stresses may not necessarily be different in their effect on metabolism. Certainly the formation of visible lesions on the leaf is a good distinction between one type of pollutant effect and another (e.g., that exhibiting no visible lesions). Indeed as will be described in a later section, visible injury formation is often linked with ethylene production and a defined mechanism. Metabolic changes without visible lesions occur but there is little agreement as to why they occur (e.g., no obvious mechanism, but again see later). Probably the majority of the differences between chronic and acute stress are due simply to the progress of developmental age of leaves through the longer period of the application of chronic stress (over days and even weeks). We do not know how to link mechanisms found for acute stress to changes induced by chronic stress–the plant simply changes and we seem not to be able to normalize the developmental stage of the plant.

10.1.1.2 Overview

During this article, some readers will be undoubtedly disappointed in our lack of discussion of forests per se; we have come to the conclusion that we do not understand the basics yet and so to generalize to full ecosystems would be premature. There have been some attempts to weave a story regarding trees and physiology (e.g., Schmieden and Wild 1995). However we have decided to concentrate on what is known at the present time in terms of a mechanistic picture. Towards the end we attempt to apply these current ideas in suggesting how measurements within an ecosystem might be accomplished. The scaling up of physiological principles to forest growth is difficult at best.

Yet what is meant by forest decline in terms of physiology? Is a decline due to a general aging of the forest or to an increased propensity for injury by other biotic and abiotic stresses? To begin of any discussion of varied processes, we must be aware that photosynthesis is a net gain, and respiration and "defense/repair" is a net loss. What sort of compensation mechanisms exist which can eliminate some leaf losses? Thus, we are confronted with how productivity is best measured spatially and temporally. In fact, the induction of the decline may only be an indirect, but interactive, consequence of total environmental exposure.

Several major physiological hypotheses are currently being discussed which define mechanisms in terms of air quality leading to detrimental effects on plants. What is difficult to understand is how physiological studies relate to ecological events. By clearly organizing the concept of how individual plants react to ozone, future studies of whole communities may be better defined. We must begin with the events induced in a single plant (generally a crop plant), because this system has been best studied from biochemical and physiological perspectives. Although we obtain data only for individual plants, the understanding reached for one system should help to define studies in more complex systems, closer to the natural ecosystem. Clearly this integrative transition is an important future area of research (Field et al. 1992; Elheringer and Field 1994; Taylor et al. 1994).

10.2 The Assault of Air Pollutants upon Plants

If ozone exposure can be considered a stress upon a plant, what effects can that stress generate within that plant? Stress is a force exerted on an object, which is at steady state before that stress (Levitt 1980). The object reacts and counters that stress with an internal strain. However, since the environment surrounding a plant is constantly changing, what is the beginning state or "zero point" of the plant? We are forced to ask: how can we measure an ozone stress (by a measure of the dose of ozone to the sensitive tissue); what is the initial state of the plant before the stress (under what conditions has the plant grown and for how long); and how does the plant respond to that stress (what is the strain within the plant)?

Plant strategies to resist ozone stress include exclusion and tolerance of ozone or repair and compensation after the exposure. Air pollutants can be excluded from the tissues via stomata closure or from cells by impermeability of their membrane. To a certain extent, plants can tolerate pollutants by storing them in a different tissue or organelle (such as for NO_x) or through detoxification, such as a chemical modification to a less harmful compound that can enter the normal metabolic pathways (see Wellburn 1990). Past observations of ozone injury have indicated physiological and metabolic changes (reviewed by Heath 1988; Reddy et al. 1993; Harris and Bailey-Serres 1994). Many of these events, especially if operative over a long period of time, are likely to have been initiated at the level of gene expression.

Response to air pollutants may also represent physiological accommodation or adaptation to a stress. There is clear evidence that plants subjected to repeated exposures at low levels will "harden off" and become more resistant to later exposures (McCool et al. 1988), a further indication that changes in gene expression may play a role. Yet it is not known at what expense to plant growth this occurs. It is also not known how rapidly a plant recovers after a single exposure nor how repeated exposures affect normal metabolism. It would seem that a permanent shift in energy allocation with continuous ozone exposure may be less detrimental than long periods of normal allocation between temporary but more dramatic allocation shifts.

10.2.1 Sequence of Events Following Air Pollutant Exposure

In order to understand any mechanism operating within a biological system, we need to postulate a process. The beginning of the mechanism for air pollution injury is the presence of the pollutant near the plant or plant canopy. The level and type of pollutant is critical for the mechanism, but for this section we assume that the pollutant is ozone[1]. The sequence from that point forward is (1) entry of the pollutant into the canopy and ultimately the leaf; (2) movement of product(s) and transformations within the leaf; and (3) biochemical and physiological reactions of the gases or their products within the leaf. There is no single process un-

[1]Oxides of nitrogen (NO_2, NO, and HNO_3) travel with ozone since they are generated by the processes which produce ozone within urban air basins.

derlying the response to ozone exposure (Guzy and Heath 1993). This conclusion was reached earlier by studies with a variety of cultivars and species (see Heagle 1989 for a summary).

10.2.1.1 Entry of the Pollutant into the Leaf

Although ozone measurements for air quality monitoring purposes are taken in the free troposphere, the pollutant's site of action lies at biochemical sites of reaction within cells of the leaf interior. The linkage between ozone sources and the pollutant's site of action is physically circuitous and chemically complex, and, as a consequence a simple empirical relationship between the concentration in the atmosphere and that at sensitive reaction sites within the leaf interior is unlikely.

Gas-phase transport begins in the free troposphere; it includes diffusion through the stomate and extends into the intercellular space of the leaf interior. In the free troposphere, ozone transport is largely governed by turbulent processes so that all gas species are transported collectively in bulk flow. Once ozone enters the laminar boundary layer of the leaf surface, transport is progressively dictated by molecular processes that reflect the unique properties of ozone. At this point along the transport pathway, the pathway and fate of all gases begins to diverge. Interestingly, the transport pathway in the free troposphere is meters in length whereas the transport pathway on the leaf surface and within the leaf interior is at least six orders of magnitude shorter. However, control over the ozone uptake is largely determined by the latter segment of the pathway since diffusion is rate limiting.

Currently the external concentration of ozone in the air surrounding the plant is used to give an indication of exposure. This exposure is often used as a measure of "dose". Clearly this is incorrect (Taylor and Hanson 1992; Heath 1994b). As ozone has no easily measured isotope, virtually no measurements have been done on an actual dose of ozone, that is, the amount of ozone which reacts with individual biochemicals in the leaf.

The calculation process to find a measure of dose should include the gas entering and passing through the canopy. There is a volume of literature regarding this, but it is much too broad to be included here (for more details, see Lovett 1994). Much work has been done on the latter stages of the gas-phase transport process which involves the diffusive bulk flow of the gas up to the leaf, including gaseous diffusion through the leaf boundary layer and stomate into the substomatal cavity (Levett 1994; Taylor and Constable 1994). The cuticle seems to be a near-absolute barrier for ozone (Kerstiens and Lendzian 1989). For our discussion, we will ignore the cuticle as a point of ozone interaction with the cell. The entry of gases into a leaf depends upon physical and chemical properties of gas phase and leaf surface and has been described for the path of CO_2 movement, which approximately follows a linear flux law of

$$j = g(Co - Ci) \ , \tag{10.1}$$

where the flux (j) into the substomal cavity of a leaf is a function of the conductance[2] (g) through the boundary layer and stomate and the gradient of concentration of gas from the outside (Co) inwards (Ci). For both water and CO_2, this formulation has been used for years (Farquhar and Sharkey 1982; Ball 1987).

For ozone, the internal concentration, estimated by traditional methods used for CO_2, is very close to zero (Laisk et al. 1989). It is felt that a zero internal concentration of ozone is due to its extreme reactivity with cellular chemicals (see later sections and Bailey 1978; 1982). Since ozone is reactive, the site of its reactions would be predicted to be close to its entry point (Moldau et al. 1990; Heath 1994a).

The effective delivery rate of ozone to the tissue within the leaf (R_D)[3] is

$$R_D = g \times Co \ , \tag{10.2}$$

with stomata conductance being the major regulatory control (Tingey and Taylor 1981; Amiro et al. 1984; Taylor and Hanson 1992). Thus, the simplest calculation of dose depends upon the leaf conductance and the external concentration. The expression of this dose is as a rate of delivery to a surface area (mole/m^2 sec). A major question exists: is the dose (total accumulation of ozone given as mole/m^2; the rate is integrated over exposure time) or the rate of delivery most critical for the development of injury? Other questions include whether the index of exposure, given as delivery rate or accumulation, has a threshold (a level below which no observable effect is noted) and linearly generates a measured effect.

The diffusive flow of gaseous pollutants is from the substomatal cavity into the cell through the wall. An equilibrium between the gas and aqueous phase must occur at the interface where the gaseous species dissolve in the water, according to Henry's law (Heath 1980, 1987; Wellburn 1990). As discussed later, once ozone enters an aqueous phase, many products are possible including organic, inorganic, and free radical species (see Hoigné and Bader 1983a,b; Hoigne et al. 1985; Bailey 1978, 1982).

---

[2]The conductance of ozone is derived from the conductance for H_2O with appropriate corrections for the mass difference between ozone and H_2O.

[3]Dose is measured in moles per area of leaf surface per unit time; so Co is calculated in moles per volume of air rather than $\mu l \ l^{-1}$.

One would expect that one molecule of ozone would change one molecule (e.g., oxidize one sulfhydryl) and thus the measured ratio would be one SH lost one O_3 reacted (see later section). Yet the cell has methods to maintain the sulfhydryls largely reduced, so the ratio itself would be lowered, depending upon the rate of reduction and ability of the cell to move the reducing agent to the affected sulfhydryl. This would mean that both the accumulated dose (over a time period) should be matched to the rate of delivery of ozone to the affected site (the highest rate would overwhelm the cell's ability to regenerate the sulfhydryl). The line of reasoning depends upon knowing the amount of ozone at the critical site within the leaf tissue. We often use only empirical measures, which relate the loss of productivity to the amount of external ozone to which the leaf/needle has been exposed integrated over time. Such empirical determinations do not give any measure of the biochemical effectiveness of ozone.

10.2.1.2 Reactions of the Gas Within the Leaf

The first reactions of ozone within the water phase at the cell's surface (apoplastic space) are critical. These potentially diverse reactions generate many dissimilar species within the wall region of the cell and these species have new reaction potentials and travel different paths. Presently, these chemical reactions and the potential new species are poorly understood although there have been some speculations (Heath 1987, 1988; Chameides 1989; Wellburn 1990). It is critical, however, to realize just what sorts of products may be formed even if our knowledge is incomplete.

As is described in Cacace and Speranza (1994), ozone has four major chemical reactivity characteristics: (1) a nucleophile due to an unshared electron pair; (2) an electrophile through an insertion into the π bonds of alkenes; (3) a Lewis acid due to its proton affinity; and (4) a Brønsted base through an insertion into the σ bond of alkanes. Many have believed that ozone reacts only as a radical; but the classical studies of Criegee (Criegee 1975; Mudd 1996) are based upon the second case in which oxidation products of the carbons next to double bonds of alkenes are formed. It has been shown that the solvent in which the reaction occurs has much to do with ozone's ultimate reactions and the products' fate (Heath and Tappel 1976). In water, hydrogen peroxide is formed as ozone attacks the double bond, while alkene peroxides are formed by reactions in polar alcohols.

Although not exact, a comparison of rate coefficients of chemical reactivity with ozone is instructive in order to develop some comprehension of ozone's reactivity with biochemicals in vivo. Giamalva et al. (1986) determined the rate coefficients for reactions of tocopherol and methyl

Table 10.1. Rate coefficients for varied biochemicals in two different solvents. Data for α-Tocopherol, Me-oleate, and Na-ascorbate from Giamalva et al. (1986) and that for sulfhydryls and tryptophan from Pryor and Uppu 1993. There is a sizable pH dependence for the reaction of tocopherol in water, declining with lower pH which indicates that the unprotonated hydroxyl may be important. No such pH depedence exists for oleate's reaction. Sulfhydryl in the form of cysteine undergoes a loss of a H^+ in water with a pK of about 9.3 (depending upon its microenvironment)

Species	Rate Coefficient (M^{-1} s^{-1})	
	In CCl_4	In water
α-Tocopherol	5.5×10^3	1×10^6 at pH 7
Me-oleate	5.0×10^5	1×10^6
Na-ascorbate		6×10^7
Sulfhydryl$^-$ H^+		4×10^4
Sulfhydryl$^-$		2.4×10^{10}
Tryptophan		2×10^7

oleate with ozone in two different solvents (water and organic, carbon tetrachloride). Their results are given in Table 10.1 and show that the reactivity of tocopherol is increased some 200-fold in water, while the reactivity for oleate in water is increased but by much less (two-fold). For comparison ascorbate in water has a 60-fold increase in reactivity compared with oleate or tocopherol. This change is to be compared with an increase in the rate constant for unprotonated thiols and tryptophan. Thus, from the chemical rate coefficients alone it can be predicted that ozone will react very rapidly with unprotonated thiols and will react rapidly with ascorbate and tryptophan. Only after the concentration of those species have been diminished by ozone, will double bonds of hydrocarbons be attacked.

Earlier work has suggested that the structures in which the double bonds are "hidden" (within the micelle) serve to protect the double bonds from ozone attack (Bakalik and Thomas 1977; Matheson and King 1978). It was hypothesized that ozone did not penetrate well into micelles (see Heath 1987). Recently, Uppu and Pryor (1994) generated reverse micelles in which the head groups of lipids form a water pocket core with the acyl units extending outward into the isooctane solvent. Under these conditions sodium oleate protected lysozyme, contained within the water pocket. The authors suggested that the double bond of oleate was therefore more reactive than the tryptophan residues of lysozyme. Unfortunately, as the authors realized, the structural packing of the acyl units of the fatty acid in a reverse micelle was much less than in a bilayer.

Pryor (1992) calculated the distance which ozone may be expected to penetrate into the pulmonary air/tissue boundary. His conclusion was that even though the fluid layer of the lung lining was from 0.1 to 20 μm thick, ozone could reach the cell membrane. While the fluid layer in plant walls contains lipid and protein materials with high reactivity towards ozone, the wall also possess a more tortuous pathway. Pryor's calculation indicates that the time required for an ozone molecule to pass through the membrane by diffusion is about 500×10^{-9} s. He then calculated the half-life of ozone within the bilayer based upon its reactivity with fatty acid double bonds and found that it could only be about 70×10^{-9}, too short to cross the membrane unreacted. For movement within a region containing reduced glutathione he calculates a half-life of about 700×10^{-9}s. These calculations suggest that ozone cannot penetrate very far into a cell unreacted.

Biologically, ozone generally reacts with water-soluble organic molecules at double bonds to form carbonyl groups and with sulfhydryls to form disulfide bridges or sulfones (Mudd et al. 1969; Mudd and Kozlowski 1975). In water, the reactions become more confusing, but some products have been described by Pryor et al. (1985) and Heath (1987). Oxidative products of ozone which have been linked to potential metabolic changes are H_2O_2, hydroxyl radical ($HO^{\bullet}$), and superoxide (O_2^-) (Tingey and Taylor 1981; Möller 1989).

For these species, effective detoxification reactions occur via antioxidant metabolites and enzymes (see Larson 1988), such as ascorbate (Castillo et al. 1987; Chameides 1989; Luwe et al. 1993), glutathione (Matters and Scandalios 1987; Castillo et al. 1987), and superoxide dismutase (Lee and Bennett 1982; but see Polle and Rennenberg 1991). These antioxidants must be present at high enough concentrations to intercept the oxidative intermediates, but their concentrations vary with the leaf's developmental stage (Price et al. 1990). If the prefumigation concentrations of these antioxidants are low, in some cases their production can be stimulated upon exposure to ozone, albeit the response time is relatively slow (Harris and Bailey-Serres 1994). The chemical modification of wall-space-specific biochemicals is also possible (Castillo et al. 1987). An increase in potential antioxidants has been demonstrated for peroxidases (Castillo et al. 1984), superoxide dismutase (Matters and Scandalios 1987; Lee and Bennett 1982), and ascorbate (Castillo et al. 1987). These results are in contrast to the loss of activities of enzymes such as glucan synthase (Ordin et al. 1969) and diamine oxidase (Peters et al. 1988) due to ozone exposure.

10.2.1.3 Movement of Product(s) and Transformations

It is believed that the initial site of ozone injury is the plasma membrane. Certainly, membrane functions such as membrane fluidity (Pauls and Thompson 1980), permeability (Elkiey and Ormrod 1979), K^+-exchange via ATPase reactions (Dominy and Heath 1985), and Ca^{2+} exclusion (Castillo and Heath 1990) are rapidly lost upon exposure. If ozone reacts with components of the cell wall, those alterations may connect to the membrane through the wall. On the other hand, if ozone reacts directly with the membrane, those alterations may affect membrane-specific proteins not directly linked to transport, e.g., signal transducers (Cohen 1992) as well as transport systems and ion channels (Tester 1990). The mean free path of ozone is very short and, thus, only its oxidative products can diffuse into the cell and react with *internal* metabolites and structural biochemicals. The presence of internal cellular antioxidants would be expected to eliminate these internal diffusing oxidants.

The events of this pathological process which disrupts normal cell and tissue homeostasis is hardest to unravel and a complete sequence of events is difficult to elucidate (see Schulte-Hostede et al. 1988). Homeostasis is largely governed by the genetic makeup of the plant and its interplay with the environment. The principal differences between initial and secondary injury can be defined only by cellular location and time scale. This division is useful, however, because an initial injury is often dependent upon the chemical species involved, while secondary reactions seem to be more related to those involved in many, quite differently induced, stress responses, e.g., wounding.

Under chronic ozone stress, growth and morphological changes occur (Pell and Weissberger 1976; Masuch and Kettrup 1985). With ozone exposure the plant's developmental sequence is altered and the plant appears different to one grown in unpolluted air. One would predict that the appearance of certain proteins and the activation of some genes should be delayed, and even their order of appearance changed. These events usually suggest that transcription factors which control sets of important genes have been affected. One would predict that chronic stress would be linked to developmentally controlled genes.

10.2.2 Amount of Exposure at the Physiological Level

10.2.2.1 Initial Site of Direct Ozone "Attack"

In order to understand what processes may protect the cell from ozone exposure, we need an understanding of possible injury processes at initial sites of action of ozone within the cell proper. It may well be that there is not a single initial site, but multiple ones depending upon the external conditions and the genetic make-up of the plant. A good review regarding the initial action within a plant which is induced by ozone was written recently by Kangasjärvi et al. (1994).

Fatty Acids. There is a still a widespread belief that lipids are the primary point of ozone attack and this seems to fixate researchers upon lipid chemistry. While a very small amount of critical lipid molecules [Heath (1987) estimates that it would be less than 2%] may be changed, it would seem that the use of Ockham's razor would redirect the focus of these researchers until our methods of lipid determinations markedly improve. Reactive amino acids of proteins seem to be a more obvious site for ozone attack (Mudd 1996).

There are little direct data to show that lipids are attacked by ozone in any living system, which has not been previously *severely* damaged by ozone. Most of the data suggesting lipid attack by ozone were on plants subjected to ozone concentrations of 500–1000 nl l^{-1} for several hours (summarized by Heath 1984; Heath 1996b). Under these conditions the plant usually shows leaf wilting and collapse of tissue, suggesting a high rate of water loss (Bobrov 1952; Evans and Miller 1972; Pell and Weissberger 1976). Often after several days most of the interveinal region has gross chlorosis and necrosis. With such severe damage, it is not surprising that lipid and protein alterations are observed. While useful in the past, those reports are not especially insightful now where even peak levels of ozone are rarely above 300 nl l^{-1}. We are becoming more concerned with levels of 100 nl l^{-1} or less daily for several hours over many weeks.

Lipid peroxidation is often suggested as a major cause of membrane injury induced by ozone. However, the basic reactions of lipid peroxidation are quite different from the reactions in which ozone in involved. Once a free radical of a methylene-interrupted double bond systems is generated, the progress and products of lipid peroxidation are well defined (Tappel 1975). However, ozone attacks double bonds directly in a very different manner (Criegee 1975). Menzel (1980) suggested that the breakdown of the ozonide formed by ozone's reaction with double bonds of lipids could generate free radicals which then in turn could initiate lipid peroxidation itself. While this sequence of events seems reasonable, the levels of ozonide necessary to accomplish this initiation would have to be high. The stability and reactivity of the ozonide suggests that it is more like to react with itself forming the Criegee products or with water forming H_2O_2 (Heath and Tappel 1976). Any lipid peroxidation, if measured properly, would be indicative of the generation of a free radical in polyunsaturated fatty acids[4]. Ozone, as often shown in chemical systems, can add to a double bond and causes a rearrangement (Heath and Tappel 1976; Heath 1987). This mechanism leads to the generation of diverse carbonyl groups and can act upon only one double bond.

In any case, in the past, the detection of "ozone-induced peroxidation" has relied upon the production of malondialdehyde and then upon the reaction of malondialdehyde with thiobarbituric acid (TBA), yielding a characteristic orange color (Heath 1978). Yet the production of malondialdehyde is a low-yield process and many types of polyunsaturated fatty acid oxidation can produce it, with further reactions of the ozone products (rearrangements of the cyclic ozonide, see Heath and Tappel 1976). Also, unfortunately, sucrose alone can give a very high positive reaction with the TBA test (Schmidt 1959). Thus, in leaf tissue the level of sucrose should be measured at the same time as any potential peroxidation. To the authors' knowledge no one has successfully measured any of the other potential products of ozonolysis in plants (see Heath 1987; Heath 1996b).

In some reports regarding ozone-induced "peroxidation" unicellular organisms such as *Chlorella* (Frederick and Heath 1975) and *Euglena* (Chevrier et al. 1990) were subjected to very high levels of ozone. Heath's group determined that malondialdehyde production in *Chlorella* was observed only after there was little differential membrane permeability left and most metabolic pathways were shut down (summarized by Heath 1987). In fact, malondialdehyde production could be observed only after direct ozone uptake by the cells was greatly increased (indicative of complete opening of the cell to the environment) and the loss of the ability of cells to grow on a glucose media (indicative of cell death). Heath (1987) attempted to measure oxidative products induced by ozone but discovered that no products of ozonation of pure lipids could be found using standard lipid isolation methods. It was thought that low molecular weight products with oxygen groups were water soluble so they would not partition into the organic phase for separation.

Free Radicals and Primary Toxicants. Mehlhorn et al. (1990) investigated free radical formation with ozone exposure found no evidence of free radicals induced by ozone *only*; the authors believed that NO_X radicals in

[4]The initiation of lipid peroxidation is typically by the removal of a hydrogen atom from the carbon within a methylene-interrupted double bond system (R-CH=CH-CH_2-CH=CH-R′). The further autocatalytic reaction is due to oxygen reacting with the carbon free radical followed by a further abstraction by the hydroperoxyl radical of a hydrogen atom on another double bond system, forming a new free radical, completing the hydroperoxide, and inducing a cyclic reaction.

polluted air were most probably responsible for generating the observed signals. Living tissues have many free radical signals and so it is difficult to show changes if they are small.

Grimes et al. (1983) looked for the presence of free radicals due to ozone exposure in living tissues. No radical signals were detected unless certain organic acids (e.g., caffeic acid) were also added. They proposed that the radicals were produced only via a concurrent mechanism with the acid. In addition, reaction rates reported in Finkelstein et al. (1980) show that while DMPO or TMPO (spin traps of free radicals) could easily trap OH radicals (4×10^9 M^{-1} s^{-1}), they could not easily trap superoxide (depending upon its reaction with other compounds; 10 M^{-1} s^{-1}) or other slowly reacting oxidizing compounds.

Many products are possible in chemical solutions (especially with organic solutions) exposed to ozone (Atkinson 1990; Bailey 1978, 1982). Depending upon the organic compound, the products may be especially nasty for plants. Yet again no one has been able to demonstrate any product. It is hypothesized that ozone will form H_2O_2 and OH radicals, but that is true only in chemical solutions. In fact, the formation of these species is generally very pH dependent and these species may not be able to form in the acidic cell wall region (see Heath 1987). It is hard to believe that a little more H_2O_2 within the wall could cause major problems since it is normally present above 100 μM in many tissues (Schopfer 1994). A hydroxyl radical is much more reactive and could cause problems; however without its detection, one simply does not know how much injury it can cause.

One possibility for the production of a toxicant involves the reactions of ozone with the ethylene formed via wounding induced by ozone. Initially the production of ethylene after an ozone exposure (Craker 1971) was thought to be a typical wounding and injury response. The studies of Mehlhorn and Wellburn (1987) in which an inhibitor of ethylene formation [AVG (aminoethoxyvinyl glycine); an inhibitor of ACC (1-aminocyclopropane-carboxylic acid) synthase] blocked ethylene formation and inhibited visible injury implied that cessation of the ethylene release prevented the formation of visible injury. This, in turn, suggested that ethylene (with ozone) generated a chemical which may play a primary role in ozone-induced visible injury.

Early work by Tingey's group (Tingey et al. 1973, 1976a) showed that phenylalanine ammonia lyase (PAL) and some other enzymes which have been shown to be induced by pathogen attack, such as polyamine oxidase (Ward et al. 1991), were stimulated by ozone exposure. Further, Tingey's group demonstrated that ethylene release was stimulated by ozone fumigation (Tingey et al. 1976b). The amount of ethylene produced after 24 h was very species dependent and also depended upon the external concentration of ozone. Much of this work was ignored until Gunderson and Taylor (1988, 1991) restudied ethylene release and showed that it could be responsible, to some degree, for partial stomata closure induced by ozone. Confusing this story is the observation that some isozymes of peroxidase are stimulated by wounding (Zheng and van Huystee 1992) while other isozymes of peroxidase are stimulated by ethylene (Ishige et al. 1992).

This wounding concept was expanded by Langebartels et al. (1991) by cleverly using two tobacco cultivars (Bel B and W3) with differential sensitivities to ozone. The exposure level was chosen such that the sensitive cultivar was injured while the tolerant one was not, thereby creating a wonderful control. They followed a time sequence to show that an increase in the activities of various metabolic systems followed the same order as for the pathogen attack (for a summary see Fangmeier et al. 1994; Heath 1994a).

Ethylene is derived from the amino acid, methionine, (Smith 1985; Fig. 10.1). The first committed step of this pathway is the formation of S-adenosylmethionine (SAM). SAM, however, is used in the formation of higher molecular weight homologues of polyamines, spermidine, and spermine, using the backbone of putrescine[5]. In the ethylene pathway, SAM is converted to ACC by the synthase which is then converted to ethylene by the oxidase. It is believed that the oxidase present in low amounts, but its level is stimulated by wounding or other triggers (Avni et al. 1994). On the other hand, the ACC synthase is not present constitutively, but its production undergoes a rapid increase immediately upon wounding (with a half-time of 50 min; Kim and Yang 1992). The regulation of the synthase seems to be under three sets of controls: (1) the production of its mRNA; (2) a complex cycle of protein kinase and protein phosphatase (Spanu et al. 1994); and (3) a feed-back/feed-forward regulation by the substrate/products. The kinase/phosphatase cycle seems to be regulated by Ca^{2+} (Schulman and Low 1989).

The linkage of all these pathways with ozone is interesting (see Heath 1988 for a summary; also Galston and Sawhney 1990) but somewhat tenuous at the present time. For example, polyamines have been shown to be able to scavenge free radicals in a chemical system (Drolet et al. 1986; Bors et al. 1989) and feeding polyamines through the transpirational stream to tomato leaves prevents visible injury (Ormrod and Beckerson 1986). Further, ozone has been shown to stimulate one of the enzymes leading to the formation of putrescine, i.e., arginine decarboxylase

[5]It is of interest that putrescine is a derivative of the phenylalanine ammonia lyase pathway which will be mentioned later in Section 10.2.3.1.

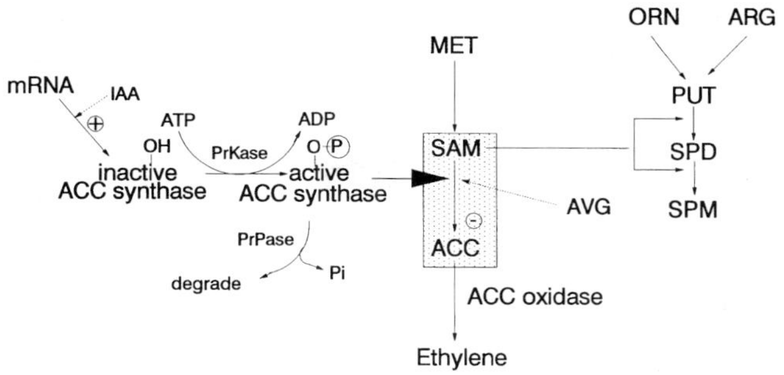

Fig. 10.1. Biochemical pathways of ethylene and polyamine formation. The pathway for the biosynthesis of ethylene is from methionine (*MET*) through S-adenosylmethionine (*SAM*) and 1-aminocyclopropane-1-carboxylate (*ACC*). Shown here is the production of an inactive form of the *ACC synthase*, which is converted into the active form by phosphorylation via a protein kinase (*PrKase*). Dephosphorylation, by a protein phosphatase (*PrPase*), seems to lead to an enzyme which is more easily degraded rather than merely an inactive one. SAM is used to add carbon and nitrogen to putrescine (*PUT*) forming the polyamines, spermidine (*SPD*) and spermine (*SPM*). Both amino acids, ornithine (*ORN*) and arginine (*ARG*) can produce PUT, but using different enzymes. The site of *AVG* inhibition is also shown (for details, see text). (Adapted from Smith 1985)

(Rowland-Bamford et al. 1989). However, as Lee and Chu (1992) have shown in rice, ethylene production (and hence its accumulation within the tissue) leads to the production of putrescine but not to the further formation of spermidine and spermine. One could argue that increased putrescine production was due to the depletion of SAM via the ethylene pathway, but the addition of ACC (at 1 m*M*) to stimulate the amount of spermidine/spermine precursor only raised the polyamine levels slightly.

The question arises whether this linkage of ethylene to ozone injury was limited to visible injury or whether the chemicals used to prevent ethylene release were also closing the stomata. Taylor et al. (1988) showed that AVG did not necessarily close stomata nor inhibit carbon assimilation per se in *Glycine max. L.* It may be that ozone generates a general wounding response much like that for a pathogen attack through a loss of membrane integrity. Loss of membrane semipermeability would cause a Ca^{2+} increase within the cell, triggering a wide variety of wound responses including the formation of ethylene. The question arises of how ethylene formation induces the visible injury with ozone.

Can the concentration of internal ethylene react efficiently with ozone? The studies of Gunderson and Taylor (1988) showed that about 1–2 ppm of ethylene induce a stomata closure (threshold of 0.12 ppm). The rates of reaction of ethylene with various components of an urban atmosphere has been measured (Finlayson-Pitts and Pitts 1986). Using these reaction coefficients and typical concentrations for measured reactive species

Table 10.2. Rates of reaction of various urban pollutant species with ethylene in the gaseous phase. k is the second-order reaction coefficient between the species and ethylene in a gas phase. The concentration of the species is a typical value in urban air, although great variation from place to place is apparent. It is assumed that the concentration of the species within the leaf is the same as in the urban air, i.e., no breakdown, and that of ethylene is 2 ppm as calculated in the text. From these numbers the rate of production of the first reaction product (see Fig. 10.2) can be calculated for a typical leaf section in terms of *M*, mol/l of leaf water. (Finlayson-Pitts and Pitts 1986)

Species	k (cm^3 mol^{-1} s^{-1})	Ref. page	Concentration (mol cm^{-3})	Ref. page number	Production (M h^{-1})
O_3	1.1×10^6	443	8.3×10^{-12}		2.6×10^{-9}
OH	5.1×10^{12}	432	1.7×10^{-18}	480	2.6×10^{-9}
NO_3	4.3×10^7	470	8.3×10^{-15}	344	1.1×10^{-10}
$O(^3P)$	4.6×10^{11}	466	1.3×10^{-19}	531	1.8×10^{-11}
$O_2(^1\Delta_g)$	1.1×10^6	480	1.7×10^{-14}	480	5.4×10^{-14}

within the atmosphere, one can calculate roughly the rates of formation of products of ethylene with each species[6]. This does not take into account other further reactions or interrelated reactions, but it does provide an estimation of the rate of formation of possible toxic compounds (see Table 10.2).

These calculations suggest that the maximum rate of formation within the gas phase of any toxicant is less than 3 nM h^{-1}. However, rates of product formation from ozone, NO_3 radicals, and hydroxyl radicals are all very similar. With the approximations used here all species could be potential reactive species. However at this rate of production, most tissue antioxidants would be expected to stop the formation of these products. Unfortunately, near the hydrated area of the surface of the cell wall, these reaction rate coefficients in the gas phase are probably not correct.

The initial products of the three potential reactive species are as shown in Fig. 10.2. All are radicals and could potentially react with many biochemicals within the cell. Most of the possible reactions end with the formation of an aldehyde (generally formaldehyde).

Polle and Junkermann (1994) focused upon one of the possible toxicants which seems inhibitory to peroxidases – hydromethyl hydroperoxide (called HMHP with a structure of $HOCH_2$–OOH). This compound could be formed from the precursor radical produced by ozone and ethylene (see Fig. 10.2). The breakdown of this radical produces formaldehyde (H_2CO) within the cell wall. Hydrogen peroxide is produced in

[6]For these calculations, the amount of internal ethylene under the wounding response is assumed to be 2 ppm for gaseous phase reactions, higher than the amount which generated a stomata response (see Gunderson and Taylor 1988).

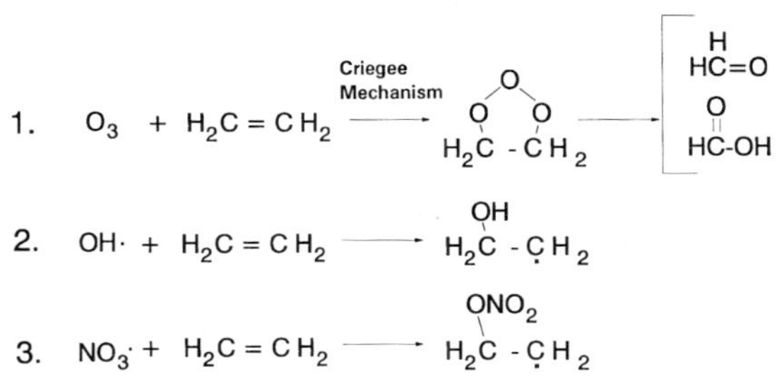

Fig. 10.2. Reactions of common air oxidants with ethylene. The formation of each radical leads to very diverse reactions sequences. The Criegee Mechanism for ozone attack on a double bond is the *top reaction* and a typical breakdown resulting from a cleavage of the double bond and the addition of oxygen to the carbons is shown. For more details see the text. For the kinetics of the reactive species in urban air, see Table 10.2. (Finlayson-Pitts and Pitts 1986)

the wall by biochemical processes (e.g., lignin production). H_2O_2+H_2CO are the reactants which produce HMHP. Thus, the possibility exists of the production of such a molecule in a region near sensitive peroxidases.

Again how much can be produced with the amounts of precursors present within the cell? It can be calculated that the production rate of the precursor radical would be about 3×10^{-9} *M* h^{-1} within the tissue. If that was the rate of formaldehyde production and the amount of H_2O_2 was in excess, the production of HMHP would be expected to be about 3 n*M* h^{-1}. For a 50% reduction in peroxidase, the level of HMHP must be about 200 μM (Polle and Junkermann 1994). To produce this amount would require a very long time.

10.2.2.2 Ionic Flows Maintained by the Plasma Membrane

The plasma membrane plays a critical role in the production of ozone injury (Heath et al. 1974; Heath 1980, 1988, 1994a; Heath and Castillo 1988). There have been two principle types of material used for these studies – alga and green plants. The algae were exposed to very high levels of ozone (within the gas stream); however, as Heath (1996a) argues, the membrane surface of algae were exposed to approximately the same amount of ozone as the membrane surface of the cells within the leaf. Early work by Evans and Ting (1973, 1974) showed a decline in metabolite and ion movements into the leaf disc (postfumigation), but they used tissue that often had a small amount of visible injury (10–20% of water-logging spots). The later work from Heath's laboratory used tissue that demonstrated no visible injury, even though it was subjected to an acute exposure at high levels of ozone as used in some field studies.

The principal features of the argument is that ozone reacts with a few critical sulfhydryl or ring amino acids on proteins (since they are so reactive in a water environment; Mudd, 1996). One such sulfhydryl on K^+-stimulated ATPase on the plasma membrane has been demonstrated to be sensitive to ozone in vivo (Dominy and Heath 1985) and in vitro (Heath and Castillo 1988). More importantly Castillo and Heath (1990) demonstrated that the plasma membrane Ca^{2+}-transport system (efflux pump and influx permeability) were altered by ozone exposure in vivo (the pump was inhibited and influx increased). Heath (1987) also argued that the breakdown of ozone under some conditions can produce H^+ and so also alter the wall space pH. The loss of K^+ from and the gain of Ca^{2+} into the cell's interior are coupled with an increased amount of K^+ outside the cell[7], and a shift in apoplastic pH. These ionic changes increase the loss of a wide range of metabolites, induce enzyme activation, and alter normal gene transcription.

Also the stomata function would be greatly altered by the shifts in ionic balance (Gilroy et al. 1991). Atkinson et al. (1989, 1990) have shown most convincingly that Ca^{2+} plays an important role with respect to the sensitivity of guard cells to ABA (abscisic acid). There is a dynamic interaction between the water potential Ca^{2+} concentrations, and ABA level of guard and subsidiary cells. The actual conductance is given by an integrated response of all these parameters. The closure (and opening) of the stomata observed within different environmental systems with ozone exposure may simply represent a shift in sensitivity of the guard cells to ABA with the existing water potential and Ca^{2+} concentrations. There would be no change in the level of any of the parameters, merely the movement of some of the apoplastic Ca^{2+} into the cytoplasm, altering how the cells respond to the existing level of ABA (see Davies et al. 1994).

Thus, ozone interactions at the plasma membrane even without the production of toxic intermediates should be able to change the ionic balances within the cell. Certainly a raise in internal Ca^{2+} induced by the changes in the Ca^{2+} transport system and a pH change in the wall space would trigger a conductance change (Paterson et al. 1989) and a series of responses by the cell (Evans 1988; Alkon and Rasmussen 1988). Unfortunately, no one has yet measured changes in the cellular concentration of these ions upon ozone exposure except in algal cells (summarized in Heath 1987).

[7]The amount of Ca^{2+} is high in the apoplastic space and even a relatively large increase in influx would not be expected to alter the net amount externally.

10.2.3 Responses of Metabolic Processes

The activity of enzymes control the speed of the transformation of various processes[8]. Those enzymes are degraded by several known processes (Ciechanover and Schwartz 1994; Hershko 1991). To maintain activity, they must be continually synthesized through the transcription of DNA to produce mRNA, the translation of which produces the enzymes. Transcription is tightly controlled by transcription factors (Karin and Smeal, 1992). Each of these steps is controlled by a combination of environmental and hormonal signals and by the levels of other metabolites.

10.2.3.1 Wounding

A common factor in ozone stress, mechanical wounding, and pathogen attack is the rapid production of ethylene. In fact, much of the more recently discovered biochemical symptomatology of ozone exposure resembles those events following pathogen attack. We are still not sure of how far to carry the analogy since pathogen attack generates hypersensitivity and molecules which seem to be carried throughout the plant to signal the attack in other tissues, but the sequence is striking (Ecker and Davis 1987; Eyal and Fluhr 1991; Ward et al. 1991; Schraudner et al. 1992). As discussed in the chapter by Langebartels et al. (this Vol.) the production of many of these enzymes upon ozone exposure occurs in a wide variety of plant material. Wounding, by mechanical injury or pathogens, and ethylene production are linked by these defensive enzymes.

Diverse metabolic processes can be altered through a cascade effect which begins at a simple fundamental level. The best example is in animal systems in which the epinephrine triggers a series of reactions which activates more steps at each level; this series involves cyclic-AMP and a protein kinase/phosphatase[9]. There are several newly found kinase/phosphatase activation systems in plants (Cohen 1992), which may be responsible for some of the metabolic imbalances which seem to be caused by ozone exposure (described above for ethylene, see Fig. 10.1). There are several superfamilies of protein kinase proteins (Stone and

[8]We often think of pathways rather than processes. One process which is critical for normal operation of the cell is the efflux pump for Ca^{2+} of the plasma membrane. Although not a pathway, the operation of the pump is controlled by metabolic regulators.

[9]A protein kinase (PrKase) is simply an enzyme which phosphorylates proteins, including enzymes, thereby activating or deactivating the latter. The related phosphatase (PrPase) is the kinase's counterpart; the phosphate group on the enzyme is removed by the phosphatase thereby deactivating or activating it.

Walker 1995), several of which have been implicated in ethylene production and wounding in plants. The two which are directly related to ethylene are ETR1 (an ethylene sensing element, which is a histidine PrKase) and CTR1 (which is related to Raf family of PrKases whose structure has seven transmembrane domains and acts to inhibit the ethylene signal transduction). The third type is the SNF-1 (which is a member of the calcium/calmodulin-dependent PrKase, CaMK family of serine/threonine PrKases). SNF-1 has a role in carbon metabolism and seems to be regulated by environmental stimuli, including light, ABA, and water stress.

Mehdy (1994) recently described a model of how an elicitor produced by a pathogen attack activates a G-protein, which opens the inward flowing Ca^{2+} channel. The flow of Ca^{2+} into the cytoplasm raises the internal level (at the μM level) and so activates a PrKase, which increases the activity of the plasma membrane NAD(P)H oxidase generating superoxide (O_2^-). Superoxide dismutase converts this into H_2O_2; this is responsible for the active oxygen species response as a defense mechanism. A branched system acts at the secondary message level to change the production of defense gene products through transcription factors.

It must be noted that any increase in the passive inward flow of Ca^{2+} could lead to the same result as activation of the G-protein. The level of cytoplasmic Ca^{2+} would rise and all else would follow. Since pathogenic wounding is believed to be caused by the elicitor altering the exclusion of Ca^{2+} from the cell, the loss of Ca^{2+} exclusion by any other means including ozone attack of the membrane would trigger the same responses (Langebartels et al. 1991; Schraudner et al. 1992). It remains a difficult problem to show that Ca^{2+} loss is the first event (but see Castillo and Heath 1990). In retrospect, it is surprising that an earlier correlation was not made between ozone-induced wounding and pathogen attack.

10.2.3.2 Stress-Induced Alterations in Gene Expression

Many abiotic stresses have been described by techniques of molecular biology. In most cases stress causes alterations in protein synthesis and a class of proteins appears. However, when the levels of certain proteins change only slightly, examination of mRNA more readily indicates an alteration of gene expression (Sachs and Ho 1986). In a number of stress responses, changes in protein synthesis can be traced back to regulation of DNA transcription.

In recent years, there has been a great deal of progress in identifying genes that are induced by several environmental stresses, including water deficit (Dure 1993), high temperature (Vierling 1991), and hypoxic stress (Bailey-Serres et al. 1989). The results from these investigations indicate that the different types of environmental stresses produce changes in gene expression which are directed towards repair of a particular injury.

Air Pollution-Induced Alterations in Gene Expression. In comparison to other abiotic stresses, relatively few studies have addressed the qualitative and quantitative effects of air pollution on protein metabolism (summarized in Harris and Bailey-Serres 1994). The few reports that have appeared, however, suggest that the physiological and metabolic consequences of exposure to pollutants may, in part, be mediated by altered gene expression. Air pollution-induced alterations in protein patterns have been reported for spruce (*Picea abies L.*) chronically exposed to ozone and acid mist (Schmitt and Sandermann 1990), pea (*Pisum sativum L.*) acutely exposed to ozone (Beckett et al. 1989, 1990), wheat (*Triticum aestivum*) exposed to ozone (Farage et al. 1991; Nie et al. 1993), and spruce exposed to ozone (Kärenlampi et al. 1994). The altered expression of these proteins may be of adaptive significance for plant accommodation to ozone stress (McCool et al. 1988).

The use of *Arabidopsis thaliana* as the plant of choice for molecular biology has spread in most fields. Two major groups have begun to study *A. thaliana* under conditions of ozone exposure. While the plant has no commercial value and requires a high level of ozone to induce changes, it possesses a wide range of mutants and nucleic acid probes for other physiological processes. Thus, it may be an excellent model to investigate the mechanisms of ozone injury. Rao et al. (1996) found ozone-enhanced Cu-Zn superoxide dismutase (SOD), peroxidase, glutathione reductase, and ascorbate peroxidase activities (at 200 nl l^{-1} ozone for 6 h/day for 8 days). Sharma and Davis (1994) described similar effects in addition to the production of mRNA for glutathione S-transferase and PAL (at 300 nl l^{-1} ozone for 6 h/day for 14 days). In addition Sharma and Davis (1995) found a novel 8.6 kDa basic protein with a structure that suggested a signal peptide after ozone exposure. These very active investigations are being continued. (K. Davis, pers. comm.).

Antioxidant Changes. One of the other vigorous, current hypotheses regarding ozone resistance/sensitivity concerns the use of antioxidants (SOD, ascorbate, tocopherol) to prevent formation of noxious oxidative products or to regenerate oxidized groups on critical membrane components (e.g., glutathione, GSH). Although these compounds are often present naturally, ozone exposure increases their level in some cases.

Willekens et al. (1994) have looked at Northern blots of varied messages for antioxidant enzymes in tobacco after exposure to ozone (maximum of 0.12 ppm) for 14 h/day. Initially there was no visible injury, but CAT[10] 1, 2, and 3 changed (CAT1 was inhibited after 9 h; CAT 2 was increased after 9 h, while CAT 3 took several days to fully increase). The authors suggested that the inhibition of CAT1 was due to a decline in photorespiration. However, it is clear that most of these CATs represent cytoplasmic catalase rather than apoplastic space activities. The message for glutathione peroxidase rose only after a day, but there was little apparent change in SOD (although it was argued that the Fe-SOD[11] declined while the cytoplasmic Cu/Zn-SOD increased upon the appearance of visible injury). Glutathione peroxidase activity is thought to be linked to the repair of lipid peroxides; yet given the uncertainty of peroxidases within plants it may be premature to classify this. Certainly, other enzymes involved in the glutathione pathway are stimulated (Sharma and Davis 1994; Rao et al. 1996; for a summary, see Hess 1994). The interesting point is the small, if any, change noted in SOD gene activity. The level of the messages seem to be altered only after long exposure times and the formation of visible injury. In this case only a very small role of increased production of SOD is suggested for ozone exposure.

Along these same lines when Willekens et al. (1994) elevated SOD levels by generating transgenic tobacco plants, only a greatly elevated level of the chloroplast SOD induced a significant reduction of visible injury. The location of this useful insertion suggests that it is not critical at the initial site of injury but rather that the initial mechanism of injury generates conditions that make photooxidation within the chloroplasts more probable against which excess antioxidants are useful (Karpinski et al. 1993).

Although developmental changes are recognized, they tend to be overlooked. How developmental stage may affect ozone-induced visible injury has been summarized by Heath (1994a). The leaf is most sensitive to ozone at a particular age (often in that period just after the maximum rate of leaf expansion). That biochemical events may be influenced by ozone in different manners depending upon the tissues' developmental stage seems obvious. Casano et al. (1994) have shown that the rate of production of messages for SOD isozymes depends upon leaf age in barley. The response of the leaf to oxidative stress also varies; the younger have a greater capacity to increase their antioxidant reactions to a stress. Yet Casano et al.

[10]CAT stands for the mRNA of an isozyme of catalase.

[11]Varied SOD possess varied metals at their active site; thus, Fe-SOD has iron at its activity site.

(1994) found that the capacity to produce Fe-SOD within the chloroplast seems to remain high throughout all developmental stages.

The role of antioxidants is not especially clear for an ozone exposure; the results of research in this area are conflicting. It may well be that in some cases the existing levels of antioxidants are sufficient to protect. In this regard it is interesting to examine in more detail three papers.

Polle et al. (1995) quantified the amount of ozone which entered Norway spruce needles and the loss of ascorbate within the apoplastic space. Their data showed quite clearly that if either the total cellular ascorbate or thiols (within the full needle sample) were measured, there was very little change in the amount of ascorbate or thiols up to a dose of about 1.6 μmol ozone g^{-1} FW (fresh weight). After this threshold was passed, the amount of total ascorbate rose from 12 to about 19 μmol g^{-1}FW (about 13%) for a doubling of the amount of effective dose. In the apoplastic space, the ascorbate levels nearly doubled. However the amount of total cellular thiols fell to about 0.14 μmol g^{-1} FW (from about 0.55 μmol g^{-1}FW).

Gupta et al. (1991) found that poplar seedlings exposed to 180 nl l^{-1} ozone had no change in the SOD level (all species) but showed a rise in cellular levels of total glutathione, with a pronounced increase in reduced to oxidize ratio. The change began after about 15 min of exposure. Yet there was little change in the reduced/oxidized ratios or concentration of nucleotides (NAD and NADP). In these experiments no stomatal conductance change was noted, but the rate of photosynthesis declined (to about 45–50%). Rao et al. (1996) likewise found that ozone exposure (120 nl l^{-1}, 8 h/day for 3 weeks) increased SOD, peroxidases, glutathione reductase, and ascorbate peroxidase initially in wheat, but longer exposures (5 weeks) caused a general decrease in all soluble proteins.

On the other hand, Benes et al. (1995) found little change of SOD, ascorbate peroxidase, or guaiacol-dependent peroxidase (an extracellular peroxidase) when ponderosa pine trees were grown in the continuous presence of low amounts of ozone. Their data are detailed, but it is of interest to examine one set on needles harvested in September (Table 10.3). Here the level of SOD is slightly (30%) lower in current-year needles of a sensitive clone. The level of ascorbate peroxidase drops in the previous-year needles of a sensitive clone, while the extracellular peroxidase rises in these needles. While the results seem unclear, the levels of the enzymes are very high and could be sufficient to detoxify the high amounts of ozone entering into the needles (peak heights of 2×70 nl l^{-1} for a very high conductance of 0.01 m s^{-1} yields an entry rate of about 20 nmol g^{-1}FW s^{-1}). Thus, it may be that the amount of total enzyme within the cell was in excess of that needed to detoxify ozone and its products.

Table 10.3. Cellular levels of antioxidant enzymes found within needles of ponderosa pines during September. Tolerant and Sensitive refer to clones of pines which vary in their sensitivity to ozone exposure as judged by visible needle injury and early loss of the needles. (Benes et al. 1995)

Enzyme	Needle Class	Tolerant	Sensitive
Superoxide Dismutase	Current-year	180	120
(Units g^{-1} FW)	Previous-year	180	185
Ascorbate Peroxidase	Current-year	0.18	0.26
(μmol min^{-1} g^{-1} FW)	Previous-year	0.80	0.18
Guaiacol-Peroxidase	Current-year	20	22
(ΔA_{450} min^{-1} g^{-1} FW)	Previous-year	85	40

Finally the penetration of ozone into the leaf tissue and the rate of ascorbate loss has been measured by Luwe et al. (1993) in spinach leaves. They measured ozone uptake on the leaf surface of $8.8 \pm 0.3 \times 10^{-13}$ mol cm^{-1} s^{-1}. Yet they found a loss of ascorbate of about 6.0×10^{-8} mol cm^{-3} h^{-1}, within the apoplastic space of the leaf tissue. With some approximations for the apoplastic space per leaf area, this corresponds to 4.5×10^{-14} mol cm^{-1} s^{-1}, about 5% of the amount of ozone which is exposed to the leaf surface. This is a good correlation for a series of measurements and calculations based upon several difficult but reasonable estimations of leaf structure. They also found a decline in glutathione, but only after 6 h following fumigation with 300 nl l^{-1} ozone; this decline occurred after the loss of extracellular ascorbate. Interestingly, the concentration of total ascorbate varied inversely with the size of the leaf (a measure of leaf age). The older leaves had two to three-fold less ascorbate than younger leaves. Finally they found that the development of necrosis varied inversely with the ratio of ascorbate to total ascorbate species (ascorbate and dehydroascorbate).

This would indicate that the amount of reducing power contained in the extracellular ascorbate can eliminate the ozone in that space. The ability for reducing power to maintain ascorbate in the reduced form is sufficient. Once that effective dose is exceeded, then the level of external ascorbate rises along with the internal ascorbate, yet the reduction state of thiol inside the cells begins to suffer. Further it is critical when the tissue is sampled. The relative amounts of varied reducing powers in varied spaces is dependent upon the effective dose delivered to the tissue.

Alteration of mRNA for the Coding of RUBISCO. Pell et al. (1994) recently summarized the findings on how ozone exposure leads to a decline in RUBISCO (ribulose 1,5-bisphosphate carboxylase/oxygenase). Fundamentally, an ozone exposure of several days induces a decline in the total RUBISCO concentration. However, exposure does not seem to alter the activation of the RUBISCO present. This decline was much more obvious in older leaves (Dann and Pell 1989). Eckardt and Pell (1994) traced this to a loss of mRNA of both subunits (*rbc*S and *rbc*L). The decline in the mRNAs was rapid but somewhat reversible over a day or so. While the inhibition of mRNA production by ozone exposure was rapid, RUBISCO is present in such large amounts within leaf tissue that it required days to clearly see the loss of the enzyme. The largest effect seemed to be upon older foliage.

The level of messages can be reduced by a loss of transcription factors which are necessary for transcription or by a more rapid degradation of the message. Since *rbc*L (encoded by the chloroplast DNA) seems to be closely coordinated with *rbc*S (encoded by the nuclear DNA), a loss of *rbc*S would tend to lead to a loss of *rbc*L and so the transcription of *rbc*S is currently an active research area. There are at least three to six different DNA regions which code for *rbc*S and, unfortunately, each seems to be regulated by different events (Peters and Silverthorne 1995). Recently, Pell (unpublished) has suggested that the stability of the message, rather than its transcription, is altered by the exposure.

How could ozone which does not penetrate deeply into the cell alter any message? Three distinct possible events may partially answer this (1) one of ozone's decomposition or reaction products diffuses to the nucleus to interfere with DNA transcription events or the stability of the message; (2) the ozone-induced loss of membrane integrity leading to ionic imbalances causes the triggering of new cellular events, including a reduction in DNA transcription; (3) ozone reacts with a membrane-linked sensor complex, generating a signal at the plasma membrane which then gives rise to a series of events leading to a decline in transcription. Yet this series of reactions is not due to the sensor complex reacting with its proper message and so the loss of transcription, while metabolically proper from the complex to the DNA, is falsely triggered.

There is an influence of the level of carbohydrate within the leaf upon the amount of *rbc*S mRNA. Krapp et al. (1993) describe four methods which have been used to alter the carbohydrate level (1) the removal of a sink organ leads to an inhibition of photosynthesis at the source; (2) an increased level of CO_2 leads to a transitory increase in photosynthesis, after which the assimilation rate returns to that measured previously; (3) using transformed cells to prevent sucrose export (the addition of invertase to the cell wall) which leads to an increase in sugar content and a decline in photosynthesis; (4) feed sugars through the transpirational stream which leads to a decline in photosynthesis. While this decline may be due to the production of control metabolites to alter important sugar production pathways, such as fructose 2,6-bisphosphate, there are indications that the RUBISCO level is directly altered by a reduction in *rbc*S.

Krapp et al. (1993) also measured the half-life for *rbc*S (about 2 days[12] when 50 mM glucose is added to a cell suspension of *Chenopodium*. In 2 days the amounts of sugars were greatly increased (fructose, seven times; glucose 24 times; sucrose 2.4 times) from a total carbohydrate concentration of 25 to nearly 150 μmol g^{-1}FW. The starch level indicated a similar increase (10 to 70 μmol g^{-1}FW). Interestingly the levels of fructose bis-phosphatase and glyceraldehyde phosphate dehydrogenase were also depressed by these feedings, suggesting an integrated control. Similar results were found using less glucose with a higher concentration of cells, but the rate of increase of carbohydrate was much faster (reaching a peak at about 20 h) and the *rbcS* message declined faster (half-life of about 4 h).

Using intact plants the carbohydrate level was increased by cold girdling the petioles of tobacco and potato. The levels of carbohydrate nearly doubled in 5 days, but showed most of the increase within 12 h. The levels of *rbcS* declined rapidly (reaching 25% after 12 h); however, the half-life was not determined. A reduction in RUBISCO followed, but more slowly (with an estimated half-life of about 108 h after a lag period of at least 12 h). This of course is what would be expected – the level of the enzyme would decline slowly with a lag after a loss of the message.

A better estimation of the half-life of *rbcS* mRNA can be found in the paper of Jiang et al. (1993) who studied the destabilization of the message antisense message. The wild type message in tobacco had a half-life of about 5 h compared to that in the mutant with the antisense RNA (half-life of about 1 h). It was argued that the antisense RNA caused an increase in the degradation of the normal *rbc*S mRNA. An estimation of the half-life of *rbcS* mRNA upon ozone fumigation seems to be in the order of 1 h (Eckardt and Pell 1994; Heath, unpubl.). Although comparisons of these biological systems cannot be easily made, it seems evident that the normal half-life of *rbcS* mRNA may be closer to 5–10 h and that ozone is not simply stopping the transcription of DNA but also alters the rate of mRNA degradation.

Williams et al. (1994) have found a correlation between the levels of ABA after water stress in *Arabidopsis thaliana* leaves and the loss of *rbcS*. Although their data were not quantitative, upon the onset of water stress, the level of ABA increased by 50% in about 1–2 h and the level of *rbcS* underwent a 50% decrease in about 2–4 h. These data agree with the half-

[12]The amount of RUBISCO drops from 0.12 to 0.04 μmol g^{-1}FW s^{-1} in 6 days.

lives mentioned above. This work suggests that water stress may alter metabolic processes much more than merely closing the stomata. If ABA can lower *rbcS* by whatever mechanism, *rbcS* may not be very valuable as a marker of ozone changes except under very controlled conditions.

The current hypothesis regarding chloroplast protein changes is that ozone alters the properties of the plasma membrane leading to the synthesis of specific polypeptides. Specific receptor molecules or channels on the membrane could be affected, thus signaling the transcription or translation of the genes controlling those polypeptides. Once this membrane disruption occurs, the cell mobilizes its repair systems to overcome this injury; thus, carbon and energy sources destined for productivity are used for rebuilding. Some of these repairs are expected to result from and in the induction of specific genes. At the very least, the decrease in photosynthesis is a result of an ozone-induced decrease in *rbcS* mRNA.

10.2.4 Tissue Level

10.2.4.1 Photosynthesis

We began by arguing that the production of fixed carbon is critical to plant's productivity and resistance to other stresses. The question remains of how to rank the various events that may lower photosynthetic response. Visible injury results in the loss of photosynthetically active tissue, but much of that loss can be compensated for by the increased light absorption by other leaves and by preventing senescence of the functioning leaves. Loss of RUBISCO in older leaves due to a loss of rbcS (mRNA) plays a role, but the relation of loss of message to loss of productivity is hard to predict. Also there are other events induced by ozone (e.g., wounding) which may play a role in the loss of productivity of a plant, and they could be more important in the long run.

Saxe (1991) summarized what is known about changes in photosynthesis upon exposure to various levels of air pollutants. For ozone, the effect on photosynthesis of long-term exposure at low levels is not very clear. One can find increases, no effects, and decreases for various conditions. The majority of the data suggest that photosynthesis can be lowered without a concurrent fall in stomata conductance. Most data suggest that there is a reduced ability to fix CO_2 and the simplest explanation is a lowered amount of RUBISCO. Although Reich and Lassoie (1984) suggest that the stomata do not respond normally and become "sluggish", Heath (1994a) has suggested that the guard cells, being closest to the stomata, are subject to the maximum concentration of ozone and so should be most affected, which is similar to what Moldau et al. (1990) have observed.

First of all, the photosynthetic productivity of a full plant should be measured, not just several individual leaves. Also, while the measurement of photosynthesis in a single leaf is technically easy, it is difficult to evaluate the actual mechanistic events leading to changes in the rate. The ability to assimilate carbon via the Calvin cycle (given as A) is linked to the availability of both energy and reducing power through the light reactions and the internal concentration of carbon dioxide (given as C_i) maintained by its movement through stomata (given as g_s). The A-C_i curve indicates the problem with various types of measurements which must be made (see Heath 1996a). The sequence of events begins with the leaf at a stable state in which both A and g_s have been balanced for that particular environmental state. It must be noted that A is a typical enzyme-dependent element; the maximum rate of A is proportional to the amount of RUBISCO present (and activated) linked to the energy produced by electron transport and K_m is measured for C_i.

If g_s is not directly altered but A declines due to an ozone effect, the trajectory on the A-C_i curve is one in which as A declines, the C_i is kept constant by a forced lowering of g_s, according to the model by Farquhar and Sharkey (1982). On the other hand, if g_s cannot adjust, C_i rises with the decline in A. If the decline in A is due to only a closing of the stomata (lowered g_s with no changing RUBISCO), the trajectory follows the A-C_i curve forcing a decline in A since C_i is maintained by the smaller g_s. Thus, a simple measurement of A or g_s will not yield a meaningful description of a mechanism. The experimenter must find all three values for each case (A, g_s, and C_i) and how they change after exposure.

Interestingly, all three cases can be found in the literature. Heath (1994b) showed that when light was limiting photosynthesis, the stomata conductance would decline with ozone fumigation of bean plants by nearly 50% before any limitation of CO_2 fixation would occur since the conductance was not limiting. Farage et al. (1991) showed that the C_i (given in relation to the external concentration) rose slightly as the stomata conductance declined in wheat subject to high levels of ozone. However, they also found that the A_{sat} (maximum assimilation rate proportional to RUBISCO) also declined. For lower levels of exposure in Thompson seedless grapes, Roper and Williams (1989) showed that ozone induced a clear lowering of the full A-C_i curve, suggesting an increase in respiration and a lowering of A_{sat}.

Less data are available for tress; however Tjoelker et al. (1995) examined the gradient of ozone and light within a forest canopy and made several interesting observations. Long periods of ozone exposure decoupled assimilation from stomata conductance, in that, A varied linearly with g_s after 18 days of exposure, but A did not change as g_s rose after 68 days of exposure (see Tables 10.4 and 10.5). Further, we calculated the accumulated dose in an 8-h day, taking the 1-h mean of external ozone, and multiplying it by the stomata conductance. The average internal ozone rate of delivery is about the same for each exposure time, with that for high light intensity being only slightly higher than that for low light intensity. Using their average inhibition of CO_2 assimilation data, we tabulated the inhibition of assimilation due to accumulated dose, as shown in Fig. 10.3. The same curve shape is observed with the daily 1-h maximum, except that the accumulated dose is about 41% higher. The curve shows that the inhibition of CO_2 assimilation is directly proportional to the total accumulated dose, with a 50% inhibition being reached at 17 mmol m^{-2} (corresponding to approximately to 50 μmol g^{-1} FW of tissue, over 50–60 days).

Table 10.4. Calculated rates of ozone flux into the leaves of sugar maple. Data were calculated as described in the text from data of Tjoelker et al. (1995). Low light average was for 0.67 to 0.93 mol m^{-2} s^{-1} and the high light average was for 6.9 to 14.5 mol m^{-2} s^{-1}

Light intensity	Internal rate of ozone delivery (nmol $m^{-2}s^{-1}$)		
	18days	43 days	68 days
Low	3.39 ± 0.41	4.34 ± 0.52	3.49 ± 1.06
High	4.46 ± 1.23	4.95 ± 0.83	3.52 ± 0.23

Table 10.5. Carbon assimilation and stomata conductance of maple exposed to ozone for various time intervals. As little or no change was noted in the average conductance over all light intensities, the conductances were averaged. The assimilation rates were taken from data regarding near light saturation (i.e., high light)

	18 days	43 days	68 days
Conductance (mmol m^{-2} s^{-1})			
Control	68.7 ± 16.6	99.9 ± 14.9	60.6 ± 20.5
Ozone	58.2 ± 18.1	76.0 ± 13.3	56.3 ± 16.8
Assimilation (μmol m^{-2} s^{-1})			
Control	3.91 ± 0.93	4.26 ± 1.42	3.46 ± 0.80
Ozone	2.99 ± 1.10	2.28 ± 1.07	1.60 ± 0.64

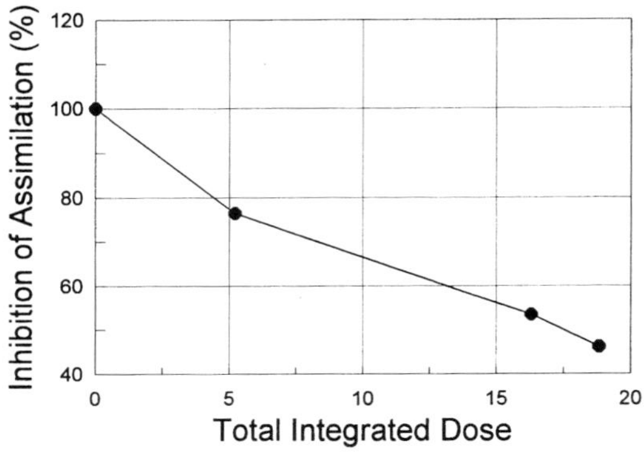

Fig. 10.3. Dose response of the inhibition of CO_2 assimilation in sugar maple. The accumulated dose of ozone (measured in mmol ozone absorbed by a m^2 leaf area during each exposure period) was calculated as described in the text and in Table 10.4. The control level of assimilation (100%) is given in Table 10.5. (Tjoelker et al. 1995)

10.2.4.2 Photoinhibition

Photoinhibition occurs when the rate of CO_2 fixation is reduced without reducing light capture (Osmond 1981; see Fig. 10.4). In most cases investigated, the link with the photosystems is through the rise in reduced plastoquinone (PQ) between photosystem I and II (see Barber and Andersson 1992). This lowers the amount of Q_A, which is the primary acceptor of reaction center (RC) II electron. The lack of Q_A then leads to a build up of equivalents within RC II [notably, pheophytin and the chlorophyll trap ($P_{680}{}^+$)]. $P_{680}{}^+$ can generate the triplet state of this chlorophyll which can give rise to singlet O_2 if the ß-carotenes cannot quench the triplet. This condition, however, can be interrupted by the lost of D_1 protein, which acts as a "fuse" for the continued electron transfer (in a pathological state), by allowing the oxidation of PQ. D_1 is a 32 kDa protein, which rapidly turns over during photoinhibition, producing 23, 16, and 10 kDa fragments[13].

Ozone injury which leads to chlorosis fits into this sequence of events well, although with some problems. (1) Much of visible injury is initially due to the loss of chlorophyll. Even without a total bleaching, chlorophyll loss is a sensitive indicator of ozone exposure. (2) While much of the chlorophyll fluorescence studies have been done under conditions of high

[13] Interestingly, Nie et al. (1993) have shown that there are some new peptides formed after ozone exposure of the sensitive tip regions of wheat. Although it is difficult to reliably obtain molecular weights from their 2-D gels, there are three new peptides in the range of 10–25 kDa.

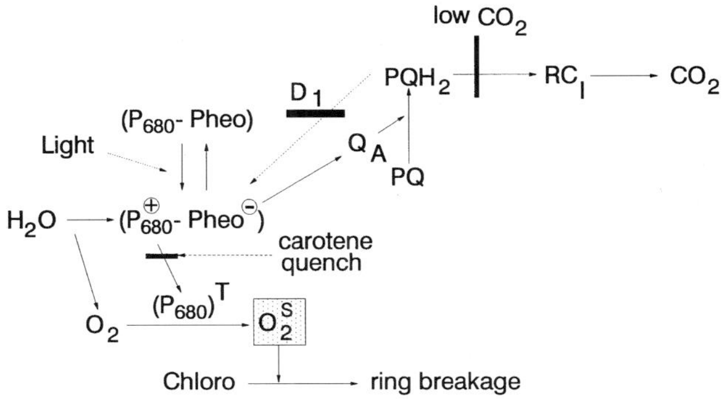

Fig. 10.4. The postulated mechanism of photoinhibition and photobleaching within the light reactions of photosynthesis. The reaction center of photosystem II is made up of a dimer of P_{680} (a specialized chlorophyll molecule) and pheophytin (*Pheo*). The electron to reduce the oxidized P_{680} molecule is ultimately derived from water oxidation (which generates molecular oxygen). The excess electron on Pheo is transfered to plastoquinone (*PQ*) through the primary acceptor (Q_A); from there it moves to reaction center I and ultimately reduces NADP. The 32 kDa protein (D_1) acts to prevent over reduction of PQ if normal electron flow is reduced, observed in general by a rapid turn over. It is believed that with overreduction of PQ,P_{680} is transformed into a triplet state which converts O_2 to the singlet O_2 (O_2^S). It is singlet O_2 which chemically attacks bulk chlorophylls, breaking their ring structure and bleaching their absorption. Also normally β-carotene acts as an antioxidant and chemically quenches the triplet of P_{680}. (Barber and Andersson 1992)

ozone levels (see Heath, 1994, 1996, for a summary), there is a pronounced loss of variable fluorescence observed under a variety of conditions. The current terminology is a decline in F_v/F_m (where F_m is the maximum peak of fluorescence induced by a pulse of light after a period of dark adaptation and F_v is the maximum peak less the fluorescence without any light pulse (Aro et al. 1993; Nie et al. 1993). (3) Furthermore while Nie et al. (1993) did not observe a loss of D_1 like protein in 2-D gels upon exposure of wheat to ozone, Pino et al. (1995) did in corn. They used 1-D gels to show a loss of a 32-kDa protein, and subsequently using antibodies to D_1 and a pulse labeling experiment they convincingly showed an increased turnover of D_1 after fumigation.

Although more work needs to be done regarding the time period and exposure indices of ozone required to trigger photoinhibition, it looks like this may be the initial cause for chlorophyll loss by ozone exposure. While chlorophyll is a sensitive molecule in vitro, when it is structurally arranged in the thylakoid structure it is very resistant to oxidation and other chemical changes (Heath 1989). It is of interest that few researchers pondered over why chlorophyll should be lost after ozone exposure. It would seem that the best current hypothesis is that a lack of ability to fix CO_2 would lead to an over-reduction of PQ which triggers photoinhibition. In the absence of effective antioxidant-like behavior (e.g., functioning β-carotenes and D_1 turnover), the triplet state of the RCII attacks the bulk chlorophyll, leading to a ring breakage and bleaching of the chlorophyll molecule. Furthermore, chlorosis is not a primary result of ozone exposure, but rather a secondary effect due to impaired photosynthetic capacity. In other words, no toxicant flows into the chloroplasts to induce injury.

10.2.4.3 Ozone Injury and Carbohydrate Status

Studies on carbohydrates were begun early in the investigations of ozone effects on plants (summarized by Dugger and Ting 1970a). Sucrose applied to leaves resulted in some protection, yet leaves stored in the dark were not susceptible to ozone but could be made so by application of sugar. Ozone susceptibility was evident when the level of sugars was between 1 and 4 mg g^{-1} FW for pinto bean, and between 5 and 14 mg g^{-1} DW for tobacco; higher or lower levels removed the susceptibility. It was believed that higher concentrations of sugars induced an osmotic effect which closed the stomata. It was not known why lower levels protected. Interestingly, the ozone-resistant variety of tobacco (Bel-B) contained less sugar than the susceptible variety (Bel-W3).

In citrus after several weeks of ozone exposure, the soluble sugars increased within the leaf while the starch declined. In pine, the levels of ethanol-soluble sugars and polysaccharides declined after long periods of ozone fumigation. In pinto bean the levels of starch and sugars roughly directly and inversely correlated with leaf damage using leaf age as the variable. The changes in sugars after fumigation or levels of sugars before fumigation have been correlated with visible injury (results summarized in Dugger and Ting 1970b).

Stomata conductance and CO_2 assimilation are affected by ozone exposure which lessens the ability to efficiently assimilate carbon (Lee 1965; Ehleringer 1991). It follows that the levels of sugars (both soluble and starch) should be reduced. The questions arise of (1) whether the ability to move the carbohydrates within the cell is altered (e.g., inability of moving triose phosphate out of the chloroplasts so that more starch is made in the chloroplast, but less sucrose can then be made in the cytoplasm), and/or [2] whether translocation of the sugars out of the leaf is inhibited, and by what mechanisms do these events occur.

10.2.4.4 Translocation

The loss of productivity is often associated with a loss of photosynthetic assimilation. While that may be true, often the photosynthetic decline is less than the decline in productivity. Certainly the accumulation of carbon is only the first step to productivity. There are many levels at which the energy accumulated may be used before primary production. Indeed the first step of carbon dioxide fixation results in the formation of triose phosphate (TP) which can be transformed into glucose phosphate and then into starch to be stored within the chloroplast. In order for the carbon to be transported to the cytoplasm, the TP must be exchanged through the envelope by a transporter with cytoplasmic inorganic phosphate (P_i).

This step is well regulated; early in the morning most of the TP is transported to the cytoplasm. Only later in the day does accumulation of starch within the chloroplast occur. Stored starch can be used for respiration within the cell which had earlier produced the fixed carbon. The TP in the cytoplasm can be used for normal metabolic events or transformed into sucrose (through a series of highly regulated steps). In fact, the use of transgenic tobacco plants which have altered amounts of the translocator has shown that the distribution of assimilates between starch and sugar can be easily affected (Barnes et al. 1994). Sucrose (or a related carbohydrate) is transported to other tissues which require carbon (the growing points or sinks, such as flowers, roots, and newly developed leaves).

As seen in Table 10.6, the inhibition of the photosynthetic rate by ozone is greater in the primary leaves of bean than in the trifoliate. However, the primary leaf retains more assimilate (as ^{14}C) while less is retained in the trifoliate upon ozone exposure. Yet the plants' total productivity is lowered by ozone exposure. It is not clear how we can describe these very diverse events which vary with leaf age.

Another interesting development along this same line involves the use of EDU to protect plants from the development of visible injury (see Kostka-Rick and Manning 1993, for an excellent review). It is not known how this compound induces protection, but its use has been found to be associated with a rise in soluble sugars, with the leaves indicating a change in translocation. It has been observed that the level of superoxide dismutase is not affected by EDU (Pitcher et al. 1992). However, EDU does change the amount of ethylene production (Zilinskas et al. 1990), again suggesting a tight linkage between ethylene production and visible injury. More important is the linkage of this observation with the earlier work of Dugger and Ting (1970b) who showed that feeding leaves with

Table 10.6. Carbon fixation and allocation due to ozone fumigation. *Phaseolus vulgaris* L. (bush bean cultivars) were exposed to filtered air (control) or 0.2 ppm ozone for 1 week. The analysis given below is from the first day of the second week of exposure. The data given here are significantly different (at a 5 or 1% level) between ozone and control. The original paper has much data at a lower level of ozone. (McLaughlin and McConathy 1983)

Allocation indices	Control (%)	Ozone (%)
Photosynthesis:		
Primary leaves	100	41
Trifoliate leaves	100	61
Leaf retention of ^{14}C:		
Primary	0.34	0.45
Trifoliate	0.36	0.25
Total plant dry weight at harvest (g)	18.0	13.7

sugars lead to a protection from visible injury. One might postulate that EDU leads to a decline in translocation and thus a rise in the level of soluble carbohydrate. Once the level of soluble carbohydrate rose, the visible injury was inhibited. It is still unknown how soluble carbohydrates within a tissue protect the cells from injury.

10.3 Summary of Physiological and Biochemical Events

In terms of physiology, the "leading" hypotheses for an ozone-induced alteration of plant productivity are: (1) loss of membrane integrity and transport abilities coupled to ionic imbalances; (2) production of antioxidants as a line of defense; (3) production of a signal which leads to a wounding response similar to that observed after a pathogen attack; (4) loss of photosynthetic efficiency involving both stomata closure and carbon assimilation decline due to RUBISCO loss; (e) decline in the ability of photosynthetic tissues to move carbohydrates out into the translocation stream to developing leaves and roots.

With respect to biochemistry, the membrane is altered, principally via protein changes, such that permeability, transport, and triggering mechanisms are no longer correct for the state of the cell, e.g., the rate of ion movements or sensitivities to messages are abnormal. The cellular pools of Ca^{2+}, K^{+}, and H^{+} are the primary suspects for distortion. It is doubtful that membrane fatty acids are attacked in the initial stages of exposure (to almost ambient levels); however, after cells begin to die or at extreme high levels of ozone (500 nl l^{-1}), the membranes will be at risk and such changes are observed.

The varied antioxidants (metabolites or enzymes) can eliminate the oxidant or its products if maintained in the correct abundance. Yet a too rapid oxidant entry can overwhelm the antioxidant response, as there simply is not time to resynthesize the antioxidants themselves or to increase the enzyme levels to produce them.

A very interesting hypothesis is that visible injury is caused by the interaction of ozone with stress-induced ethylene either by direct chemical transformation of ethylene to a toxic product or by alteration of the ethylene/binding site relations. This concept would place wounding (and events associated with it) in the same category as visible injury. The sequence would be no visible injury until the wounding level reaches a critical point. Wounding might be expected to produce biochemical conditions (ethylene production) that would lead to early senescence (which is observed at low exposure levels).

A product of ozone (less probably ozone itself) enters the cell and causes a decline in the mRNA for RUBISCO (*rbc*S and *rbc*L) such that the level of RUBISCO declines within the chloroplast, leading to a lowered rate of CO_2 fixation and lowered productivity. Yet ozone alters the normal relations of guard cells and subsidiary cells such that the stomata close (or open under very mild exposure) and so limit the CO_2 fixation patterns by a different mechanism which is controlled by very different environmental conditions (e.g., water stress at the root level).

A sensitive site of ozone-induced changes seem to be the translocation of sugars such that even a mild exposure will alter the ability to transform carbohydrate within the cell and to transport sugars out of the cell to "sinks". Although the mechanisms for this effect have not been well studied, the loss of translocation would place sinks such as developing seeds, fruits, and roots at great risk and would have pronounced long-term effects on any long-lived species.

Naturally these ideas are not independent but are linked to each other. Some are concerned with primary sites of interaction and others, secondary phenomena linked to the primary site. These concepts from biochemistry may allow us to build a "model" upwards to show what the critical measurements are as the level of integration of these biological systems rises.

10.3.1 O_3-Induced Changes in the Plant's Ability to Respond to Other Stresses

The effects of ozone on the physiology and growth of forest trees include both direct and indirect responses. Direct effects alter leaf processes, including changes in protein synthesis and degradation, CO_2 assimilation and photosynthate allocation, transpiration, stomata physiology, and the development of visible foliar lesions. Increasing attention from a scientific and regulatory perspective is focusing on indirect effects of ozone, whereby the pollutant induces a change(s) in the plant's physiological state which, in turn, influences its ability to either compete for limited resources (e.g., light, nutrients, and water) or withstand the ecological mixture of biotic and abiotic stresses that determine survival (Taylor et al. 1994). Notable examples of indirect effects of ozone on forest trees are changes in drought resistance, winter hardiness, susceptibility to pathogens, intraspecific and interspecific competition, and whole-plant imbalances in dry matter partitioning. A lack of awareness of multiple stress interactions underlies the shortcomings of traditional, single factor dose-response methodologies used to characterize the role of a single air pollutant on a tree's growth, physiology, and development. Accordingly, more mechanistically based studies in conjunction with process-level modeling activities are being formulated to address how air pollution in general and ozone specifically interact with the host of biotic and abiotic factors to influence a tree's physiology and growth.

Whereas the direct effects of ozone are well recognized, the potential for interactive effects of ozone with other stresses of natural and anthropogenic origin are also noteworthy. An example is that of drought, one of the major recurrent stresses in forest landscapes and one that clearly influences physiology and growth. In conjunction with intensive investigations of stomata physiology and CO_2 assimilation, the exposure of sensitive forest trees under field conditions to ozone concentrations greater than or equal to ambient levels can result in a more efficient use of water by the foliage. A corollary to this observation is that seedlings grown in ambient or charcoal-filtered atmospheres are more sensitive to recurrent drought stress than plants grown in atmospheres with elevated levels of ozone. These conclusions are supported by independent studies showing a direct effect of ozone on stomata physiology in both forest trees (Reich and Amundson 1985) and agricultural crops (Temple 1986).

10.4 Role of Models

Even though ozone is documented to be a chronic-level stress, is regionally distributed, and appears to affect the physiology and growth of a variety of sensitive species, the significance of ozone exposures on regional forest resources is poorly documented. The assessment of the pollutant's effects on forests is complicated by the wide variation in sensitivity at the interspecific and intraspecific level, mixed species composition and multiple-age class structure of forests, temporal and spatial variation in pollutant exposures, site-specific interaction of air pollutants with other stresses of natural or anthropogenic origin, and limited applicability of many site-specific experimental studies.

Pollution effects are driven by biochemical dysfunction induced at the cellular level and subsequently manifest as physiological responses at the whole-tree level. These ozone-induced effects are subsequently followed by alterations of competitive relationships at the community level, reflecting variability in ozone exposure and other stress factors across broad geographical regions. Each level of integration affects vegetation through the different forest microenvironment processes, beginning at the level of cellular biochemistry and individual leaf physiology and ranging to that at the level of the whole tree, community, and ecosystem. The large number of variables that can be directly or indirectly affected by ozone constrains the assessment of pollution effects by experimentation alone (Johnson and Taylor 1989). An alternative is to use models of forest trees and stands to predict how forest resources will respond to existing or projected pollution scenarios.

Modeling has become an essential tool in the study of air pollution effects on forests (Taylor et al. 1994). Empirical or descriptive models have been used for decades to describe the growth of a species under a defined range of conditions on a given site. These models are useful for describing growth over time and, when averaged over long time periods, can be used to predict tree growth under a specific set of site and climatic conditions.

Process-based models are potentially more useful in the study of forest responses to stress given that they are based on the mechanisms underlying the responses and incorporate processes by which stresses can be allowed to interact. Consequently, process models are useful for simulating conditions that could not be studied (and subsequently modeled). Unlike empirical models, process-based models generally include environmental variables which drive plant growth, allowing prediction of plant response to stress under a set of untested conditions.

Most of the physiologically based process models are modular in form and have varying degrees of detail. In general, most have relatively detailed photosynthetic modules, allowing the simulation of stresses that directly influence the acquisition of carbon. Many of the models use source–sink and phenological relationships to stimulate carbon and nutrient transport throughout the plant and to drive growth (Weinstein et al. 1991). Less developed in most models is the below-ground component, due in part to our relatively poor understanding of below-ground processes and our relative lack of understanding of the effects of anthropogenic factors on the rhizosphere. At the individual tree level, the most commonly used model for addressing the effects of ozone is TREGRO (Weinstein et al. 1991).

One of the most promising methodologies for investigating air pollution effects on regional forest resources relies on a hierarchical linkage among the output of models based on individual forest tree responses to air pollution with a forest stand simulator; the product of this model then serves as an input variable to a geographic information system (GIS). This approach is particularly valuable since it relies heavily on the wealth of physiological and growth data collected in controlled environments for a variety of forest trees. For example, the most probable effect of a regionally distributed pollutant (e.g., ozone chronic levels) is to reduce a tree's growth so that neighboring trees are more competitive in acquiring otherwise limiting resources (e.g., sunlight, water, nutrients), which in turn alters community dynamics. It is proposed that physiological effects predicted by individual tree, process level models (e.g., TREGRO) could serve as inputs to the stand-level models to evaluate the influence of changes in growth patterns of plants competing with other individuals under a range of environmental conditions. Finally, the stand level output could serve as input into a GIS, allowing for a matrix of interactive effects to be developed depending on a suite of natural and anthropogenic factors that the GIS provides. This methodology allows regional forest effects due to air pollution to be simulated under a range of environmental and climatic conditions that would not be amenable to experimentation.

TREGRO has been used to simulate growth and carbon allocation patterns in several different tree species at risk from ozone exposure (Weinstein et al. 1991; Taylor and Hanson 1992; Taylor and Constable 1994; Weinstein and Yanai 1994;). Tree growth is controlled in the model by two data sets: a parameter file defining the physiological and biochemical characteristics of the tree species (e.g., rates of growth, CO_2 assimilation characteristics, respiration) and a meteorological file defining the environment in hourly time steps (e.g., air temperature, rainfall, and light). The model uses biochemical-based photosynthetic equations to calculate carbon gain, subtracts respiratory carbon loss, and allocates the remaining carbon to the growth of different tree compartments (e.g., foliage, branch, stem, and fine roots).

10.4.1 Is an Effect Real?

Many experimental analyses of an air pollution-induced change of a certain physiological process have been carried out. Often, however, these studies do not exactly correspond with one another. Possible causes are variations in species, growing conditions, and exposure indices between each experimental protocol. Although the effects are expected to be very similar, the same physiological processes do not show the same reactions to the pollutant – the process is activated, inhibited or remains the same. This effect leads to uncertainty, especially for policy makers.

Typically, a statistical analysis is done on the experimental data; that is, a measure of "effect size" is tabulated whereby the results observed with a certain dose of ozone is compared to what is found with a carbon-filtered (or zero dose) exposure. The effect size is analysed statistically, which rejects effect sizes near zero. The test is highly dependent upon the sample size; one must use a very large sample in order to see small effects. However, these small effects may have a pronounced effect when a long-lived tree is involved. Are we forced to continuously carry out experiments over and over, because we do not have the support or ability to handle a large number of experimental treatments?

There is another approach (meta-analysis) to determine if there is any effect. "Meta-analysis, by contrast, looks at the distribution of all effect sizes, significant or not. If they are randomly clustered around zero, meta-analysis suggests that the treatment has no effect. But if they cluster off to one side, meta-analysis shows that something is going on...." (Mann 1994). Rather than having each study offset another in "vote-counting", each study seems to "add" to the total and amplify the effect. It seems that a combination of good physiological modeling with the use of meta-analysis may allow a better understanding of how air quality affects plant health.

10.5 Conclusion

We have attempted to (1) present a coherent view of an integrated biochemical and physiological view of ozone effects; (2) define a functional ozone standard for plants; and (3) present a judgement on biomarker techniques to check for ozone effects in the absence of visible symptoms. Presently these goals are difficult to reach. We don't believe even for crop plants that objective (1) is possible in terms of biochemistry, although we have made large strides in the last decade.

The question of what sort of inputs from a lower scale of integration (e.g., cell) must be carried upwards to the next level (e.g., tissue or whole plant) is not answerable at the highest level of the ecosystem. Furthermore, there is little agreement on objective (2) in terms of duration, rates, and peaks of ozone exposure. In fact, much of the ozone standard is based upon visible injury in the field which is generally reached at a higher level of ozone than that which induces a productivity decline. Must we base our "dose" responses upon necrosis/chlorosis? This question leads to objective (3). Currently there seems to be no specific biomarkers for ozone alone. Many of the physiological events seem to point to wounding responses as the best markers, but they are also markers for other stresses.

Control of air quality ultimately depends upon the pollutant levels society will tolerate. All political arguments rest upon the quandary that tighter controls are costly, but result in cleaner air. A political choice must be based upon sound science in order to counter-balance economic obstacles.

On the other hand, from the political and economic point of view, plant productivity must be related to some measure of ozone level within the atmosphere. From this there are three distinct measures: (1) the total loss in productivity at any actual level of ozone[14]; (2) the slope of the curve at the actual level which is the marginal cost of increasing or decreasing the ozone level; and (3) the lag of the initial decline of productivity indicating the threshold of action below which economic, and hence political, arguments are unimportant. Experimental designs may not be especially useful for all possible ozone levels for the above measures; each measure should be determined using a specific protocol. Indeed any one measure is difficult to describe in an ecological setting because the baseline level of ozone is unknown (and hence the level from which we measure productivity is unknown). We simply do not know how to judge the "cost"[15]

[14]Of course it depends upon what is taken as the "zero" ozone level of productivity. If there is a lag in the productivity/ozone curve, then a small but finite level of ozone would not change the 100% level. For a curve with no lag, one must agree upon what is an "acceptable but non-zero" level of ozone for the 100% value of productivity.

[15]The concept of cost is used in a rough manner. In general it means that the plant's productivity is not used for making more plant but rather used to prevent or repair air pollutant-induced injuries. In some cases the air quality could merely lower the ability of the plant to obtain that productivity in the first case, e.g., lowering carbon assimilation by closing the stomata.

of changing the level of ozone except by arguing on the basis of controlled field studies.

There is generally debate regarding how well various indices fit the field data. It is a retrospective argument. The index deemed to be best is then used to predict the marginal cost. However, often any given index is not linked to any particular mechanistic model and thus, one cannot describe how that index may change with another environmental parameter, e.g., average field temperature or relative humidity. In the final analysis we must attempt to link physiology/biochemical data to ecological field data to develop a full mechanistic model upon which upon ultimate political decisions can be based. Thus, it seems to us that the day of pure empirical field testing is past.

References

Alkon DL, Rasmussen H (1988) A spatial-temporal model of cell activation. Science 239:998–1004

Amiro BD, Gillespie TJ, Thurtell GW (1984) Injury response of *Phaseolus vulgaris* to ozone flux density. Atmos Environ 18:1207–1215

Aro E-M, McCaffery S, Anderson JM (1993) Photoinhibition and D1 protein degradation in peas acclimated to different growth irradiances. Plant Physiol 103:835–843

Atkinson CJ, Mansfield TA, Kean AM, Davies WJ (1989) Control of stomatal aperture by calcium in isolated epidermal tissue and whole leaves of *Commelina communis* L. New Phytol 111:9–17

Atkinson CJ, Mansfield TA, McAinsh MR, Brownless C, Hetherington AM (1990) Interactions of calcium with abscisic acid in the control of stomatal aperature. Biochem Physiol Pflanz 186:333–339

Atkinson R (1990) Gas phase tropospheric chemistry of organic compounds: A review. Atmos Environ 24A:1–41

Avni A, Bailey BA, Matto AK, Anderson JD (1994) Induction of ethylene biosynthesis in *Niotiana tabacum* by a *Trichoderma viride* xylanase is correlated to the accumulation of 1-aminocyclopropane-1-carboxylic acid (ACC) synthase and ACC oxidase transcripts. Plant Physiol 106:1049–1055

Bailey PS (1978) Ozonation of olefins: routes to peroxide products. In: Ozonation in organic chemistry, vol II. Olefinic compounds. Academic Press, New York, Chapt 6

Bailey PS (1982) Ozonation of nucleophiles. In: Ozonation in organic chemistry, vol II. Nonolefinic compounds. Academic Press New York Chapt 7

Bailey-Serres J, Kloeckener-Gruissem B, Freeling M (1989) Genetic and molecular approaches to the study of the anaerobic response and tissue specific gene expression in maize. Plant Cell Environ 11:351–357

Bakalik DP, Thomas JK (1977) Miceller catalysis of radical reactions. A spin trapping study. J Phys Chem 81:1905–1908

Ball JT (1987) Calculations related to gas exchange. In: Zeiger E, Farquhar GD, Cowan IR (eds) Stomatal function. Stanford Press, Stanford, pp 445–476

Barber J, Andersson B (1992) Too much of a good thing: light can be bad for photosynthesis. TIBS 17:61–66

Barnes SA, Knight JS, Gray JC (1994) Alteration of the amount of the chloroplast phosphate translocator in transgenic tobacco affects the distribution of assimilate between starch and sugar. Plant Physiol 106:1123–1129

Beckett P, Hozley D, Price A, Heatherington A, Lea P (1989) Ozone stress proteins in pea plants. Plant Physiol 89S:128

Beckett PL, Heatherington AM, Lea PJ (1990) Ozone stress proteins in plants. J Exp Bot 40:10–1 (Abstr)

Benes SE, Murphy TM, Anderson PD, Houpis JLJ (1995) Relationship of antioxidants enzymes to ozone tolerance in branches of mature ponderosa pine (*Pinus ponderosa*) trees exposed to long-term, low concentration, ozone fumigation and acid precipitation. Physiol Plant 94:123–134

Bobrov RA (1952) The effect of smog on the anatomy of oat leaves. Phytopathology 42:558–563

Bors W, Langebartels C, Michel C, Sandermann H Jr (1989) Polyamines as radical scavengers and protectants against ozone damage. Phytochemistry 28:1589–1595

Cacace F, Speranza M (1994) Protonated ozone: Experimental detection of O_3H^+ and evaluation of the proton affinity of ozone. Science 265:208–209

Casano L. Martin M, Sabater B (1994) Sensitivity of superoxide dismutase transcript levels and activities to oxidative stress is lower in mature-senescent than in young barley leaves. Plant Physiol 106:1033–1039

Castillo FJ, Heath RL (1990) Ca^{2+} Transport in membrane vesicles from pinto bean leaves and its alteration after ozone exposure. Plant Physiol 94:788–795

Castillo FJ, Penel C. Greppin H (1984) Peroxidase release induced by ozone in *Sedum album* leaves: Involvement of Ca^{2+}. Plant Physiol 74:846–851

Castillo FJ, Miller PR, Greppin H (1987) Extracellular biochemical markers of photochemical oxidant air pollution damage to Norway spruce. Experientia 43:111–115

Chameides WL (1989) The chemistry of ozone deposition to plant leaves: Role of ascorbic acid. Environ Sci Technol 23:595–600

Chevrier N, Chung YS, Sarhan F (1990) Oxidative damages and repair in *Euglena gracilis* exposed to ozone: 2 membrane permeability and uptake of metabolites. Plant Cell Physiol 31:987–992

Ciechanover AM, Schwartz AL (1994) The ubiquitin-mediated proteolytic pathway–mechanisms of recognition of the proteolytic substrate and involvement in the degradation of native cellular proteins. FASEB J 8:182–191

Cohen P (1992) Signal integration at the level of protein kinases, protein phosphatases and their substrates. TIBS 17:408–413

Craker LE (1971) Ethylene production from ozone injured plants. Environ Pollut 1:299–304

Criegee R (1975) Mechanism of ozonolysis. Angew Chem (Int Ed) 14:745–760

Dann MS, Pell EJ (1989) Decline of activity and quantity of ribulose bisphosphate carboxylase/oxygenase and net photosynthesis in O_3-treated potato foliage. Plant Physiol 91:427–432

Darrall NM, Jäger HJ (1984) Biochemical diagnostic tests for the effect of air pollution on plants. In: Koziol MJ, Whatley FR (eds). Gaseous air pollutants and plant metabolism. Butterworths, London, pp 333–349

Davies WJ, Tardieu F, Trejo CL (1994) How do chemical signals work in plants that grow in drying soil? Plant Physiol 104:309–314

Davison AW, Barnes JD, Renner CJ (1988) Interactions between air pollutants and cold stress. In: (eds) Schulte-Hostede S, Darrall NM, Blank LW, Wellburn AR Air Pollution and Plant Metabolism. Elsevier, London, pp 307–328

Dominy PJ, Heath RL (1985) Inhibition of the K^+-stimulated ATPase of the plasmalemma of pinto bean leaves by ozone. Plant Physiol 77:43–45

Drolet G, Dumbroff EB, Legge RL, Thompson JE (1986) Radical scavenging properties of polyamines. Phytochemistry 25:367–371

Dugger WM, Ting IP (1970a) Air pollution oxidants–Their effects on metabolic processes in plants. Annu Rev Plant Physiol 21:215–234

Dugger WM, Ting IP (1970b) Physiological and biochemical effect of air pollutant oxidants on plants. Recent Adv Phytochem 3:31–58

Dure LS III (1993) Structural motifs in Lea proteins. In: Close TJ, Bray EA (eds) Plant responses to cellular dehydration during environmental stress. Current Topics in Plant Physiology, vol 10. American Society of Plant Physiologists, Rockville, pp 91–103

Eckardt NA, Pell EJ (1994) O_3-induced degradation of Rubisco protein and loss of Rubisco mRNA in relation to leaf age in *Solanum tuberosum* L. New Phytol 127:741–748

Ecker JR, Davis RW (1987) Plant defense genes are regulated by ethylene. Proc Natl Acad Sci USA 84:5202–5206

Ehleringer JR (1991) Estimating the cost of air pollution resistance. In: Taylor GE, Jr, Ditelka LF, Clegg MT (eds) Ecological Genetics and Air Pollution. Springer, Berlin Heidelberg New york, pp 203–208

Ehleringer JR, Field CB (eds) (1994) Scaling physiological processes: leaf to globe. Academic Press, San Diego

Elkiey T, Ormrod DP (1979) Ozone and/or sulphur dioxide effects on tissue permeability of petunia leaves. Atmos Environ 13:1165–1168

Elmore CD (1980) The paradox of no correlation between leaf photosynthetic rates and crop yields. In: Hesketh JD, Jones JW (eds) Predicting Photosynthesis for Ecosystem Models, vol 1. CRC Press, Boca Raton, pp 155–168

Evans DE (1988) Regulation of cytoplasmic free calcium by plant cell membranes. Cell Biol Int Rep 12:383–396

Evans LS, Miller PR (1972) Ozone damage to ponderosa pine: A histological and histochemical appraisal. Am J Bot 59:297–304

Evans LS, Ting IP (1973) Ozone induced membrane permeability changes. Am J Bot 60:155–162

Evans LS, Ting IP (1974) Effect of ozone on ^{86} Rb-labeled potassium transport in leaves of *Phaseolus vulgaris* L. Atmos Environ 8:855–861

Eyal Y, Fluhr R (1991) Cellular and molecular biology of pathogenesis related proteins. Oxford Surv Plant Mole Cell Biol 7:223–254

Fangmeier A, Brunschön HJ, Jäger HJ (1994) Time course of oxidant stress biomarkers in flag leaves of wheat exposed to ozone and drought stress. New Phytol 125:63–69

Farage PK, Long SP, Lechner EG, Baker NR (1991) The sequence of changes within the photosynthetic apparatus of wheat following short term exposure to ozone. Plant Physiol 95:529–535

Farquhar GD, Sharkey TD (1982) Stomatal conductance and photosynthesis. Annu Rev Plant Physiol 33:317–345

Field CB, Chapin FS III, Matson PA, Mooney HA (1992) Responses of terrestrial ecosystems to the changing atmosphere: a resource based approach. Annu Rev Ecol Syst 23:201–235

Finkelstein E, Rosen GM, Rauckman EJ (1980) Spin trapping of superoxide and hydroxyl radical: practical aspects. Arch Biochem Biophys 200:1–16

Finlayson-Pitts BJ, Pitts JN Jr (1986) Atmospheric chemistry: fundamentals and experimental techniques. John Wiley, New York 890 pp

Frederick P, Heath RL (1975) Ozone-induced fatty acid and viability changes in *Chlorella*. Plant Physiol 55:15–19

Galston AW, Sawhney RK (1990) Polyamines in plant physiology. Plant Physiol 94:406–410

Giamalva DH, Church DF, Pryor WA (1986) Kinetics of ozonation 4: Reactions of ozone with α-tocopherol and oleate and linoleate esters in carbon tetrachloride and in aqueous micellar solvents. J Am Chem Soc 108:6646–6651

Gilroy S, Fricker MD, Read ND, Trewavas AJ (1991) Role of calcium in signal transduction of *Commelina* guard cells. Plant Cell 3:333–344

Grimes HD, Perkins KK, Boss WF (1983) Ozone degrades into hydroxyl radical under physiological conditions. Plant Physiol 72:1016–1020

Gunderson CA, Taylor GE Jr (1988) Kinetics of inhibition of foliar gas exchange by exogenous ethylene: An ultrasensitive response. New Phytol 110:517–524

Gunderson CA, Taylor GE Jr (1991) Ethylene directly inhibits foliar exchange *in Glycine max*. Plant Physiol 95:337–339

Gupta AS, Alscher RG, McCune D (1991) Response of photosynthesis and cellular antioxidants to ozone in Populus leaves. Plant Physiol 96:650–655

Guzy MR, Heath RL (1993) Response to ozone of varieties of common bean (*Phaseolus vulgaris* L.) New Phytol 124:617–625

Harris MJ, Bailey-Serres JN (1994) Ozone effects on gene expression and molecular approaches to breeding for air pollution resistance. In: Basra AS (ed) Stress induced gene expression in plants. Hardwood Academic Press, Chur, pp 185–207

Heagle AS (1989) Ozone and crop yield. Annu Rev Phytopathol 27:397–423

Heath RL (1978) The reaction stoichiometry between ozone and unsaturated fatty acids in an aqueous environment. Chem Phys Lipids 22:25–37

Heath RL (1980) Initial events in injury to plants by air pollutants. Annu Rev Plant Physiol 31:395–431

Heath RL (1984) Air pollutant effects on biochemicals derived from metabolism: Organic, fatty and amino acids. In: Koziol MJ, Whatley FR (eds) Gaseous air pollutants and plant metabolism. Butterworths, London, pp 275–290

Heath RL (1987) The biochemistry of ozone attack on the plasma membrane of plant cells. Adv Phytochem 21:29–54

Heath RL (1988) Biochemical mechanisms of pollutant stress. In: Heck WW, Taylor OC, Tingey DT (eds) Assessment of crop loss from air pollutants. Elsevier Applied Science, London, pp 259–286

Heath RL (1989) Alteration of chlorophyll in plants upon air pollutant exposure. In: Woodwell GM, Cook ER, Cowling EB, Johnson AH, Kimmerer TW, Matson PA, McLaughlin SS, Raynal DJ, Swank WT, Waring RH, Winner WE, Woodman JN (eds) Biologic Markers of air-pollution stress and damage in forests. US National Academy Press, Washington, DC, pp 347–356

Heath RL (1994a) Alterations of plant metabolism by ozone exposure. In: Alscher RG, Wellburn AR (eds) Plant responses to the gaseous environment. Chapman & Hall, London, pp 121–146

Heath RL (1994b) Possible mechanisms for the inhibition of photosynthesis by ozone. Photosynth Res 39:439–451

Heath RL (1996a) The modification of photosynthetic capacity induced by ozone exposure. In: Baker NR (ed) Photosynthesis and environment, Chap 17. Adv Photosynthesis. Kluwer, Dordrecht, 25 pp

Heath RL (1996b) Lipid metabolism and oxidant air pollutants. In: Iqbal M, Yunus M (eds) Plant response to air pollution. John Wiley, Chichester, pp 353–373

Heath RL, Castillo FJ (1988) Membrane disturbances in response to air pollutants. In: Schulte-Hostede S, Darrall NM, Blank LW, Wellburn AR (eds) Air pollution and plant metabolism. Elsevier, London, pp 53–75

Heath RL, Tappel AL (1976) A new, sensitive assay for the measurement of hydroperoxides. Anal Biochem 76:184–191

Heath RL, Chimiklis PE, Frederick P (1974) The role of potassium and lipids in ozone injury to plant membranes. ACS symp ser 3. In: Dugger WM (ed) Air pollution effects on plant growth. American Chemical Society, Washington, DC, pp 58–75

Heck WW, Taylor OC, Tingey DT (1988) Assessment of Crop Loss from Air Pollutants. Elsevier, London, 552pp

Hershko A (1991) The ubiquitin pathway for protein degradation. TIBS 16:265–267

Hess JL (1994) Plant response to the gaseous environment. In: Alscher RG, Wellburn AR (eds) Free radical scavenging. Chapman & Hall, London, pp 99–120

Hogsett WE, Tingey DT, Lee EH (1988) Ozone exposure indices: concepts for development and evaluation of their use. In: Heck WW, Taylor OC, Tingey DT, (eds) Assessment of Crop Loss from Air Pollutants. Elsevier, London, pp 107–138

Hoigné J, Bader H (1983a) Rate constants of reactions of ozone with organic and inorganic compounds in water I: Non-dissociating organic compounds. Water Res 17:173–183

Hoigné J, Bader H (1983b) Rate constants of reactions of ozone with organic and inorganic compounds in water II: Dissociating organic compounds. Water Res 17:185–194

Hoigné J, Bader H, Hoog WR, Staehelin J (1985) Rate constants of reactions of ozone with organic and inorganic compounds in water III: Inorganic compounds and radicals. Water Res 19:993–1004

Ishige F, Mori H, Yamazaki K, Imaseki H (1992) Identification of a basic glycoprotein induced by ethylene in primary leaves of Azuki bean as a cationic peroxidase. Plant Physiol 101:193–199

Jacobson JS, Hill AC (1970) Recognition of air pollution injury to vegetation: a pictorial atlas. Air Pollution Control Assoc, Pittsburg, 150 pp

Jiang C-Z, Rodermel SR, Shibles RM (1993) Photosynthesis, Rubisco acitivity and amount, and their regulation by transcription in senescing Soybean leaves. Plant Physiol 101:105–112

Johnson DW, Taylor GE Jr (1989) Role of air pollution in forest decline in eastern North America. Water Air Soil Pollut 48:21–43

Kangasjärvi J, Talvinen J, Ultriainen M, Karjalainen R (1994) Plant defense systems induced by ozone. Plant Cell Environ 17:783–794

Kärenlampi SO, Airaksinen K, Miettinen ATE, Kokko NI, Holopainen JK, Kärenlampi LV, Karjalainen RO (1994) Pathogenesis-related proteins in ozone exposed Norway spruce [*Pincea abies* (Karst) L.]. New Phytol 126:81–89

Karin M, Smeal T (1992) Control of transcription factors by signal transduction pathways: the beginning of the end. TIBS 17:418–422

Karpinski S, Wingsle G, Karpinski B, Hällgren J-E (1993) Molecular responses to photoxidative stress in *Pinus sylvestris* (L.) Plant Physiol 103:1385–1391

Kerstiens G, Lendzian KJ (1989) Interaction between ozone and plant cuticles. I. ozone deposition and permeability. New Phytol 112:13–19

Kim WT, Yang SF (1992) Turnover of 1-aminocyclopropane-1-carboxylic acid synthase protein in wounded tomato fruit tissue. Plant Physiol 100:1126–1131

Kostka-Rick R, Manning WJ (1993) Dose-response studies with the antiozonant ethylenediurea (EDU), applied as a soil drench to two growth substrates, on greenhouse-grown varieties of *Phaseolus vulgaris* L. Environ Pollut 82:63–72

Koukol J, Dugger WM Jr (1967) Anthocyanin formation as a response to ozone and smog treatment in *Rumex crispus* L. Plant Physiol 42:1023–1024

Krapp A, B Hofmann, Schafer C, Stitt M (1993) Regulation of the expression of *rbc*S and other photosynthetic genes by carbohydrates: A mechanism for the sink regulation of photosynthesis. Plant J 3:817–828

Laisk A, Kull O, Moldau H (1989) Ozone concentration in leaf intercellular air space is close to zero. Plant Physiol 90:1163–1167

Langebartels C, Kerner K, Leonardi S, Schraudner M, Trost M, Heller W, Sandermann H Jr (1991) Biochemical plant responses to ozone I. Differential induction of polyamine and ethylene biosynthesis in tobacco. Plant Physiol 95:882–889

Larson RA (1988) The antioxidants of higher plants. Phytochemistry 27:969–978

Lee EH, Bennett JH (1982) Superoxide dismutase: A possible protective enzyme against ozone injury in snap beans (*Phaseolus vulgaris* L.). Plant Physiol 69:1444–1449

Lee T-M, Chu C (1992) Ethylene induced polyamine accumulation in rice (*Oryza sativa* L) coleoptiles. Plant Physiol 99:238–245

Lee TT (1965) Sugar content and stomatal width as related to ozone injury in tobacco leaves. Can J Bot 43:677–685

Levitt J (1980) Responses of plants to environmental stress, 2nd edn. Academic Press, New York, 697 pp

Lovett GM (1994) Atmospheric deposition of nutrients and pollutants in North America: an ecological perspective. Ecol Appl 4:629–650

Luwe MWF, Takahama U, Heber U (1993) Role of ascorbate in detoxifying ozone in the apoplast of spinach (*Spinacia oleracea* L) leaves. Plant Physiol 101:969–976

Mann CC (1994) Can meta-analysis make policy? Science 266:960–962

Manning WJ, Keane KD (1988) Effects of air pollutants on interactions between plants, insects, and pathogens. In: Heck WW, Taylor OC, Tingey DT (eds) Assessment of crop loss from air pollutants. Elsevier, New York, pp 365–386

Masuch G, Kettrup A (1985) Investigation on the effect of ozone on leaves of pinto. In: Troyanowsky C (ed) Air Pollution and Plants. VCH Publishers, Deerfield Beach, pp 142–145

Matheson IBC, King AD (1978) Solubility of gases in micellar solution. J Colloid Interfacial Sci 66:464–469

Matters GL, Scandalios JG (1987) Synthesis of isozymes of superoxide dismutase in maize leaves in response to O_3, SO_2, and elevated O_2. J Exp Bot 38:842–852

McCool PM, Musselmann RC, Younglove T, Teso RR (1988) Response of kidney bean to sequential ozone exposures. Environ Exp Bot 28:307–313

McLaughlin SB, McConathy RK, (1983) Effects of SO_2 and O_3 on allocation of ^{14}C-labeled photosynthate in *Phaseolus vulgaris*. Plant Physiol 73:630–635

Mehdy MC (1994) Active oxygen species in plant defence against pathogens. Plant Physiol 105:467–472

Mehlhorn H, Wellburn AR (1987) Stress ethylene formation determines plant sensitivity to ozone. Nature 327:417–418

Mehlhorn H, Tabner BJ, Wellburn AR (1990) Electron spin resonance evidence for the formation of free radicals in plants exposed to ozone. Physiol Plant 79:377–383

Menzel DB (1980) Protection against environment toxicants. In: Machlin LJ (ed) Vitamin E: A comprehensive treatise. Marcel Dekker, New York, pp 474–494

Moldau H, Sober J, Sober A (1990) Differential sensitivity of stomata and mesophyll to sudden exposure of bean shoots to ozone. Photosynthetica 24:446–458

Möller D (1989) The possible roles of H_2O_2 in new type forest decline. Atmos Environ 23:1625–1627

Mudd JB (1996) Biochemical basis for the toxicity of ozone. In: Iqbal M, Yunus M (eds) Plant response to air pollution. John Wiley, Chichester, pp 267–283

Mudd JB, Kozlowski TT (eds) (1975) Responses of plants to air pollution. Academic Press, New York, 383pp

Mudd JB, Leavitt R, Ongun A, McManus TT (1969) Reaction of ozone with amino acids and proteins. Atmos Environ 3:669–682

Nie G-Y, Tomasevic M, Baker NR (1993) Effects of ozone on the photosynthetic apparatus and leaf proteins during leaf development in wheat. Plant Cell Environ 16:643–651

Ordin L, Hall MA, Kindinger JI (1969) Oxidant induced inhibition of enzymes involved in cell wall polysaccharide synthesis. Arch Environ Health 18:623–625

Ormrod DP, Beckerson DW (1986) Polyamines as antioxidants for tomato. Hort Sci 21:1070–1071

Osmond CB (1981) Photorespiration and photoinhibition: some implications for the energetics of photosynthesis. Biochim Biophys Acta 639:77–98

Paterson NW, Weyers JDB, A'Brook RA (1989) The effect of pH on stomatal sensitivity to abscisic acid. Plant Cell Environ 11:83–89

Patterson DT (1980) Light and temperature adaptation. In: Hesketh JD, Jones JW (eds) Predicting photosynthesis for ecosystem models. CRC Press, Boca Raton pp 205–235

Pauls KP, Thompson JE (1980) In vitro simulation of senescence related membrane damage by ozone induced lipid peroxidation. Nature 283:504–506

Pell EJ, Weissberger WC (1976) Histopathological characterization of ozone injury to soybean foliage. Phytopathology 66:856–861

Pell EJ, Eckardt N, Glick RE (1994) Biochemical and molecular basis for impairment of photosynthetic potential. Photosynth Res 39:453–462

Peters JL, Silverthorne J (1995) Organ-specific stability of two *Lemna rbc*S mRNAs is determined primarily in the nuclear compartment. Plant Cell 7:131–140

Peters JL, Castillo FJ, Heath RL (1988) Alteration of extracellular enzymes in pinto bean leaves upon exposure to air pollutants, ozone and sulfur dioxide. Plant Physiol 89:159–164

Pietila M, Lahdesmaki P, Pakonen T, Laine K, Saari E, Havas P (1990) Effect of nitrogenous air pollutants on changes in protein spectra with the onset of winter in the leaves and shoots of the bilberry (*Vaccinium myrtillus* L.). Environ Pollut 66:103–116

Pino ME, Mudd JB, Bailey-Serres J (1995) Ozone-induced alterations in the accumulation of newly synthesized proteins in leaves of maize. Plant Physiol 108:777–785

Pitcher LH, Brennan E, Zilinskas BA (1992) The antiozonant ethylenediurea does not act via superoxide dismutase induction in bean. Plant Physiol 99:1388–1392

Polle A, Junkermann W (1994) Inhibition of apoplastic and symplastic peroxidase activity from Norway Spruce by the photooxidant hydroxymethyl hydroperoxide. Plant Physiol 104:617–621

Polle A, Rennenberg H (1991) Superoxide dismutase activity in needles of Scots pine and Norway spruce under field and chamber conditions: Lack of ozone effect. New Phytol 117:335–343

Polle A, Wieser G, Havranek WM (1995) Quantification of ozone influx and apoplastic ascorbate content in needles of Norway spruce trees (*Picea abies* L. Karst) at high altitude. Plant Cell Environ 18:681–688

Price A, Lucas PW, Lea PJ (1990) Age dependent damage and glutathione metabolism in ozone fumigated barley: A leaf section approach. J Exp Bot 41:1309–1317

Pryor WA (1992) How far does ozone penetrate into the pulmonary air-tissue boundary before it reacts? Free Radical Biol Med 12:83–88

Pryor WA, Uppu RM (1993) A kinetic model for the competitive reactions of ozone with amino acid residues in proteins in reverse micelles. J Biol Chem 268:3120–3126

Pryor WA, Giamalva DH, Church DF (1985) Kinetics of Ozonation 3: Substituent Effects on the Rates of Reaction of Alkenes. J Am Chem Soc 107:2793–2796

Ranieri A, Bernardi R, Pisanelli A, Lorenzini G, Soldatini GF (1990) Changes in the free amino acid pool and in the protein pattern of maize leaves under continuous SO_2 fumigation. Plant Physiol Biochem 28:601–607

Rao MV, Plaiyath G, Ormrod DP (1996) Ultraviolet-B- and ozone-induced biochemical changes in antioxidant enzymes of *Arabidopsis thaliana*. Plant Physiol 110:125–136

Reddy GN, Arteca RN, Dai YR, Flores HE, Negm FB, Pell EJ (1993) Changes in ethylene and polyamines in relation to mRNA levels of the large and small subunits of ribulose bisphosphate carboxylase/oxygenase in ozone-stress potato foliage. Plant Cell Environ 16:819–826

Reich PB, Amundson RG (1985) Ambient levels of ozone reduce net photosynthesis in tree and crop species. Science 230:566–570

Reich PB, Lassoie JB (1984) Effects of low level O_3 exposure on leaf diffusive conductance and water-use efficiency in hybrid poplar. Plant Cell Environ 7:661–668

Rodriguez R, Sanchez Tames R, Durzan DJ, (eds) (1990) Plant Aging: Basic and Applied Approaches. NATO ASI Series, vol 186. Plenum Press, New York, 450 pp

Roper TR, Williams LE (1989) Effects of ambient and acute partial pressure of ozone on leaf net CO_2 assimilation of field grown *Vitis vinifera* L. Plant Physiol 91:1501–1506

Rowland-Bamford AJ, Borland AM, Lea PJ, Mansfield TA (1989) The role of arginine decarboxylase in modulating the sensitivity of barley to ozone. Environ Pollut 61:95–106

Sachs MM, Ho TD (1986) Alteration of gene expression during environmental stress in plants. Annu Rev Plant Physiol 37:363–376

Saxe H (1991) Photosynthesis and stomatal response to polluted air and the use of physiological and biochemical response for early detection and diagnostic tools. Adv Bot Res 18:1–128

Schmidt IH (1959) Über Thiobarbitursäure Methin-Farbstoffe. Fette Seifen Anstrichen 61:881–888

Schmieden U, Wild A (1995) The contribution of ozone to forest decline. Physiol Plant 94:371–378

Schmitt R, Sandermann H Jr (1990) Biochemical response of Norway spruce (*Picea abies* (L.) Karst.) towards 14-month exposure to ozone and acid mist: Part II Effects on protein biosynthesis. Environ Pollut 64:367–373

Schopfer P (1994) Histochemical demonstration and localization of H_2O_2 in organs of higher plants by tissue printing on nitrocellulose paper. Plant Physiol 104:1269–1275

Schraudner M, Ernst D, Langebartels C, Sandermann H Jr (1992) Biochemical plant responses to O_3 III: Activation of the defense-related proteins β-1,3 glucanase and chitinase in tobacco leaves. Plant Physiol 99:1321–1328

Schulman H, Low LL (1989) Multifunctional Ca^{2+}/calmodulin-dependent protein kinase. TIBS 14:62–66

Schulte-Hostede S, Darrall NM, Blank LW, Wellburn AR (eds) (1988) Air Pollution and Plant Metabolism. Elsevier, London, 381 pp

Sharma YK, Davis KR (1994) Ozone-induced expression of stress-related genes in *Arabidopsis thaliana*. Plant Physiol 105:1089–1096

Sharma YK, Davis KR (1995) Isolation of a novel *Arabidopsis* ozone-induced cDNA by differential display. Plant Mol Biol 29:91–98

Smith TA (1985) Polyamines. Annu Rev Plant Physiol 36:117–143

Spanu P, Grosskopf DG, Felix G, Boller T (1994) The apparent turnover of 1-aminocyclopropane-1-carboxylate synthase in tomato cells is regulated by protein phosphorylation and dephosphorylation. Plant Physiol 106:529–535

Stone JM, Walker JC (1995) Plant protein kinase families and signal transduction. Plant Physiol 108:451–457

Tappel AL (1975) Lipid peroxidation. In: Trump BA, Arstila AV (eds) Pathology of cell membrane. Academic Press, New York, pp 145–170

Taylor GE Jr, Constable JVH (1994) Modelling pollutant deposition to vegetation: scaling down from the canopy to the biochemical level. In: Percy KE, Cape JN, Garrels R, Simpson CJ (eds) Air Pollution and the Leaf Cuticle. NATO ASI Series G; Ecological Sciences, Springer, Berlin, Heidelberg, New York, pp 15–37

Taylor GE Jr, Hanson PJ (1992) Forest trees and tropospheric ozone: Role of canopy deposition and leaf uptake in developing exposure-response relationships. Agric Ecosyst Environ 42:255–273

Taylor GE Jr, Ross-Todd BM, Gunderson CA (1988) Action of ozone on gas exchange in *Glycine max*. L. Merr: A potential role for endogenous stress ethylene. New Phytol 110:301–307

Taylor GE Jr, Johnson DJ, Andersen CP (1994) Commissioned review - Air pollution and forest ecosystems: a regional to global perspective. Ecol Appl 4:662–689

Temple PJ (1986) Stomatal conductance and transpirational response of field grown cotton to ozone. Plant Cell Environ 9:315–321

Tester M (1990) Plant ion channels: Whole cell and single channel studies. New Phytol 114:305–340

Tingey DT, Taylor GE Jr (1981) Variation in plant response to ozone: A conceptual model of physiological events. In: Unsworth M, Ormrod D (eds) Effects of Gaseous Air Pollutants in Agriculture and Horticulture. Butterworth, London, pp 113–138

Tingey DT, Fites RC, Wickliff C (1973) Foliar sensitivity of soybeans to ozone as related to several leaf parameters. Environ Pollut 4:183–192

Tingey DT, Fites RC, Wickliff C (1976a) Differential foliar sensitivity of soybean cultivars to ozone associated with differential enzyme activities. Physiol Plant 37:69–72

Tingey DT, Standley C, Field RW (1976b) Stress ethylene evolution: A measure of ozone effects on plants. Atmos Environ 10:969–974

Tingey DT, Hogsett WE, Henderson S (1990) Definition of adverse effects for the purpose of establishing secondary national ambient air quality standards. J Environ Qual 19:635–639

Tjoelker MG, Volin JC, Oleksyn J, Reich PB (1995) Interaction of ozone pollution and light effects on photosynthesis in a forest canopy experiment. Plant Cell Environ 18:895–905

Uppu RM, Pryor WA (1994) The reactions of ozone with proteins and unsaturated fatty acids in reverse micelles. Chem Res Toxicol 7:47–55

US EPA (1988) Air quality criteria for ozone and other photochemical oxidants. US Environmental Protection Agency, Environmental Criteria and Assessment Office, Research Triangle Park, NC, EPA-600/8-84-020aF, vol III, chapt 6

US EPA (1992) Air quality criteria for oxides of nitrogen. US Environmental Protection Agency, Environmental Criteria and Assessment Office, Research Triangle Park, EPA-600/8-91/049aA, vol II, chapt 9

Vierling E (1991) The role of heat shock protein in plants. Annu Rev Plant Physiol Plant Mol Biol 42:579–620

Ward ER, Uknes SJ, Williams SC, Dincher SS, Wiederhold DL, Alexander DC, Ahl-Goy P, Métraux J-P, Ryals JA (1991) Coordinate gene activity in response to agents that induce systemic acquired resistance. Plant Cell 3:1085–1094

Weinstein DA, Yanai RD (1994) Integrating the effects of simultaneous stresses on plants using the simulation model TREGRO. J Environ Qual 23:418–423

Weinstein DA, Beloin RM, Yanai RD (1991) Modeling changes in red spruce: carbon balance and allocation in response to interacting ozone and nutrient stresses. Tree Physiol 9:127–146

Wellburn AR (1990) Why are atmospheric oxides of nitrogen usually phytotoxic and not alternative fertilizers? New Phytol 115:395–429

Willekens H, Van Camp W, Van Montagu M, Inzé D, Langebartels C, Sandermann H Jr (1994) Ozone, sulfur dioxide, and ultraviolet B have similar effects on mRNA accumulation of antioxidant genes in *Nicotiana plumbaginifolia* L. Plant Physiol 106:1007–1014

Williams J, Bulman MP, Neill SJ (1994) Wilt-induced ABA biosynthesis, gene expression and down-regulation of *rbc*S mRNA levels in *Arabidopsis thaliana*. Physiol Plant 91:177–182

Woodwell GM, Cook ER, Cowling EB, Johnson AH, Kimmerer TW, Matson PA, McLaughlin SS, Raynal DJ, Swank WT, Waring RH, Winner WE, Woodman JN (1989) Biologic markers of air-pollution stress and damage in forests. US National Academy Press, Washington, DC, 363pp

Zheng X, van Huystee RB (1992) Anionic peroxidase catalysed ascorbic acid and IAA oxidations in the presence of hydrogen peroxide: A defence system against peroxidative stress in peanut plants. Phytochemistry 31:1895–1898

Zilinskas NA, Greenhalgh-Weidman B, Brennan E (1990) The relationship between EDU pre-treatment and C_2H_4 evolution in ozonated pea plants. Environ Pollut 65:241–249

11 Forest Decline and Ozone: Synopsis

H. Sandermann Jr., A.R. Wellburn and R.L. Heath

11.1 Decline Definitions

There are two main definitions of forest decline. The operational definition is based on the visual damage inventory (loss and yellowing of needles or leaves, changes in morphology) that was newly introduced in the early 1980s. Appreciable damage rates in various tree species have been found by applying it (FBW 1989). The operational definition is purely descriptive and does not contain claims of specificity of the symptoms or of a trend in damage development. Furthermore, it has not been linked to rates of wood formation or any defined natural or anthropogenic causal factor. In contrast, the functional definitions of forest decline are based on a causal analysis of the various abiotic and biotic disease factors that play a role in the turnover and population shifts of forest trees (Manion 1981, Schwerdtfeger 1981). Yet this definition is very difficult to apply. Meaningful ecological conclusions depend on a transition from the operational to the functional level of forest decline studies. Interestingly, the lack of adequate data has led most of the authors of this book to use operational definitions of forest decline.

11.2 Ozone History of the Series "Ecological Studies"

The possible relationship between forest decline and ozone has been considered in previous volumes of the series "Ecological Studies". Guderian et al. (1985) in the milestone book on photochemical oxidants were hesitant to explain the **novel forest decline** in Germany in terms of ozone and placed more emphasis on soil acidification. Volume 85 of the Ecological Studies series described the mosaic cycle model of forest turnover (Remmert 1991), but the editor (now deceased) of that volume wrote in a personal letter to the first author (dated May 24, 1994) that the mosaic

Ecological Studies, Vol. 127
Sandermann et al. (eds) Forest Decline and Ozone

cycle concept was developed without any regard for environmental pollutants (*Umweltdreck*). A later volume in the Ecological Studies series (Schulze et al. 1989) dealt in detail with the needle yellowing caused by magnesium deficiency in the German Fichtelgebirge forest. A direct influence of ozone was discounted, and soil factors were exclusively held responsible for the observed decline symptoms. This conclusion was made in spite of the presence of appreciable ozone exposure (about 200 $\mu l\ l^{-1}$ h per season; Führer 1989). The yellowing symptoms induced by Mg^{2+} deficiency occurred in spruce at the time of the new flush and thereby resembled the "memory" effect of ozone (see Chap. 6). A later volume of "Ecological Studies" dealt with red spruce in eastern United States (Eagar and Adams 1992). This species was found to be ozone-tolerant in controlled exposure experiments and no direct indications for ozone damage could be observed in the field. This negative observation is in line with data from surveys carried out in eastern US, as well as in Germany, Austria and Switzerland, described in this volume. On the other hand, extensive ozone damage along a 500 km stretch from California to Nevada was described in a subsequent volume of Ecological Studies devoted to the Western Pacific Forests. Ozone caused not only visible needle and leaf symptoms on *Pinus* species and deciduous trees, but also enhanced mortality and decreased wood formation (Olson et al. 1992). This classical decline linked to ozone has been updated in the present Chapter 2. A recent volume (Fox and Mickler 1996) presents detailed evidence for ecophysiological and growth effects of ozone on southern pine forests [mainly for loblolly pine (*Pinus taeda*, L.), slash pine (*Pinus elliotti* Engelm.) and shortleaf pine (*Pinus echinata* Mill.)].

Previous volumes of Ecological Studies also clearly illustrate that ozone is a main factor of forest decline in western US, but is only a possible, contributary factor in eastern US and Europe. When the various chapters of the present book are examined for the decline/O_3 relationship, a situation resembling that in the previous volumes of the Ecological Studies series still remains (Table 11.1), in spite of all the new results obtained in recent years. The lack of positive evidence outside California could be due to three main reasons: (1) ozone is not an important factor for decline in eastern US and Europe; (2) the diagnostic methods applied so far have not proved to be sufficiently discriminating to identify O_3 as having a distinct effect in some areas; (3) O_3 only acts in concert with other environmental stresses and does not have a separate effect. The present book is not able to provide a clear answer to these possibilities although there is mounting evidence for options (2) and (3). The new results presented allow a better characterization of the potential role of ambient O_3 in forest decline, which can be used to propose a strategy for future O_3 research.

Table 11.1. Conclusions of the individual book chapters on the relationship between forest decline and ozone. Each chapter is evaluated in terms of 'yes/likely/unlikely/no evaluation' with regard to a causal link between ozone and forest decline

Chapter	Content	Evalutation
Chapter 1	Current ambient O_3 values are often above international threshold values for forest tree species.	Likely
Chapter 2	Southern California constitutes the best case study for the detrimental effects of O_3 on conifers. New results on wood formation and on a shift to O_3-tolerant, fire-sensitive tree species have been obtained. Visible needle symptoms were the only specific diagnostic parameter employed.	Yes
Chapter 3	Some correlations between O_3 levels and visible damage to or growth effects on deciduous forest trees and loblolly pine have been obtained. The situation with *Pinus strobus* remains unclear.	Likely
Chapter 4	The levels of both O_3 and decline were high at the described sites, but a causal link could not be established. The potential of spruce for acclimation was high but specific diagnostic markers were lacking.	Unlikely
Chapter 5	The levels of both O_3 and decline were high at the Wank Mountain, but a causal link could not be established although products from O_3-alkene reactions were detected. The potential of spruce for acclimation was high but specific diagnostic markers were lacking.	Likely
Chapter 6	Controlled exposures have characterized O_3 as a powerful abiotic elicitor at the gene, enzyme and metabolite levels. Abiotic or biotic damage pathways for ambient O_3 are proposed, but have not yet been examined in the field.	Likely
Chapter 7	Ozone values in the UK are near those that led to direct and indirect effects in controlled exposure studies. Causal links between controlled study and field decline are still required.	Likely
Chapter 8	Ozone values in Scandinavia are approaching those that led to effects in controlled exposure studies. Causal links between controlled study and field decline are missing, as are suitable diagnostic markers.	Likely
Chapter 9	Experimental exposure techniques have improved. There is variability in tree populations with respect to ozone effects and in published results on exposure.	No evaluation
Chapter 10	Ozone induces defined tolerance as well as damage mechanisms. Potential targets and signal chains are starting to emerge. Mechanisms defined experimentally cannot yet be extrapolated to field situations. No specific diagnostic markers are available.	No evaluation

11.3 Decline Potential of Ozone

Several chapters document that accumulated total hourly ozone concentration (SUMO) and accumulated seasonal ozone doses above concentration thresholds are in a potentially phytotoxic range over much of eastern and western USA as well as Europe. Two thresholds are commonly used: 40 nl l^{-1} (given as AOT40) or 60 nl l^{-1} (given as AOT60). This conclusion is based on limit dose values of 100–400 μl l^{-1} h for SUMO or of 10 μl l^{-1} h for AOT40. In spite of exceeding the limit values, the controlled exposure and field studies provide no clear confirmation of a role of ozone in forest decline (see Table 11.1). Yet, as often argued, phytotoxicity has to depend on O_3 dose internal to the leaf or cell rather than on external O_3; the use of an integrated dose over an entire season or tree lifetime remains another open problem.

11.4 Ozone and Plant Homeostasis

Under a given set of environmental parameters, the plant maintains a state of homeostasis that allows growth and development while allowing repair and response to moderate changes in the environment. Chapter 10 describes in detail how homeostasis can be achieved at different levels of organization. Important metabolic changes include those taking place at the plasma membrane induced by chronic ozone exposure or products derived from O_3 (Fig. 11.1). These are now thought to be due to changes in the structure of the outward-facing portions of membrane proteins, or proteins within the wall region, rather than to alterations within the lipid portions of the plasma membrane. Cellular free-radical scavenging systems comprise both antioxidants (e.g. glutathione and ascorbate) and enzyme systems that regenerate antioxidants (e.g. glutathione reductase) or remove potential sources of free radicals (e.g. superoxide dismutases or catalases). Antioxidative systems are often induced by O_3, but O_3 effects occur even in the presence of high apoplastic ascorbate (see Chaps. 5 and 6). It is possible that analytical data on total leaf extracts are of limited significance. Some exposed cells within a leaf may resist O_3, whereas other cells seem to succumb to O_3 attack, leading to patchiness or spots. There is thus a need for more histological and subcellular studies. Stress ethylene appears to cause visible injury, either by direct chemical transformation of ethylene by O_3 to toxic products or by ozone-induced alterations at binding sites for ethylene (see Chaps. 6, 7 and 10).

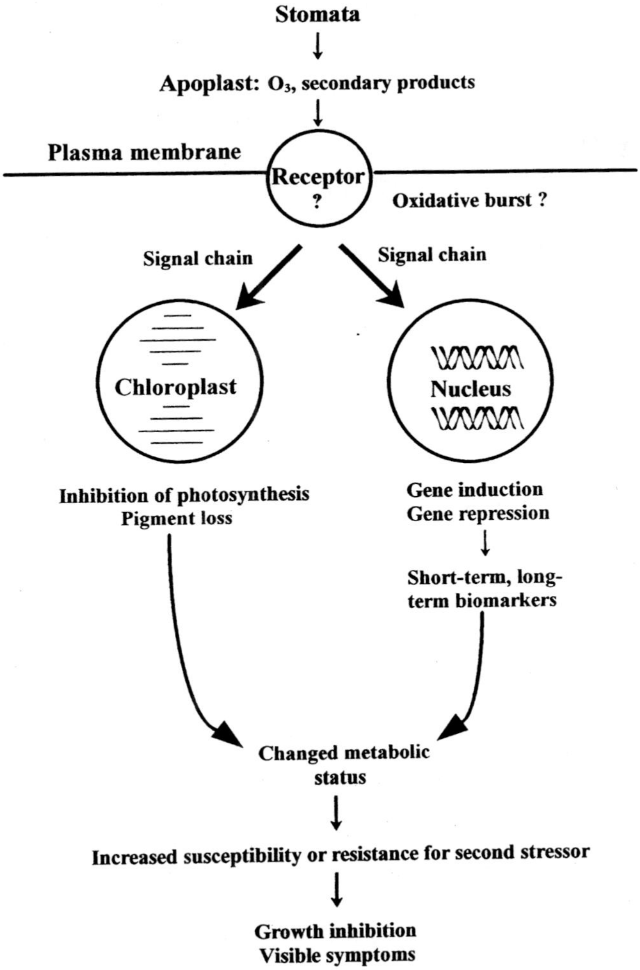

Fig. 11.1. Schematic overview of ozone effects at the cellular level (Sandermann 1996). After stomatal uptake, much of the O_3 reacts in the apoplast, with antioxidants counteracting O_3 movement. Well-documented O_3-induced effects on chloroplast functions and nuclear gene expression occur by unknown mechanisms, possibly involving O_3 receptor(s), an oxidative burst and signal chains. These effects are often linked to a second stressor and finally lead to visible effects. The *left* signal chain may be designated organelle signal chain, while the *right* signal chain may be designated gene signal transduction chain. Inhibition of photosynthesis and pigment loss comprise photo-inhibition

Ozone also alters the normal ionic and water relations of guard cells and subsidiary cells, with subsequent effects on CO_2 fixation and on control of H_2O loss, which in turn affects drought resistance. The translocation of sugar and the loading of the phloem seems to be one of the more sensitive biochemical O_3 targets. Again, the loss of carbon in some tissues of the plant has implications for water usage. Many controlled enviornment experiments have shown that exposure to O_3 causes shoot growth at the expense of root growth which renders plants more susceptible to drought (Chap. 7).

Parameters used for the expression of external O_3 doses have been considered above (Sect. 11.3). However, the problem of leaf-internal (i.e. cells plus internal air spaces) and cell-internal ozone doses has not been resolved. How much O_3 reacts in the sub-stomatal cavities with phyto-produced alkenes, how much reacts with apoplastic antioxidants, and how much reacts with critical protein moieties are unanswered questions. Also, we do not know which chemical species react in the cytoplasm and elsewhere. In addition to intercellular reactions, low O_3 levels may confuse and interfere with the plant's signals with respect to ageing and pathogen defense, or membrane transducing processes which lead to hardening against other stresses. Chapter 7 gives an overview of how summer O_3 exposure of conifers may affect subsequent winter hardening in the following autumn.

11.5 Types of Injury and Improved Ozone Diagnostics

Ozone phytotoxicity under conditions of above-standard ozone levels may be classified as follows:

- A loss of productivity occurs, but there is no visible ozone injury on ozone bioindicator species.
- Typical ozone symptoms occur that are connected for example with crown thinning, loss of older needles and growth decline.
- Non-typical symptoms may occur due to interaction of ozone with biotic and abiotic stressors. These may lead to a loss of needles and dieback of certain genotypes; this third pathway is little documented so far.

In each case, responses start at the level of cellular metabolism. Most O_3 exposure studies, including those in southern California, have not used any cellular metabolism process to monitor ozone effects, but rather have only used one or more of the following diagnostic parameters: leaf/needle injury, growth effects, pigment loss, apparent photosynthesis, photosynthetic capacity and stomatal conductance. A schematic overview of O_3 effects is given in Fig 11.1 Stomatal- or chloroplast-related parameters and the induction of additional free radical scavenging capabilities have been frequently studied. This is not true for parameters governing the oxidative

Table 11.2. Diagnostic parameters for ozone. References and further details are to be found in the indicated chapters. Visible symptoms and growth effects are routinely studied as late response parameters

Cellular compartment process	Parameter	Chapter number
Apoplast, plasma membrane	Oxidative burst, antioxidative status, oxidation products, channel and pump systems	2, 4, 6, 7, 8, 10
Signal chains	Ethylene, ACC, MACC, oxidative burst, protein kinases and phosphatases	6, 7, 10
Chloroplast	Pigment loss, antioxidative status, Rubisco level, control of the processes of photosynthesis, photosynthetic capacity, D1 protein turnover, chlorophyll fluorescence	4, 5, 6, 7, 8, 10
Nucleus	Transcript amounts for PR-proteins, secondary metabolic enzymes, antioxidative systems, general stress proteins	6, 10
Physiology of tissues	Stomatal function, gas exchange, allocation processes	2, 4, 5, 6, 7, 10
Induced biomarkers	Phytoalexins, catechin, other secondary products, PEP-carboxylase, apoplastic ascorbate, polyamines, lignins, ethylene	4, 6, 7, 10

burst similar to that caused by pathogen attack, the signal chains, specific gene induction, and O_3-induced protein formation as well as other short- and long-term biomarkers. These, along with more recently described diagnostic parameters have been listed in Table 11.2 together with the relevant chapters.

Some of the induced biomarker compounds, such as ACC and the polyamines, may act themselves as signal molecules, while the effect of O_3 on signal molecules such as jasmonate, salicylate, oligosaccharides and secondary messenger peptides is unknown, as are the response differences between tissue and cell types. The diagnostic parameters listed in Table 11.2 have been induced by O_3 exposure in controlled experiments. It is unfortunate, perhaps, that many of these parameters have so far not been tested at the classical field sites in southern California. Some of these parameters have been linked to spruce decline in certain German forest stands (Schmieden and Wild 1995) but, as already stated, the decline/O_3 link has not been established in central Europe (see Chaps. 4 and 5).

The plant as an individual. Adult plant, seedling, cell culture: gene expression, stress metabolism. Acclimation or damage of vegetative and reproductive functions, including root/shoot allocations and primary/secondary metabolism.
The plant as part of its microenvironment. Effects of O_3 on mycorrhization, nutrient uptake, and susceptibility towards pathogens, frost, heat, or other microclimatic factors.
The plant as part of an ecosystem. Population genetics and shifts; gene flow. Food chains. Interactions with non-plant organisms including herbivores, saprophytes and decomposers. Element, water and carbon cycling.

Fig. 11.2. Biological complexity of research on ozone phytotoxicity

11.6 Population and Ecosystem Responses

The question then is: What is the role of biochemical processes at the ecosystem level? Scaling-up is an integral part of ecophysiology as summarized in Fig 11.2. However, many questions remain and are actively being discussed. They include:

1. Are seedling and sapling responses a good measure of responses in older trees?
2. How does a relatively small loss of rubisco affect the overall health of an adult tree?
3. What causes a particular forest to age faster?
4. Does O_3 increase respiration much more dramatically than it lowers photosynthesis, and do these effects reduce mycorrhization and mineral uptake from soil?
5. Which metabolic *setpoints* are sensitive to O_3; in other words, which pathways are most sensitive to ozone exposure?
6. How can the health of an ecosystem and risk to a stressor such as ozone be defined?
7. If homeostasis is shifted, could the altered metabolic state cause a pathogen and/or insect attack to be a major observable response to ozone, thereby inducing the forest decline and a shift in population dynamics?'

Perhaps the transition from seedlings to stand as well as the interactions between stress factors can be best bridged by models. However, the mathematical models used should be clearly linked to the physiological processes outlined in Fig. 11.1.

To summarize, a research strategy could be as follows:

1. Establish the O_3 dose/response relationships for the early and late parameters listed in Table 11.2 in various tree species.
2. Verify or refute these relationships using only the recognized O_3-dominated field sites, e.g. those in southern California.
3. Re-examine the forest stands in eastern USA and Europe with respect to the diagnostic parameters in Table 11.2, and compare the observed effects at ambient O_3 levels with the extent of the expected responses.
4. Establish how O_3 can predispose trees to interacting abiotic or biotic stress factors. Can non-dominant O_3 in this way trigger visible symptoms and decline?
5. Establish scaling-up procedures and devise effective ecosystem modelling.
6. Redefine current O_3 standards for the research levels summarized in Fig. 11.2.

In the coming years, it is hoped that the research strategy derived from this publication will help to clarify the still largely unresolved relationship between forest decline and atmospheric O_3.

References

Eagar C, Adams MB (eds) (1992) Ecology and decline of red spruce in the eastern United States. Ecological Studies, Vol 96. Springer, Berlin Heidelberg New York

FBW (Forschungsbeirat Waldschäden / Luftverunreinigungen, 1989). 3. Bericht, Kernforschungszentrum Karlsruhe, Karlsruhe, ISSN 0931–7805

Fox S, Mickler RA (eds) (1996) Impact of air pollutants on southern pine forests. Ecological Studies, Vol 118. Springer, Berlin Heidelberg New York

Führer G (1989) Photosyntheseverhalten, Transpiration und Spaltöffnungsregulation von Fichten [*Picea abies*] unter experimenteller Schadstoffbelastung in Expositionskammern. PhD Diss, University of Würzburg, Würzburg p 176

Guderian R, Tingey DT, Rabe R (1985) Effects of photochemical oxidants on plants. In: Guderian R (ed) Air pollution by photochemical oxidants. Ecological Studies, Vol 52. Springer, Berlin Heidelberg New York pp 127–333

Manion PD (1981) Tree disease concepts. Prentice-Hall, Englewood Cliffs

Olson RK, Binkley D, Böhm M (1992) The response of western forests to air pollution. Ecological Studies, Vol 97. Springer, Berlin Heidelberg New York

Remmert H (ed) (1991) The mosaic-cycle concept of ecosystems. Ecological Studies, Vol 85. Springer, Berlin Heidelberg New York

Sandermann H (1996) Ozone and plant health. Annu Rev Phytopathol 34: (in press)

Schmieden U, Wild A (1995) The contribution of ozone to forest decline. Physiol Plant 94: 371–378

Schulze E-D, Lange OL, Oren R (eds) (1989) Forest decline and air pollution. A study of spruce (*Picea abies*) on acid soils. Ecological Studies, Vol 77. Springer, Berlin Heidelberg New York

Schwerdtfeger F (1981) Die Waldkrankheiten. Ein Lehrbuch der Forstpathologie und des Forstschutzes, 4th edn. P Parey, Berlin

Subject Index

Appendix: Color Illustrations

Fig. 1.3. Surface O_3 concentrations in the Los Angeles region for August 27, 1987, calculated through the use of the California Institute of Technology (CIT) urban airshed model (McRae et al. 1992a,b,c). The clock shows the local time and the final plot shows the daily maximum O_3. The data files used for the simulation were provided through the courtesy of Dr. Russell (Carnegie Mellon University)

Fig. 1.12. Map of the spatial distribution of AOT40 indices over Europe. Measurements during daylight hours between April and September 1992 were considered. (Hjellbrekke 1995)

Fig. 2.2. Appearances of ozone air pollution in a San Bernardino mountain setting (**a**), a stand with injured ponderosa pines (**b**), an individual injured tree (**c**), reduced needle whorl retention (**d**), and specific needle symptoms of ozone injury (**e, f**)

Fig. 3.1. Ambient ozone exposure in 1988 for the eastern half of the United States. Note the highest exposures were monitored directly within the Shenandoah National Park; the park boundaries closely coincide with the darkest shaded area. Areas of reported investigations are indicated as: *1* Acadia National Park; *2* Northcentral Pennsylvania; *3* Shenandoah National Park; *4* Great Smoky Mountains National Park; *5* Southeastern United States, i.e., from Virginia to Texas. (Map produced by the Environmental Monitoring and Assessment Program-Forests, USEPA, Research Triangle Park, North Carolina; USDA-For Serv 1992)

Fig. 3.5. A Black cherry leaves showing typical dark-colored adaxial stippling; note that stippling does not occur on the veins of the leaf. **B** Yellow poplar leaves collected from two trees at Sawmill Run, Shenandoah National Park, Virginia, 1992. Adaxial stipple is evident on the sensitive tree (*left*) while the leaf from the tolerant tree (*right*) is free of any visible symptoms. **C** White Ash showing typical upper surface stippling as induced by ozone (*left part*). Note the also typical absence of symptoms where the leaves shaded one another (*right part*)

Fig. 4.12. Symptoms of O_3 injury in leaves of *Prunus serotina* (black cherry). **A–D** Progressing leaf injury. Note that in **B** only the discolored part of the lamina was exposed to light

Fig. 5.2. View on the Wank Mountain near Garmisch-Partenkirchen in the Alps, Germany

Fig. 6.1. View of a walk-in exposure chamber with 3-year-old Norway spruce and Scots pine seedlings exposed to a montainous climate (Wank Bavarian Alps) for one growing season

Ecological Studies, Vol. 127
Sandermann et al. (eds) Forest Decline and Ozone

Fig. 6.7.A Ozone symptoms on European beech. Leaves treated with twice the ambient level of ozone. Exposure lasted from leaf emergence to leaf fall (April to October). Symptoms found in mid–August are shown. The control leaves were green. **B, C** "Memory" effect for ozone in Norway spruce and Scots pine. **B** One-year-old needles of clonal spruce plants treated for one vegetation period with 80 nl l^{-1} (24 h day^{-1}), and post-cultivated outdoors. Symptoms were assessed in June of the year following exposure. The control twig was green. **C** Needle loss of the youngest ozone-treated needles of Scots pine saplings (*right tree*) in November of the year following ozone exposure (39 days; 8 h day^{-1}; 150 nl l^{-1}). The control tree (*left*) was exposed to 10 nl l^{-1} ozone. The youngest field-grown needles did not have visible symptoms

Fig. 7.2. Spatial distribution of 24-h-mean summer (April-September) height-corrected O_3 concentrations based on the AOT40 for the protection of forests, mapped at a resolution of 1 km^2 using rural monitoring data collected at 17 sites distributed throughout the UK. The critical level for the protection of forests has been set at an accumulated exposure over a 40 nl^{-1} threshold of 10 μl l^{-1} h 24 h day^{-1} over the growing period (April-September) when trees are most sensitive (UN-ECE 1993). Courtesy of D. Fowler and R.I. Smith

Fig. 7.4. Photograph of representative 5-year-old Sitka spruce trees taken in June 1989 approximately 6 weeks after the first appearance of symptoms. The tree on the *left* came from the batch that had been exposed for the three previous summers (1986–1988) to episodes of O_3 in solardomes at Lancaster while the one on the *right* came from those that had received only clean air. (Courtsey of Longmans, Essex)

Fig. 7.5. Aerial photograph of the Liphook site showing the seven treatment plots

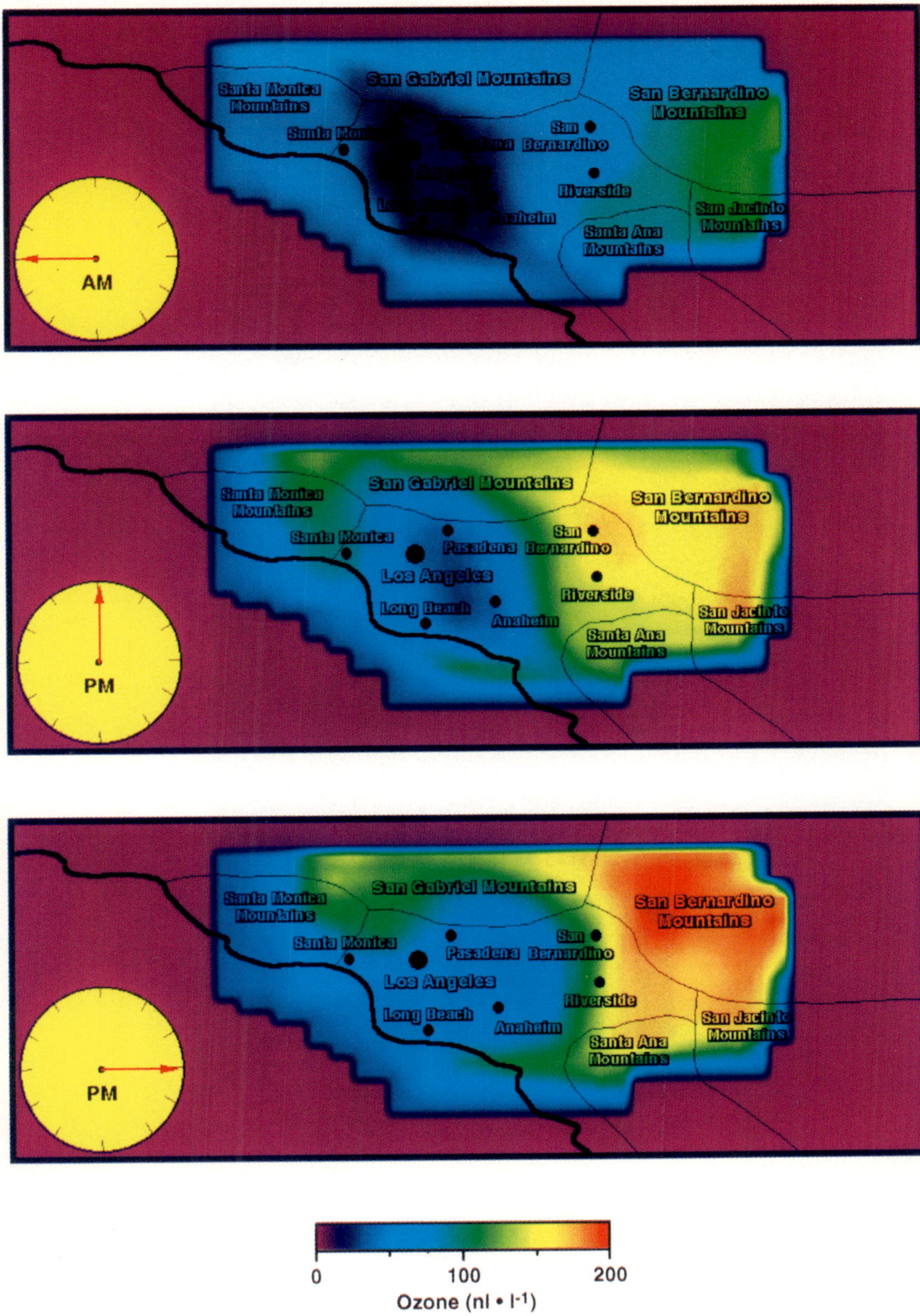
San Gabriel Mountains
Santa Monica Mountains
San Bernardino Mountains
San Bernardino
Riverside
Anaheim
San Jacinto Mountains
Santa Ana Mountains
AM
San Gabriel Mountains
Santa Monica Mountains
San Bernardino Mountains
Santa Monica
San Bernardino
Pasadena
Los Angeles
Riverside
Long Beach
Anaheim
San Jacinto Mountains
Santa Ana Mountains
PM
San Gabriel Mountains
Santa Monica Mountains
San Bernardino Mountains
Santa Monica
San Bernardino
Pasadena
Los Angeles
Riverside
Long Beach
Anaheim
San Jacinto Mountains
Santa Ana Mountains
PM
0
100
200
Ozone (nl • l-1)

Fig. 1.3

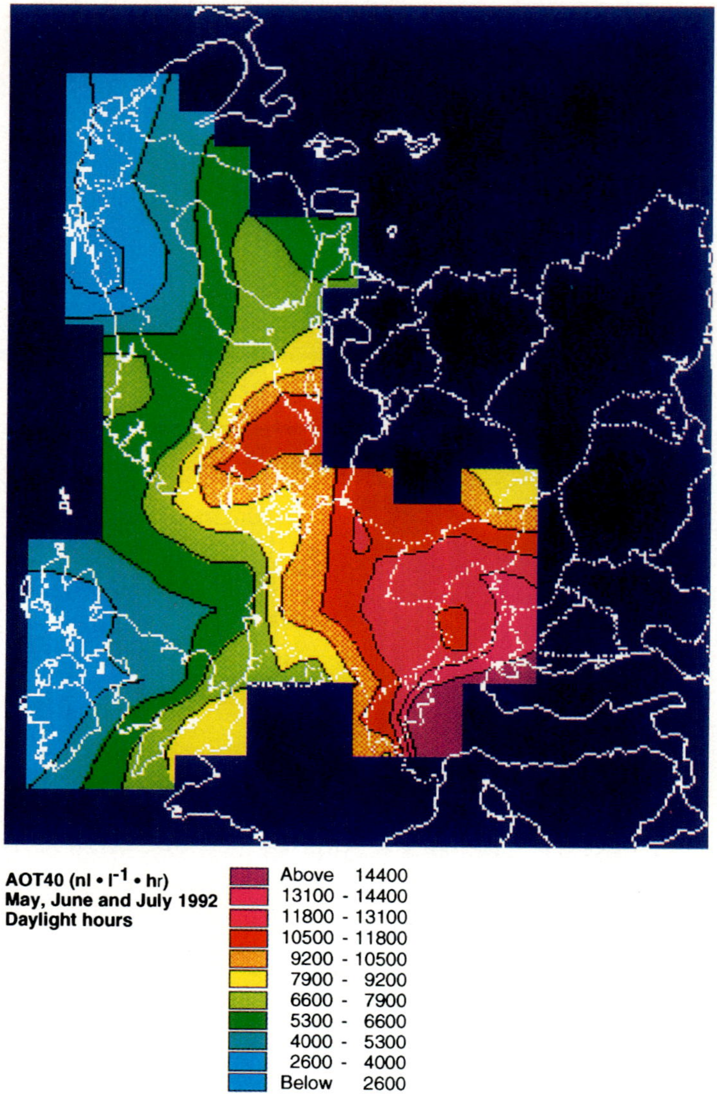
AOT40 (nl • l^{-1} • hr)
May, June and July 1992
Daylight hours
Above 14400
13100 - 14400
11800 - 13100
10500 - 11800
9200 - 10500
7900 - 9200
6600 - 7900
5300 - 6600
4000 - 5300
2600 - 4000
Below 2600

Fig. 1.12

Fig. 2.2

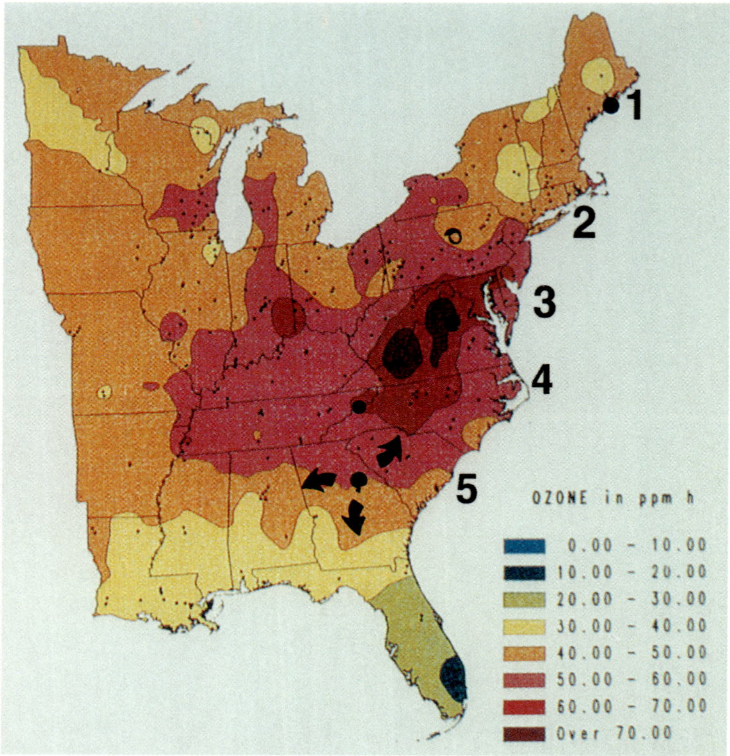
1
2
3
4
5
OZONE in ppm h
0.00 - 10.00
10.00 - 20.00
20.00 - 30.00
30.00 - 40.00
40.00 - 50.00
50.00 - 60.00
60.00 - 70.00
Over 70.00

Fig. 3.1

A
B
C

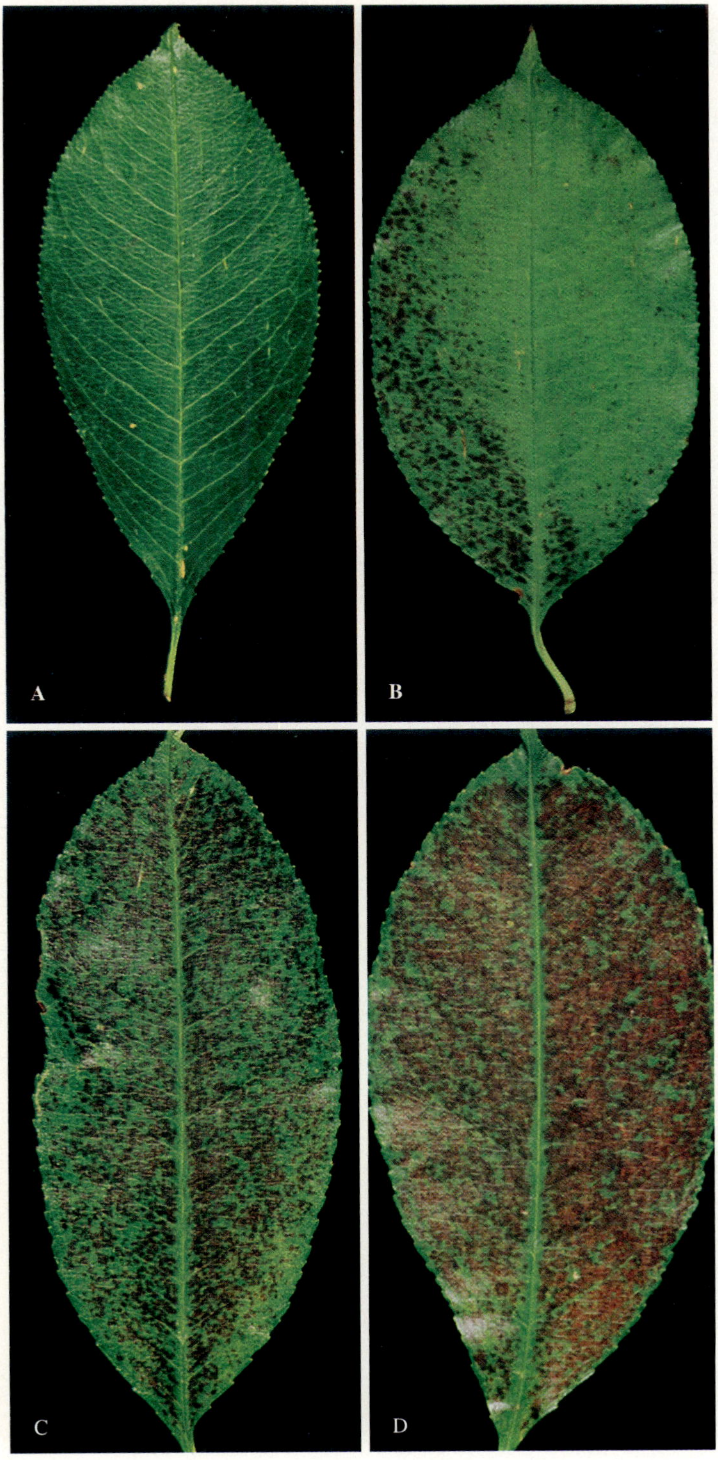

Fig. 4.12

Fig. 5.2

Fig. 6.1

Fig. 6.7

A
B
C

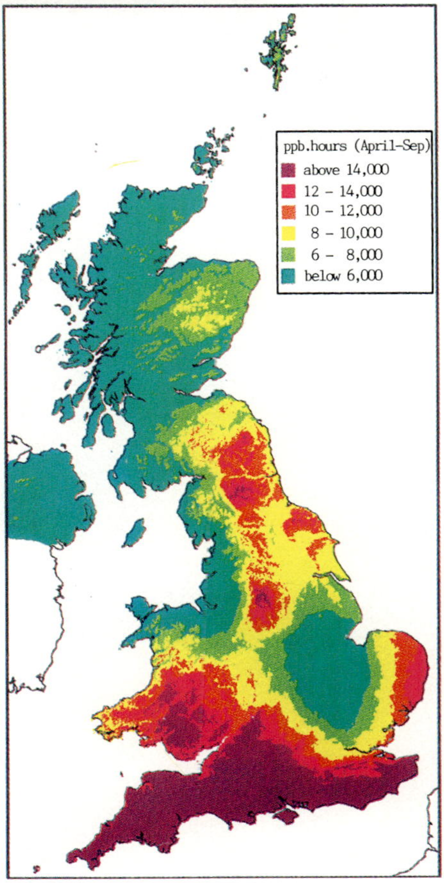
ppb.hours (April–Sep)
above 14,000
12 – 14,000
10 – 12,000
8 – 10,000
6 – 8,000
below 6,000

Fig. 7.2

Fig. 7.4

Fig. 7.5

Ecological Studies

Volumes published since 1992

Volume 89
Plantago: A Multidisciplinary Study (1992)
P.J.C. Kuiper and M. Bos (Eds.)

Volume 90
Biogeochemistry of a Subalpine Ecosystem: Loch Vale Watershead (1992)
J. Baron (Ed.)

Volume 91
Atmospheric Deposition and Forest Nutrient Cycling (1992)
D.W. Johnson and S.E. Lindberg (Eds.)

Volume 92
Landscape Boundaries: Consequences for Biotic Diversity and Ecological Flows (1992)
A.J. Hansen and F. di Castri (Eds.)

Volume 93
Fire in South African Mountain Fynbos: Ecosystem, Community, and Species Responses at Swartboskloof (1992)
B.W. van Wilgen et al. (Eds.)

Volume 94
The Ecology of Aquatic Hyphomycetes (1992)
F. Bärlocher (Ed.)

Volume 95
Palms in Forest Ecosystems of Amazonia(1992)
F. Kahn and J.-J. DeGranville

Volume 96
Ecology and Decline of Red Spruce, in the Eastern United States (1992)
C. Eagar and M.B. Adams (Eds.)

Volume 97
The Response of Western Forests to Air Pollution (1992)
R.K. Olson, D. Binkley, and M. Böhm (Eds.)

Volume 98
Plankton Regulation Dynamics (1993)
N. Walz (Ed.)

Volume 99
Biodiversity and Ecosystem Function (1993)
E.-D. Schulze and H.A. Mooney (Eds.)

Volume 100
Ecophysiology of Photosynthesis (1994)
E.-D. Schulze and M.M. Caldwell (Eds.)

Volume 101
Effects of Land Use Change on Atmospheric CO_2 Concentrations: South and South East Asia as a Case Study (1993)
V.H. Dale (Ed.)

Volume 102
Coral Reef Ecology (1993)
Y.I. Sorokin

Volume 103
Rocky Shores: Exploitation in Chile and South Africa (1993)
W.R. Siegfried (Ed.)

Volume 104
Long-Term Experiments With Acid Rain in Norwegian Forest Ecosystems (1993)
G. Abrahamsen et al. (Eds.)

Volume 105
Microbial Ecology of Lake Plußsee (1993)
J. Overbeck and R.J. Chrost (Eds.)

Volume 106
Minimum Animal Populations (1994)
H. Remmert (Ed.)

Volume 107
The Role of Fire in Mediterranean-Type Ecosystems (1994)
J.M. Moreno and W.C. Oechel

Volume 108
Ecology and Biogeography of Mediterranean Ecosystems in Chile, California and Australia (1994)
M.T.K. Arroyo, P.H. Zedler, and M.D. Fox (Eds.)

Volume 109
Mediterranean-Type Ecosystems. The Function of Biodiversity (1995)
G.W. Davis and D.M. Richardson (Eds.)

Volume 110
Tropical Montane Cloud Forests (1995)
L.S. Hamilton, J.O. Juvik, and
F.N. Scatena (Eds.)

Volume 111
Peatland Forestry, Ecology and Principles (1995)
E. Paavilainen and J. Päivänen

Volume 112
Tropical Forests: Management and Ecology (1995)
A.E. Lugo and C. Lowe (Eds.)

Volume 113
Arctic and Alpine Biodiversity. Patterns, Causes and Ecosystem Consequences (1995)
F.S. Chapin III and C. Körner (Eds.)

Volume 114
Crassulacean Acid Metabolism. Biochemistry, Ecophysiology and Evolution (1996)
K. Winter and J.A.C. Smith (Eds.)

Volume 115
Islands. Biological Diversity and Ecosystem Function (1995)
P.M. Vitousek, L.L. Loope, and
H. Adsersen (Eds.)

Volume 116
High Latitude Rainforests and Associated Ecosystems of the West Coast of the Americas: Climate, Hydrology, Ecology and Conservation (1996)
R.G. Lawford, P. Alaback, and
E.R. Fuentes (Eds.)

Volume 117
Anticipated Effects of a Changing Global Environment on Mediterranean-Type Ecosystems (1995)
J. Moreno and W.C. Oechel (Eds.)

Volume 118
Impact of Air Pollutants on Southern Pine Forests (1996)
S. Fox and R.A. Mickler (Eds.)

Volume 119
Freshwater Ecosystems of Alaska (1996)
A.M. Milner and M.W. Oswood (Eds.)

Volume 120
Landscape Function and Disturbance in Arctic Tundra (1996)
J.F. Reynolds and J.D. Tenhunen (Eds.)

Volume 121
Biodiversity and Savanna Ecosystem Processes. A Global Perspective (1996)
O.T. Solbrig, E. Medina, and J.F. Silva
(Eds.)

Volume 122
Biodiversity and Ecosystem Processes in Tropical Forests (1996)
G.H. Orians, R. Dirzo, and J.H. Cushman
(Eds.)

Volume 123
Marine Benthic Vegetation. Recent Changes and the Effects of Eutrophication (1996)
W. Schramm and P.H. Nienhuis (Eds.)

Volume 124
Global Change and Arctic Terrestrial Ecosystems (1997)
W.C. Oechel (Ed.)

Volume 125
Ecology and Conservation of Great Plains Vertebrates (1997)
F.L. Knopf and F.B. Samson (Eds.)

Volume 126
The Central Amazon Floodplain: Ecology of a Pulsing System (1997)
W.J. Junk (Ed.)

Volume 127
Forest Decline and Ozone: A Comparison of Controlled Chamber and Field Experiments (1997)
H. Sandermann, A.R. Wellburn, and R.L. Heath
(Eds.)

Volume 128
The Productivity and Sustainability of Southern Forest Ecosystems in a Changing Environment (1997)
S. Fox and R.A. Mickler (Eds.)

Springer and the environment

At Springer we firmly believe that an international science publisher has a special obligation to the environment, and our corporate policies consistently reflect this conviction.

We also expect our business partners – paper mills, printers, packaging manufacturers, etc. – to commit themselves to using materials and production processes that do not harm the environment. The paper in this book is made from low- or no-chlorine pulp and is acid free, in conformance with international standards for paper permanency.

Springer

Printing: Saladruck, Berlin
Binding: Buchbinderei Lüderitz & Bauer, Berlin